The Ultimate Regents Physics Question and Answer Book

2016 Edition

edited by Dan Fullerton

Silly Beagle Productions
Webster, NY 14580

Silly Beagle Productions
656 Maris Run
Webster, NY 14580
Internet: www.SillyBeagle.com
E-Mail: info@SillyBeagle.com

DFullerton@APlusPhysics.com

Cover Design:
Interior Illustrations by Dan Fullerton and Jupiterimages unless otherwise noted

Sales and Ordering Information
http://www.APlusPhysics.com
Sales@SillyBeagle.com
Volume discounts available
E-book editions available

Printed in the United States of America
ISBN: 978-0-9907243-3-9

1 2 3 4 5 6 7 8 9 0 9 8 7 6

Table of Contents

Name:________________________________ Period: ______________

Metric Estimation

1. What is the approximate width of a person's little finger?
   1. 1 m
   2. 0.1 m
   3. 0.01 m
   4. 0.001 m

2. What is the approximate mass of an automobile?
   1. 10^1 kg
   2. 10^2 kg
   3. 10^3 kg
   4. 10^6 kg

3. The diameter of a United States penny is closest to
   1. 10^0 m
   2. 10^{-1} m
   3. 10^{-2} m
   4. 10^{-3} m

4. An egg is dropped from a third-story window. The distance the egg falls from the window to the ground is closest to
   1. 10^0 m
   2. 10^1 m
   3. 10^2 m
   4. 10^3 m

5. The approximate height of a 12-ounce can of root beer is
   1. 1.3×10^{-3} m
   2. 1.3×10^{-1} m
   3. 1.3×10^0 m
   4. 1.3×10^1 m

6. The mass of a paper clip is approximately
   1. 1×10^6 kg
   2. 1×10^3 kg
   3. 1×10^{-3} kg
   4. 1×10^{-6} kg

7. The length of a dollar bill is approximately
   1. 1.5×10^{-2} m
   2. 1.5×10^{-1} m
   3. 1.5×10^1 m
   4. 1.5×10^2 m

8. What is the approximate length of a baseball bat?
   1. 10^{-1} m
   2. 10^0 m
   3. 10^1 m
   4. 10^2 m

9. What is the approximate diameter of an inflated basketball?
   1. 2×10^{-2} m
   2. 2×10^{-1} m
   3. 2×10^0 m
   4. 2×10^1 m

10. The length of a football field is closest to
    1. 1000 cm
    2. 1000 dm
    3. 1000 km
    4. 1000 mm

11. The approximate length of an unsharpened No. 2 pencil is
    1. 2.0×10^{-2} m
    2. 2.0×10^{-1} m
    3. 2.0×10^0 m
    4. 2.0×10^1 m

12. The height of a 30-story building is approximately
    1. 10^0 m
    2. 10^1 m
    3. 10^2 m
    4. 10^3 m

13. The diameter of an automobile tire is closest to
    1. 10^{-2} m
    2. 10^0 m
    3. 10^1 m
    4. 10^2 m

Name:________________________ Period: ____________

Kinematics-Defining Motion

1. A student on her way to school walks four blocks east, three blocks north, and another four blocks east, as shown in the diagram.

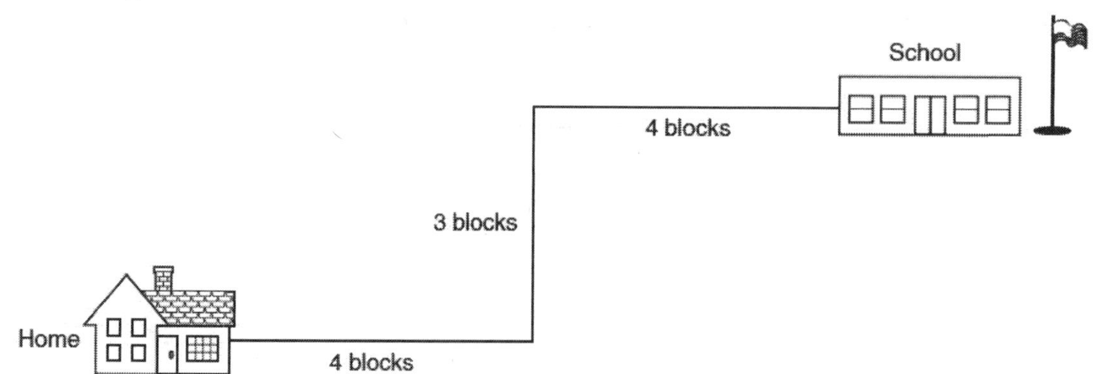

Compared to the distance she walks, the magnitude of her displacement from home to school is
   1. less
   2. greater
   3. the same

2. A motorboat, which has a speed of 5 meters per second in still water, is headed east as it crosses a river flowing south at 3.3 meters per second. What is the magnitude of the boat's resultant velocity with respect to the starting point?
   1. 3.3 m/s
   2. 5.0 m/s
   3. 6.0 m/s
   4. 8.3 m/s

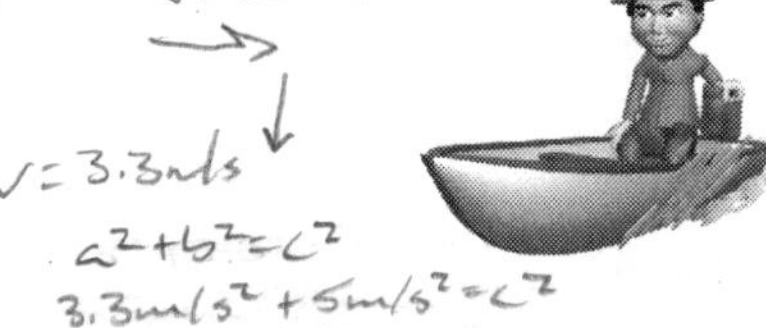

3. A speedometer in a car does *not* measure the car's velocity because velocity is a
   1. vector quantity and has a direction associated with it
   2. vector quantity and does not have a direction associated with it
   3. scalar quantity and has a direction associated with it
   4. scalar quantity and does not have a direction associated with it

4. A person observes a fireworks display from a safe distance of 0.750 kilometer. Assuming that sound travels at 340 meters per second in air, what is the time between the person seeing and hearing a fireworks explosion?
   1. 0.453 s
   2. 2.21 s
   3. 410 s
   4. 2.55×10^5 s

5. On the surface of Earth, a spacecraft has a mass of 2.00×10^4 kilograms. What is the mass of the spacecraft at a distance of one Earth radius above Earth's surface?
   1. 5.00×10^3 kg
   2. 2.00×10^4 kg
   3. 4.90×10^4 kg
   4. 1.96×10^5 kg

6. An airplane flies with a velocity of 750 kilometers per hour, 30.0° south of east. What is the magnitude of the eastward component of the plane's velocity?
   1. 866 km/h
   2. 650 km/h
   3. 433 km/h
   4. 375 km/h

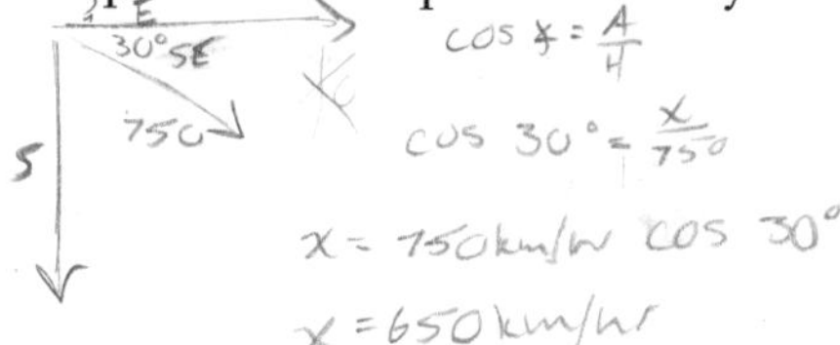

7. One car travels 40 meters due east in 5 seconds, and a second car travels 64 meters due west in 8 seconds. During their periods of travel, the cars definitely had the same
   1. average velocity
   2. total displacement
   3. change in momentum
   4. average speed

8. State the *two* general characteristics that are used to define a vector quantity.

Name:________________________________ Period: ____________

Kinematics-Defining Motion

Base your answers to questions 9 through 12 on the information and diagram below.

A model airplane heads due east at 1.50 meters per second, while the wind blows due north at 0.70 meter per second. The scaled diagram below represents these vector quantities.

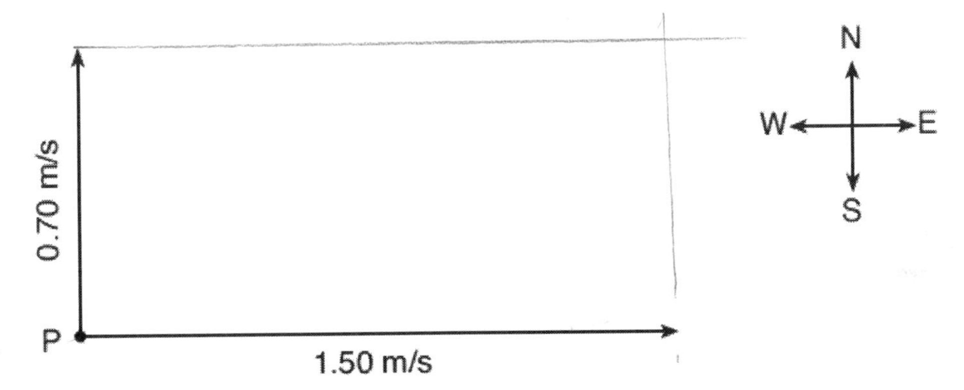

9. Using a ruler, determine the scale used in the vector diagram. **1 cm = ____________ m/s**

10. On the diagram above, use a protractor and a ruler to construct a vector to represent the resultant velocity of the airplane. Label the vector R.

11. Determine the magnitude of the resultant velocity.

12. Determine the angle between north and the resultant velocity.

13. A baseball player runs 27.4 meters from the batter's box to first base, overruns first base by 3.0 meters, and then returns to first base. Compared to the total distance traveled by the player, the magnitude of the player's total displacement from the batter's box is
    1. 3 m shorter
    2. 6 m shorter
    3. 3 m longer
    4. 6 m longer

14. In a 4-kilometer race, a runner completes the first kilometer in 5.9 minutes, the second kilometer in 6.2 minutes, the third kilometer in 6.3 minutes, and the final kilometer in 6 minutes. The average speed of the runner for the race is approximately
    1. 0.16 km/min
    2. 0.33 km/min
    3. 12 km/min
    4. 24 km/min

15. A girl leaves a history classroom and walks 10 meters north to a drinking fountain. Then she turns and walks 30 meters south to an art classroom. What is the girl's total displacement from the history classroom to the art classroom?
    1. 20 m south
    2. 20 m north
    3. 40 m south
    4. 40 m north

16. Which is a vector quantity?
    1. speed
    2. work
    3. mass
    4. displacement

17. A cart travels 4 meters east and then 4 meters north. Determine the magnitude of the cart's resultant displacement.

Name:______________________________ Period: __________

Kinematics-Defining Motion

Base your answers to questions 18 through 20 on the information and vector diagram below.

A dog walks 8 meters due north and then 6 meters due east.

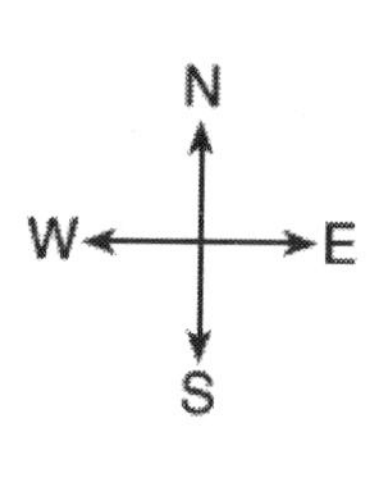

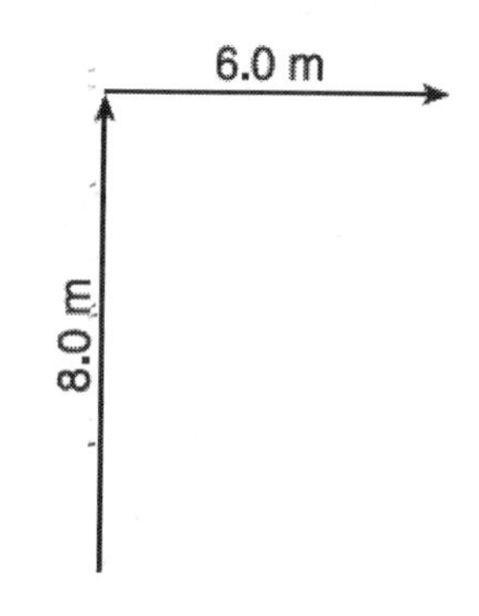

18. Using a metric ruler and the vector diagram, determine the scale used in the diagram.

    1 cm = ______________ **m**

19. On the diagram above, construct the resultant vector that represents the dog's total displacement.

20. Determine the magnitude of the dog's total displacement.

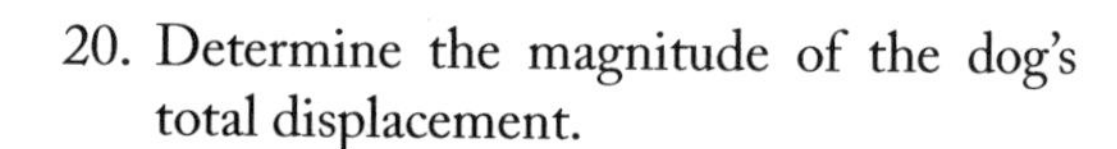

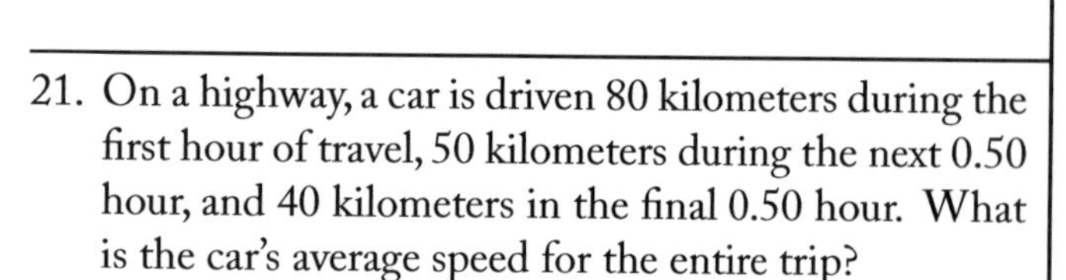

21. On a highway, a car is driven 80 kilometers during the first hour of travel, 50 kilometers during the next 0.50 hour, and 40 kilometers in the final 0.50 hour. What is the car's average speed for the entire trip?
    1. 45 km/h
    2. 60 km/h
    3. 85 km/h
    4. 170 km/h

22. Explain the difference between a scalar and a vector quantity.

23. The diagram below shows a resultant vector, R.

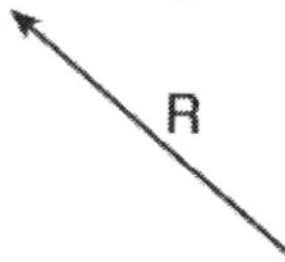

Which diagram below best represents a pair of component vectors, A and B, that would combine to form resultant vector R?

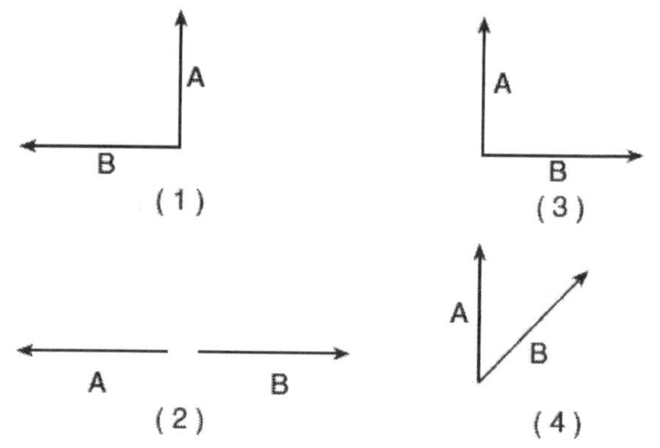

Base your answers to questions 24 and 25 on the information below.

A stream is 30 meters wide and its current flows southward at 1.5 meters per second. A toy boat is launched with a velocity of 2.0 meters per second eastward from the west bank of the stream.

24. What is the magnitude of the boat's resultant velocity as it crosses the stream?
    1. 0.5 m/s
    2. 2.5 m/s
    3. 3.0 m/s
    4. 3.5 m/s

25. How much time is required for the boat to reach the opposite bank of the stream?
    1. 8.6 s
    2. 12 s
    3. 15 s
    4. 60 s

26. A person walks 150 meters due east and then walks 30 meters due west. The entire trip takes the person 10 minutes. Determine the magnitude and direction of the person's total displacement.

Name: ____________________ Period: __________

Kinematics-Defining Motion

27. A high-speed train in Japan travels a distance of 300 kilometers in 3.60×10^3 seconds. What is the average speed of this train?
    1. 1.20×10^{-2} m/s
    2. 8.33×10^{-2} m/s
    3. 12.0 m/s
    4. 83.3 m/s

28. A child walks 5 meters north, then 4 meters east, and finally 2 meters south. What is the magnitude of the resultant displacement of the child after the entire walk?
    1. 1.0 m
    2. 5.0 m
    3. 3.0 m
    4. 11.0 m

29. Scalar is to vector as
    1. speed is to velocity
    2. displacement is to distance
    3. displacement is to velocity
    4. speed is to distance

30. A car travels 90 meters due north in 15 seconds. Then the car turns around and travels 40 meters due south in 5 seconds. What is the magnitude of the average velocity of the car during this 20-second interval?
    1. 2.5 m/s
    2. 5.0 m/s
    3. 6.5 m/s
    4. 7.0 m/s

31. Velocity is to speed as displacement is to
    1. acceleration
    2. time
    3. momentum
    4. distance

Base your answers to questions 32 and 33 on the following information.

A hiker walks 5 kilometers due north and then 7 kilometers due east.

32. What is the magnitude of her resultant displacement?

33. What total distance has she traveled?

Base your answers to questions 34 and 35 on the information below.

In a drill during basketball practice, a player runs the length of the 30-meter court and back. The player does this three times in 60 seconds.

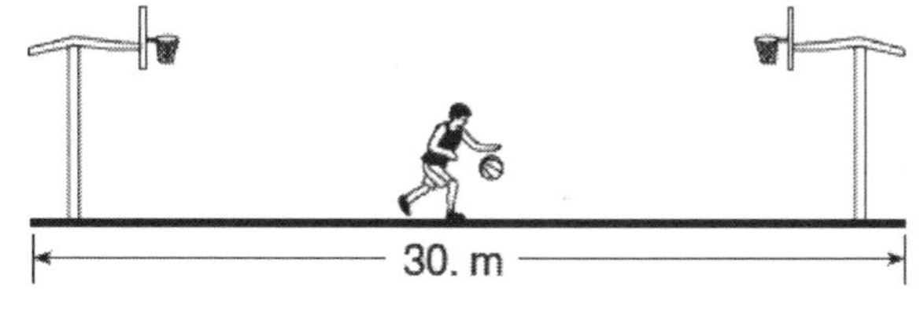

(Not drawn to scale)

34. The magnitude of the player's total displacement after running the drill is
    1. 0.0 m
    2. 30 m
    3. 60 m
    4. 180 m

35. The average speed of the player during the drill is
    1. 0.0 m/s
    2. 0.50 m/s
    3. 3.0 m/s
    4. 30 m/s

Name:______________________________ Period: ____________

Kinematics-Defining Motion

Base your answers to questions 36 through 38 on the information below.

A river has a current flowing with a velocity of 2 meters per second due east. A boat is 75 meters from the north riverbank. It travels at 3 meters per second relative to the river and is headed due north. In the diagram below, the vector starting at point P represents the velocity of the boat relative to the river water.

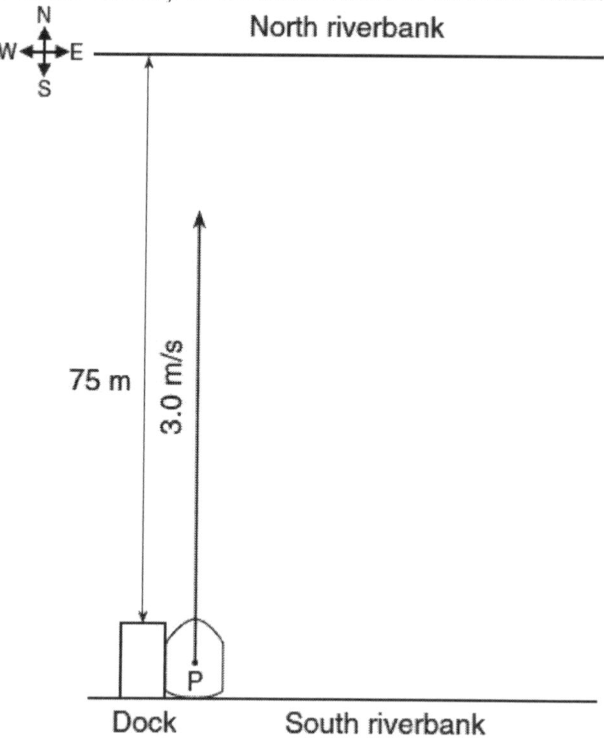

36. Calculate the time required for the boat to cross the river. [Show all work, including the equation and substitution with units.]

37. On the diagram, use a ruler and protractor to construct a vector representing the velocity of the river current. Begin the vector at point P and use a scale of 1 cm = 0.50 meter per second.

38. Calculate or find graphically the magnitude of the resultant velocity of the boat. [Show all work, including the equation and substitution with units or construct the resultant velocity vector on the diagram, using the scale given. The value of the magnitude must be written below.

Name:______________________ Period: ____________

Kinematics-Defining Motion

Base your answers to questions 39 through 42 on the information below.

A girl rides her bicycle 1.40 kilometers west, 0.70 kilometer south, and 0.30 kilometer east in 12 minutes. The vector diagram in your answer booklet represents the girl's first two displacements in sequence from point P. The scale used in the diagram is 1.0 centimeter = 0.20 kilometer.

39. On the vector diagram below, using a ruler and protractor, construct the following vectors:
    - Starting at the arrowhead of the second displacement vector, draw a vector to represent the 0.30 kilometer east displacement. Label the vector with its magnitude.
    - Draw the vector representing the resultant displacement of the girl for the entire bicycle trip and label the vector R.

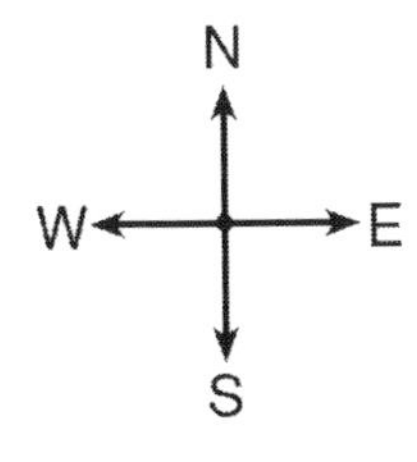

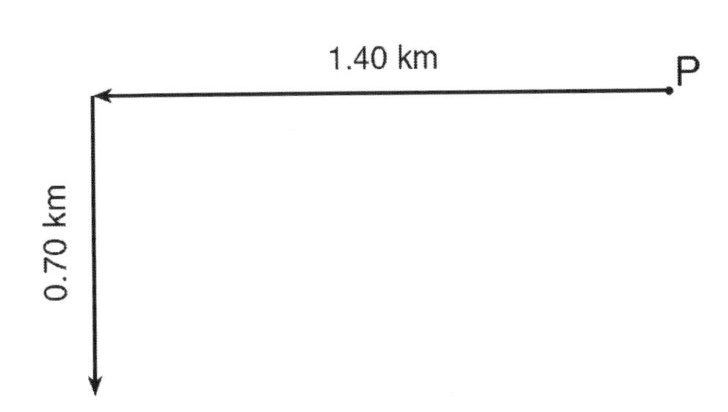

40. Calculate the girl's average speed for the entire bicycle trip. [Show all work, including the equation and substitution with units.]

41. Determine the magnitude of the girl's resultant displacement for the entire bicycle trip, in kilometers.

42. Determine the measure of the angle, in degrees, between the resultant and the 1.40-kilometer displacement vector.

Name:_______________________________ Period: ___________

Kinematics-Defining Motion

43. An airplane traveling north at 220 meters per second encounters a 50.0-meters-per-second crosswind from west to east, as represented in the diagram below.

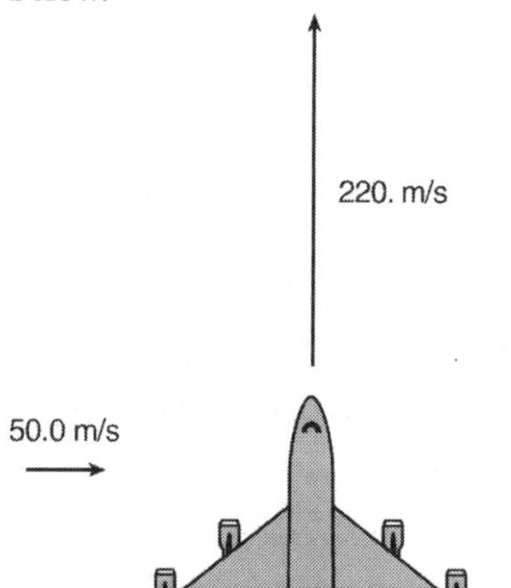

What is the resultant speed of the plane?
1. 170 m/s
2. 214 m/s
3. 226 m/s
4. 270 m/s

44. The vector diagram below represents the velocity of a car traveling 24 meters per second 35° east of north.

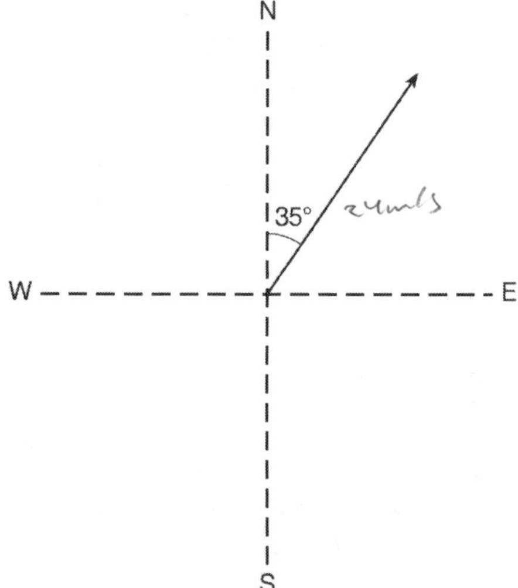

What is the magnitude of the component of the car's velocity that is directed eastward?
1. 14 m/s
2. 20 m/s
3. 29 m/s
4. 42 m/s

Name:______________________________ Period: ____________

Kinematics-Motion Graphs

1. A cart travels with a constant nonzero acceleration along a straight line. Which graph best represents the relationship between the distance the cart travels and time of travel?

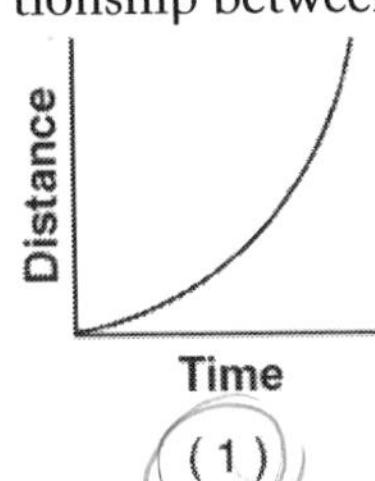

(1)

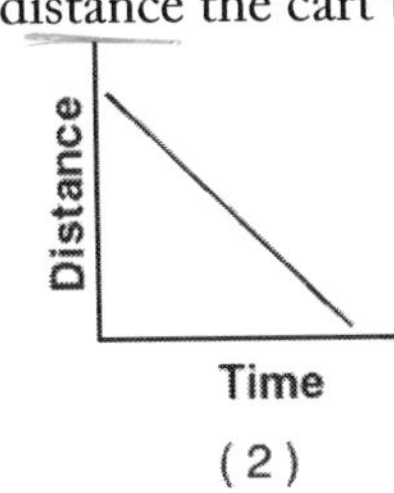

(2)

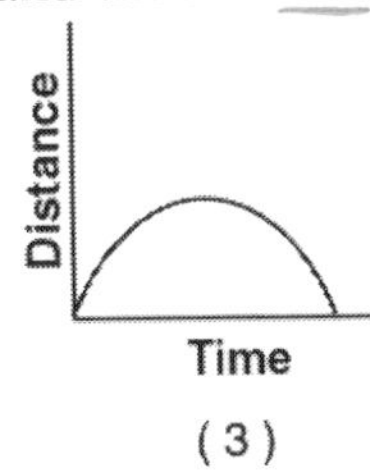

(3)

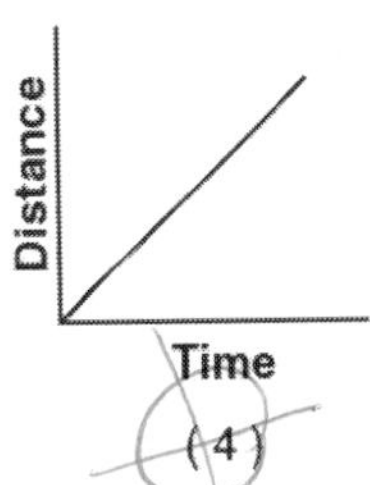

(4)

Base your answers to questions 2 through 4 on the information below.

A car on a straight road starts from rest and accelerates at 1.0 meter per second^2 for 10 seconds. Then the car continues to travel at constant speed for an additional 20 seconds.

2. Determine the speed of the car at the end of the first 10 seconds.

3. On the grid at below, use a ruler or straightedge to construct a graph of the car's speed as a function of time for the entire 30-second interval.

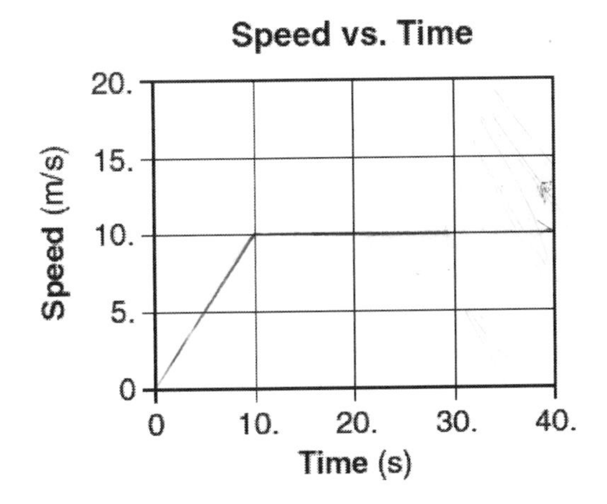

4. Calculate the distance the car travels in the first 10 seconds. [Show all work, including the equation and substitution with units.]

5. A student throws a baseball vertically upward and then catches it. If vertically upward is considered to be the positive direction, which graph best represents the relationship between velocity and time for the baseball? [Neglect friction.]

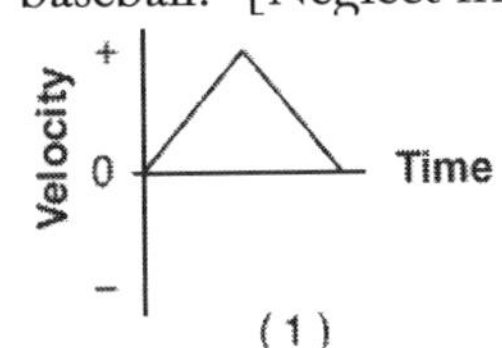

(1)

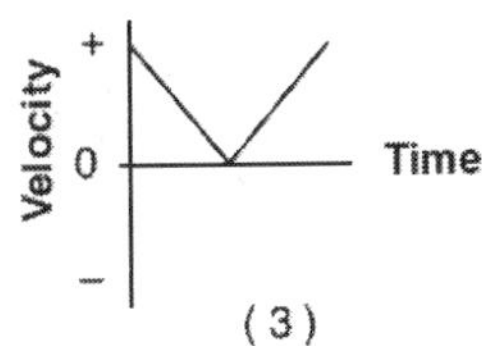

(3)

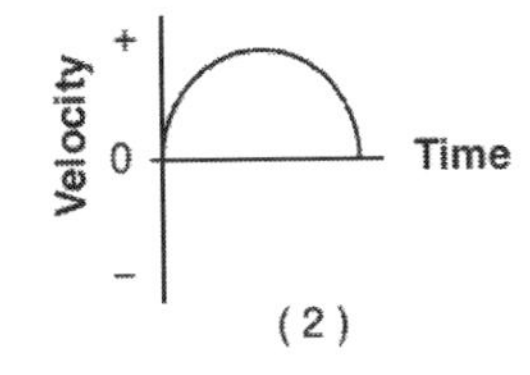

(2)

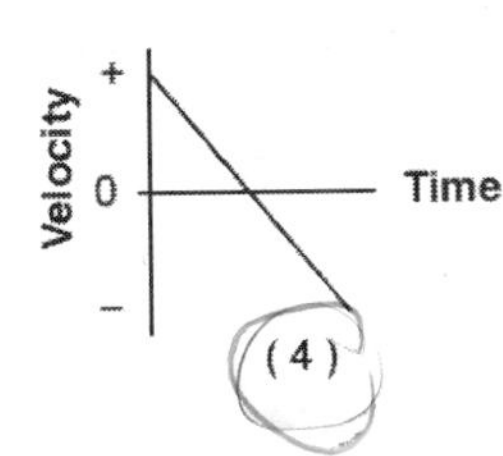

(4)

6. The graph below represents the displacement of an object moving in a straight line as a function of time.

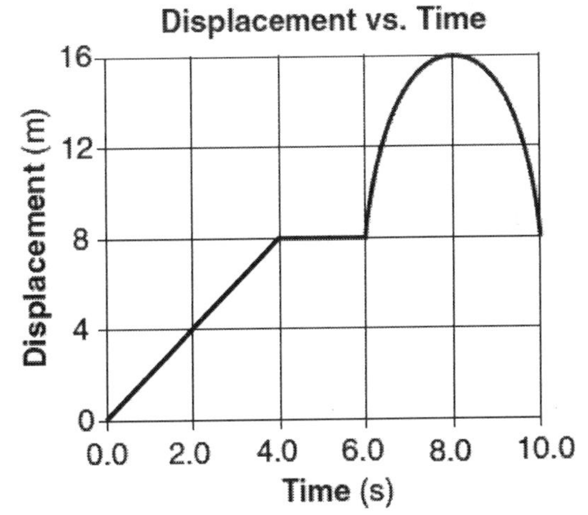

What was the total distance traveled by the object during the 10-second time interval?

1. 0 m
2. 8 m
3. 16 m
4. 24 m

Name: ______________________ Period: __________

Kinematics-Motion Graphs

7. Which graph best represents the relationship between the acceleration of an object falling freely near the surface of Earth and the time that it falls?

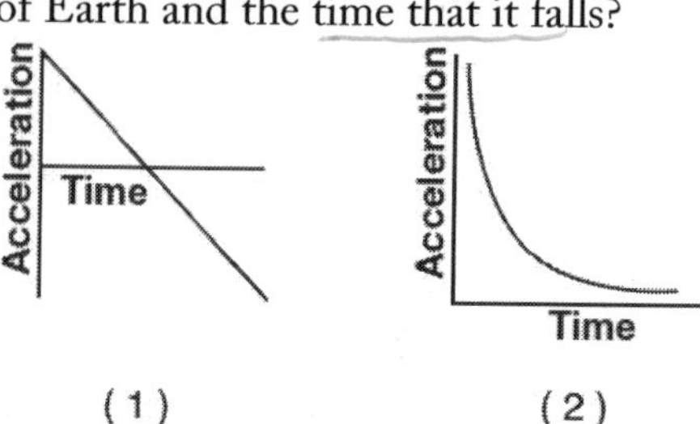

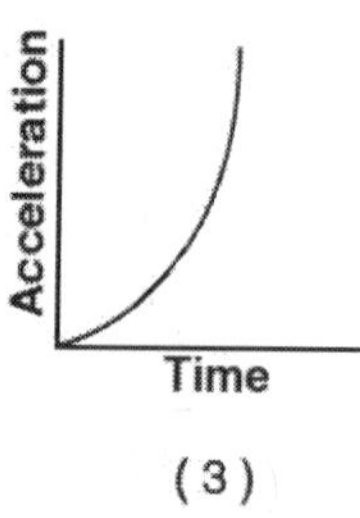

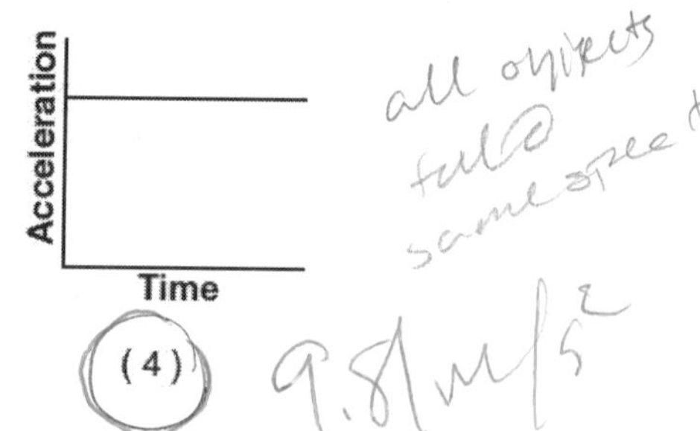

(1) (2) (3) (4)

8. Which pair of graphs represent the same motion of an object?

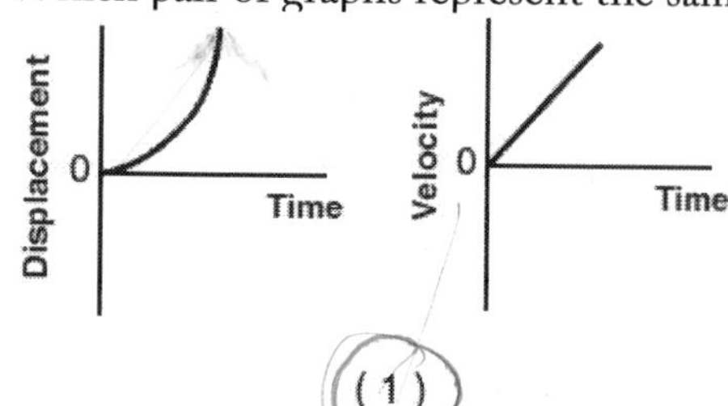

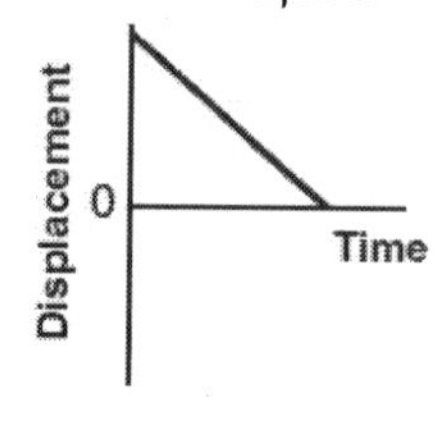

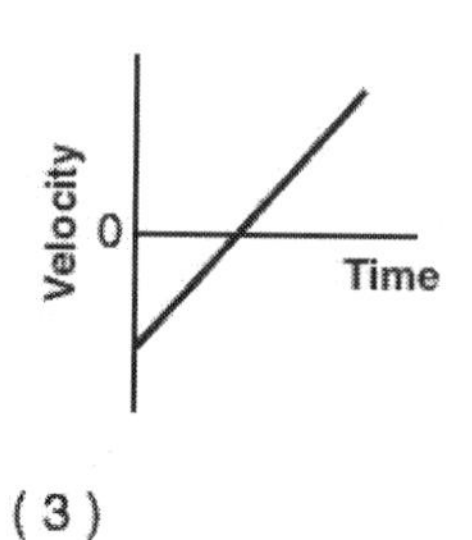

(1) (3)

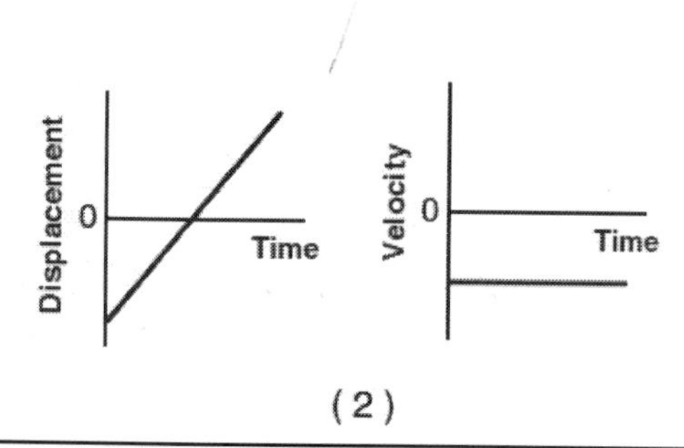

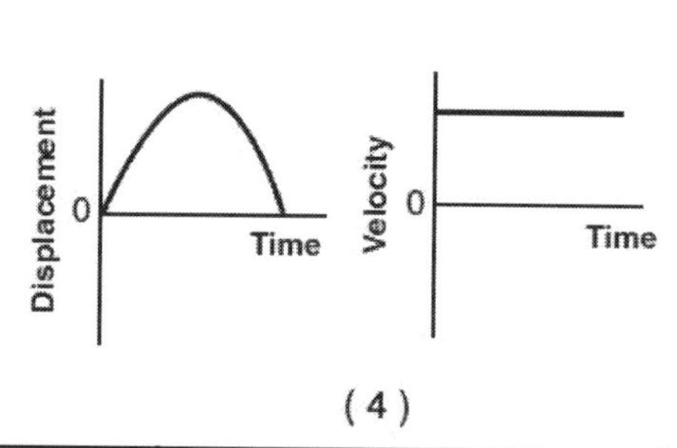

(2) (4)

9. The graph below represents the velocity of an object traveling in a straight line as a function of time.

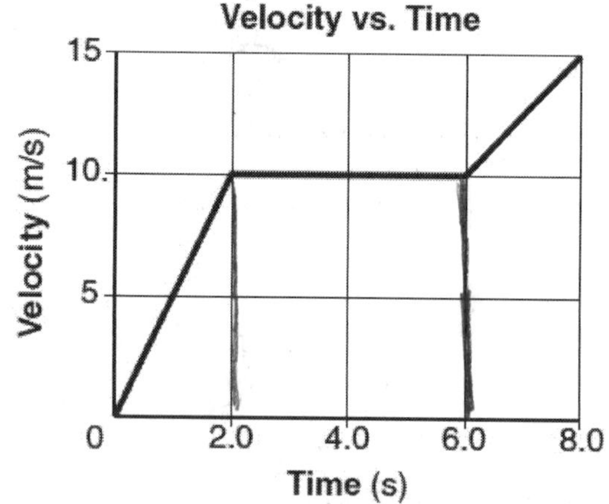

Determine the magnitude of the total displacement of the object at the end of the first 6 seconds.

10. Which graph best represents the motion of a block accelerating uniformly down an inclined plane?

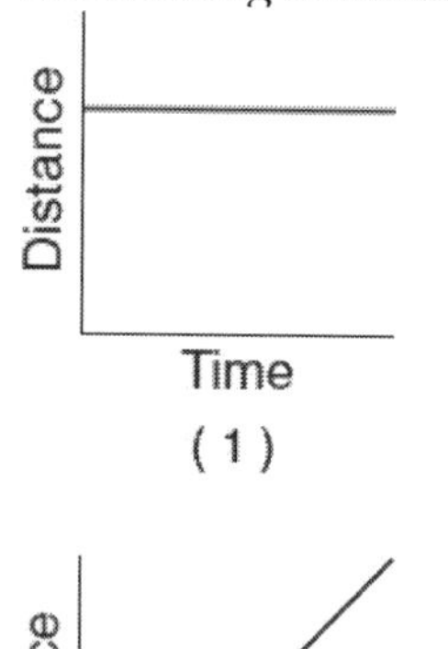

(1)

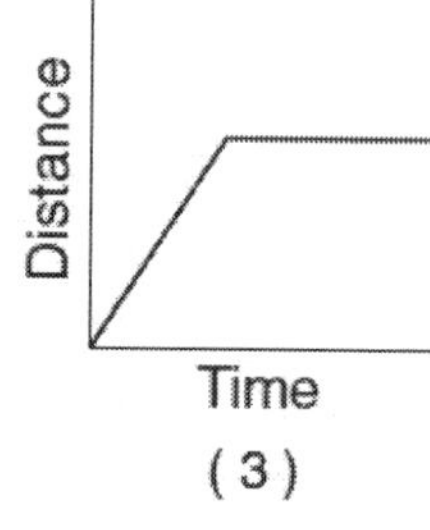

(3)

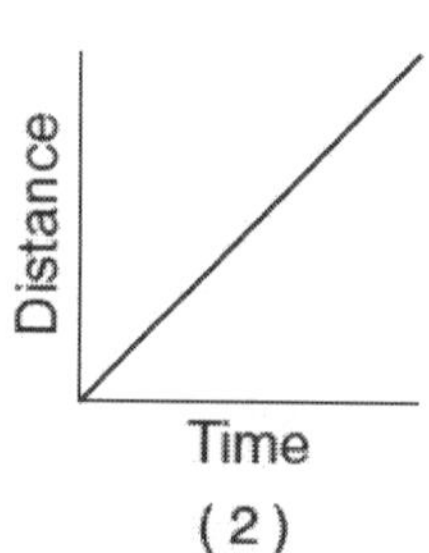

(2)

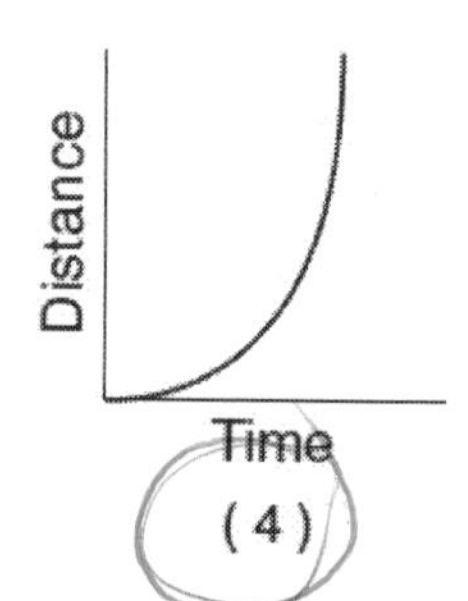

(4)

Name:______________________________ Period: __________

Kinematics-Motion Graphs

Base your answers to questions 11 and 12 on the graph below, which represents the motion of a car during a 6-second time interval.

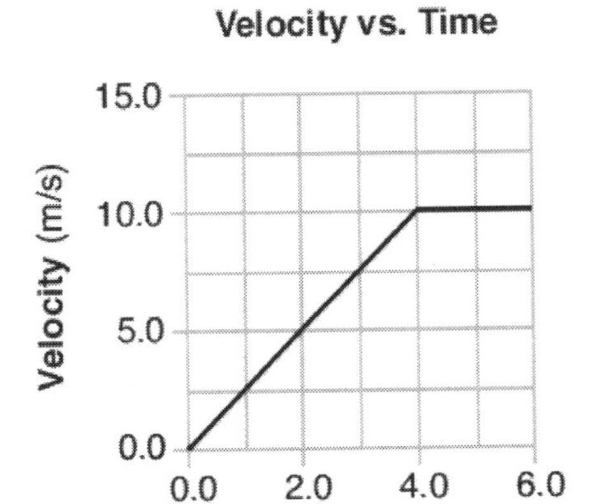

11. What is the acceleration of the car at t=5.0 seconds?
    1. 0.0 m/s^2
    2. 2.0 m/s^2
    3. 2.5 m/s^2
    4. 10 m/s^2

12. What is the total distance traveled by the car during this 6-second interval?
    1. 10 m
    2. 20 m
    3. 40 m
    4. 60 m

13. Which graph best represents the relationship between the velocity of an object thrown straight upward from Earth's surface and the time that elapses while it is in the air? [Neglect friction.]

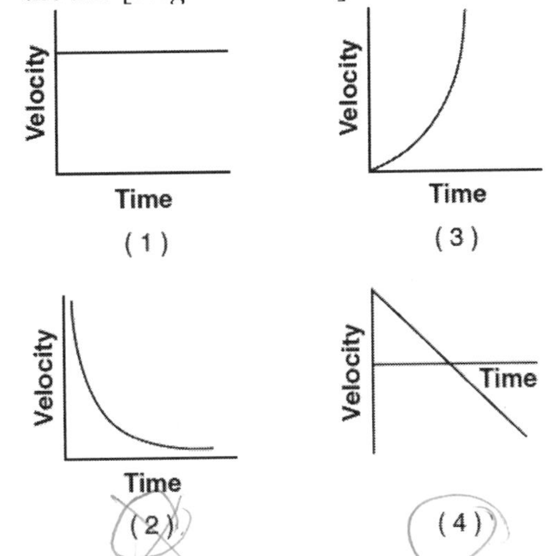

14. The graph below shows the relationship between the speed and elapsed time for an object falling freely from rest near the surface of a planet.

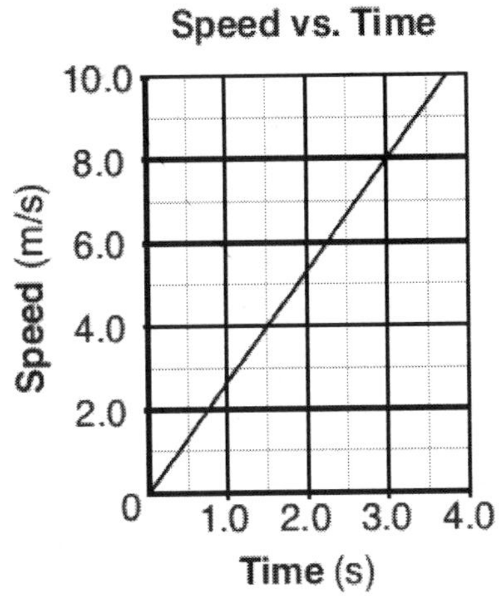

What is the total distance the object falls during the first 3 seconds?
1. 12 m
2. 24 m
3. 44 m
4. 72 m

15. The graph below represents the relationship between speed and time for an object moving along a straight line.

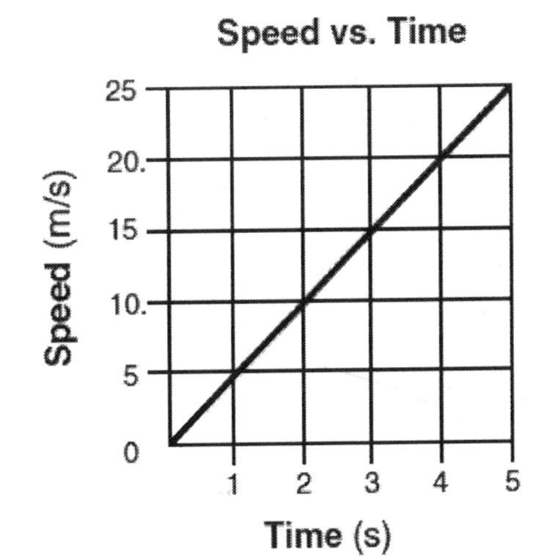

What is the total distance traveled by the object during the first 4 seconds?
1. 5 m
2. 20 m
3. 40 m
4. 80 m

Name:____ Period: _______________

Kinematics-Motion Graphs

Base your answers to questions 16 and 17 on the graph below, which shows the relationship between speed and elapsed time for a car moving in a straight line.

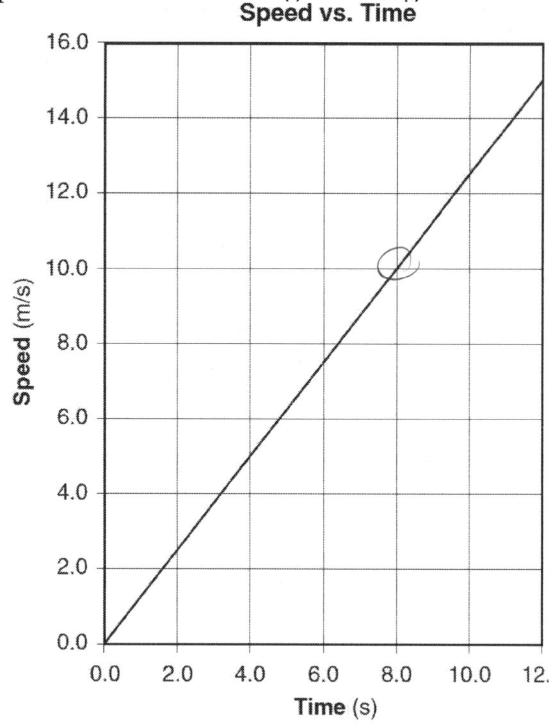

16. Determine the magnitude of the acceleration of the car.

17. Calculate the total distance the car traveled during the time interval 4.0 seconds to 8.0 seconds. [Show all work, including the equation and substitution with units.]

Base your answers to questions 18 through 20 on the graph below, which represents the relationship between velocity and time for a car moving along a straight line, and your knowledge of physics.

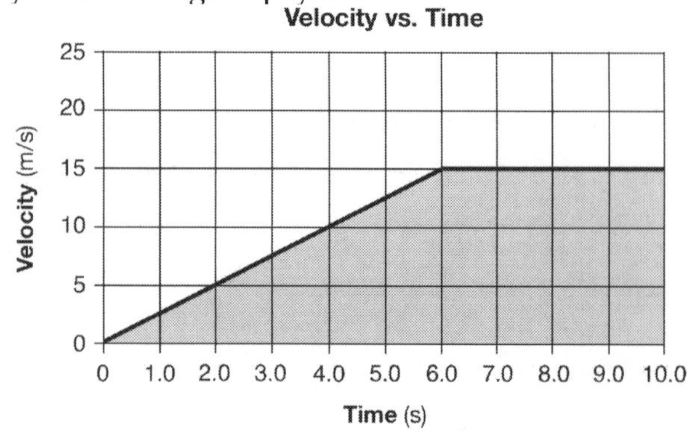

18. Determine the magnitude of the average velocity of the car from t=6.0 seconds to t=10.0 seconds.

19. Determine the magnitude of the car's acceleration during the first 6.0 seconds.

20. Identify the physical quantity represented by the shaded area on the graph.

Name: ______________________________ Period: __________

Kinematics-Horizontal Kinematics

Base your answers to questions 1 and 2 on the information below.

A 747 jet, traveling at a velocity of 70 meters per second north, touches down on a runway. The jet slows to rest at the rate of 2.0 meters per second^2.

1. Calculate the total distance the jet travels on the runway as it is brought to rest. [Show all work, including the equation and substitution with units.]

2. On the diagram below, point P represents the position of the jet on the runway. Beginning at point P, draw a vector to represent the magnitude and direction of the acceleration of the jet as it comes to rest. Use a scale of 1.0 centimeter = 0.50 meter/second^2.

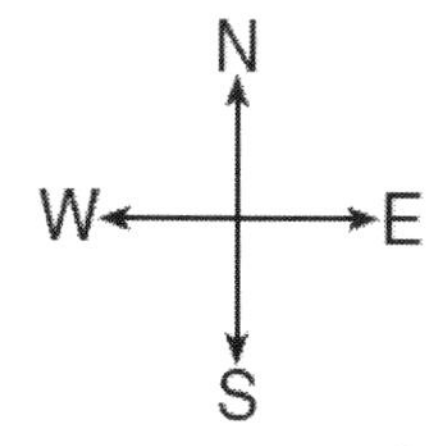

•P

3. An observer recorded the following data for the motion of a car undergoing constant acceleration.

Time (s)	Speed (m/s)
3.0	4.0
5.0	7.0
6.0	8.5

What was the magnitude of the acceleration of the car?
1. 1.3 m/s^2
2. 2.0 m/s^2
3. 1.5 m/s^2
4. 4.5 m/s^2

4. A car traveling on a straight road at 15 meters per second accelerates uniformly to a speed of 21 meters per second in 12 seconds. The total distance traveled by the car in this 12-second time interval is
1. 36 m
2. 180 m
3. 216 m
4. 252 m

5. A race car starting from rest accelerates uniformly at 4.9 m/s^2. What is the car's speed after it has traveled 200 meters?
1. 1960 m/s
2. 62.6 m/s
3. 44.3 m/s
4. 31.3 m/s

Name:____ Period: ______________

Kinematics-Horizontal Kinematics

Base your answers to questions 6 through 9 on the information and diagram below.

A spark timer is used to record the position of a lab cart accelerating uniformly from rest. Each 0.10 second, the timer marks a dot on a recording tape to indicate the position of the cart at that instant, as shown.

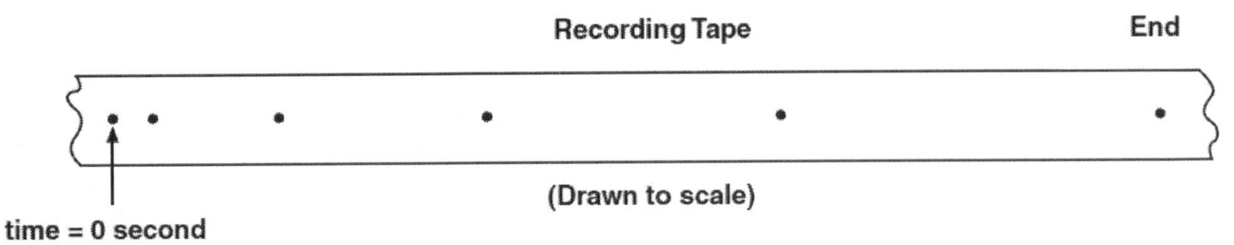

6. Using a metric ruler, measure the distance the cart traveled during the interval t=0 second to t=0.30 second. Record your answer to the nearest tenth of a centimeter.

7. Calculate the magnitude of the acceleration of the cart during the time interval t=0 second to t=0.30 second. [Show all work, including the equation and substitution with units.]

8. Calculate the average speed of the cart during the time interval t=0 second to t=0.30 second. [Show all work, including the equation and substitution with units.]

9. On the diagram below, mark at least four dots to indicate the position of a cart traveling at constant velocity.

Recording Tape

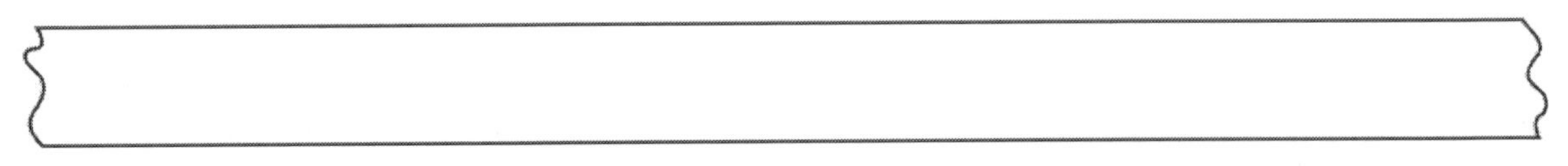

10. A car initially traveling at a speed of 16 meters per second accelerates uniformly to a speed of 20 meters per second over a distance of 36 meters. What is the magnitude of the car's acceleration?
    1. 0.11 m/s²
    2. 2.0 m/s²
    3. 0.22 m/s²
    4. 9.0 m/s²

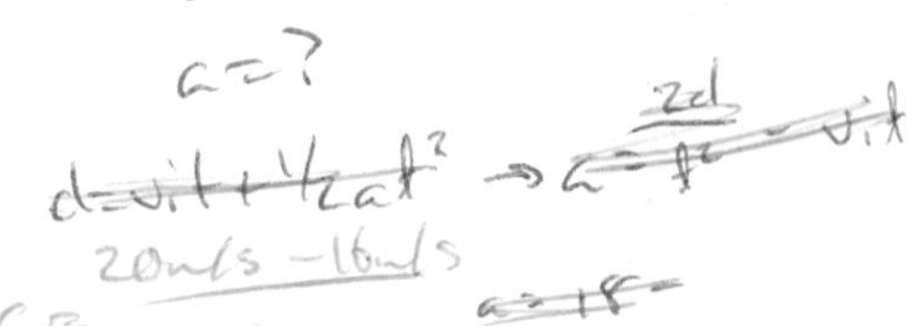

11. An object accelerates uniformly from 3 meters per second east to 8 meters per second east in 2.0 seconds. What is the magnitude of the acceleration of the object?
    1. 2.5 m/s²
    2. 5.0 m/s²
    3. 5.5 m/s²
    4. 11 m/s²

Name: ____________________ Period: __________

Kinematics-Horizontal Kinematics

Base your answers to questions 12 and 13 on the information below.

A physics class is to design an experiment to determine the acceleration of a student on inline skates coasting straight down a gentle incline. The incline has a constant slope. The students have tape measures, traffic cones, and stopwatches.

12. Describe a procedure to obtain the measurements necessary for this experiment.

13. Indicate which equation(s) they should use to determine the student's acceleration.

14. A car increases its speed from 9.6 meters per second to 11.2 meters per second in 4.0 seconds. The average acceleration of the car during this 4-second interval is
    1. 0.40 m/s^2
    2. 2.4 m/s^2
    3. 2.8 m/s^2
    4. 5.2 m/s^2

15. As a car is driven south in a straight line with decreasing speed, the acceleration of the car must be
    1. directed northward
    2. directed southward
    3. zero
    4. constant, but not zero

16. The speed of an object undergoing constant acceleration increases from 8.0 meters per second to 16.0 meters per second in 10 seconds. How far does the object travel during the 10 seconds?
    1. 3.6×10^2 m
    2. 1.6×10^2 m
    3. 1.2×10^2 m
    4. 8.0×10^1 m

Base your answers to questions 17 and 18 on the information below.

A car traveling at a speed of 13 meters per second accelerates uniformly to a speed of 25 meters per second in 5.0 seconds.

17. Calculate the magnitude of the acceleration of the car during this 5.0-second time interval. [Show all work, including the equation and substitution with units.]

18. A truck traveling at a constant speed covers the same total distance as the car in the same 5.0-second time interval. Determine the speed of the truck.

19. If a car accelerates uniformly from rest to 15 meters per second over a distance of 100 meters, the magnitude of the car's acceleration is
    1. 0.15 m/s^2
    2. 1.1 m/s^2
    3. 2.3 m/s^2
    4. 6.7 m/s^2

20. The speed of a wagon increases from 2.5 meters per second to 9.0 meters per second in 3.0 seconds as it accelerates uniformly down a hill. What is the magnitude of the acceleration of the wagon during this 3.0-second interval?
    1. 0.83 m/s^2
    2. 2.2 m/s^2
    3. 3.0 m/s^2
    4. 3.8 m/s^2

Name:______________________________ Period: ____________

Kinematics-Horizontal Kinematics

21. A skater increases her speed uniformly from 2.0 meters per second to 7.0 meters per second over a distance of 12 meters. The magnitude of her acceleration as she travels this 12 meters is
    1. 1.9 m/s^2
    2. 2.2 m/s^2
    3. 2.4 m/s^2
    4. 3.8 m/s^2

22. During a 5.0-second interval, an object's velocity changes from 25 meters per second east to 15 meters per second east. Determine the magnitude and direction of the object's acceleration.

23. A car, initially traveling east with a speed of 5 meters per second, is accelerated uniformly at 2 meters per second^2 east for 10 seconds along a straight line. During this 10-second interval, the car travels a total distance of
    1. 50 m
    2. 60 m
    3. 1.0 × 10^2 m
    4. 1.5 × 10^2 m

24. A child riding a bicycle at 15 meters per second accelerates at -3.0 meters per second^2 for 4.0 seconds. What is the child's speed at the end of this 4.0-second interval?
    1. 12 m/s
    2. 27 m/s
    3. 3.0 m/s
    4. 7.0 m/s

25. A car traveling west in a straight line on a highway decreases its speed from 30.0 meters per second to 23.0 meters per second in 2.00 seconds. The car's average acceleration during this time interval is
    1. 3.5 m/s^2 east
    2. 3.5 m/s^2 west
    3. 13 m/s^2 east
    4. 13 m/s^2 west

26. In a race, a runner traveled 12 meters in 4.0 seconds as she accelerated uniformly from rest. The magnitude of the acceleration of the runner was
    1. 0.25 m/s^2
    2. 1.5 m/s^2
    3. 3.0 m/s^2
    4. 48 m/s^2

27. What is the final speed of an object that starts from rest and accelerates uniformly at 4.0 meters per second^2 over a distance of 8.0 meters?
    1. 8.0 m/s
    2. 16 m/s
    3. 32 m/s
    4. 64 m/s

28. A truck, initially traveling at a speed of 22 meters per second, increases speed at a constant rate of 2.4 meters per second^2 for 3.2 seconds. What is the total distance traveled by the truck during this 3.2-second time interval?
    1. 12 m
    2. 58 m
    3. 70 m
    4. 83 m

29. A car is moving with a constant speed of 20 meters per second. What total distance does the car travel in 2 minutes?
    1. 10 m
    2. 40 m
    3. 1200 m
    4. 2400 m

Name: ______________________ Period: ____________

Kinematics-Horizontal Kinematics

30. A car, initially traveling at 15 m/s north, accelerates to 25 m/s north in 4 seconds. The magnitude of the average acceleration is
    1. 2.5 m/s^2
    2. 6.3 m/s^2
    3. 10 m/s^2
    4. 20 m/s^2

Name:________________________________ Period: ____________

Kinematics-Free Fall

Base your answers to questions 1 through 4 on the information and data table provided.

A 1.00-kilogram mass was dropped from rest from a height of 25 meters above Earth's surface. The speed of the mass was determined at 5-meter intervals and recorded in the data table. Using the information in the data table, construct a graph on the grid, following the directions below.

Data Table

Height Above Earth's Surface (m)	Speed (m/s)
25.0	0.0
20.0	9.9
15.0	14.0
10.0	17.1
5.0	19.8
0	22.1

1. Mark an appropriate scale on the axis labeled "Height Above Earth's Surface (m)."

2. Plot the data points for speed versus height above Earth's surface.

3. Draw the line or curve of best fit.

4. Using your graph, determine the speed of the mass after it has fallen a vertical distance of 12.5 meters.

Speed vs. Height Above Earth's Surface

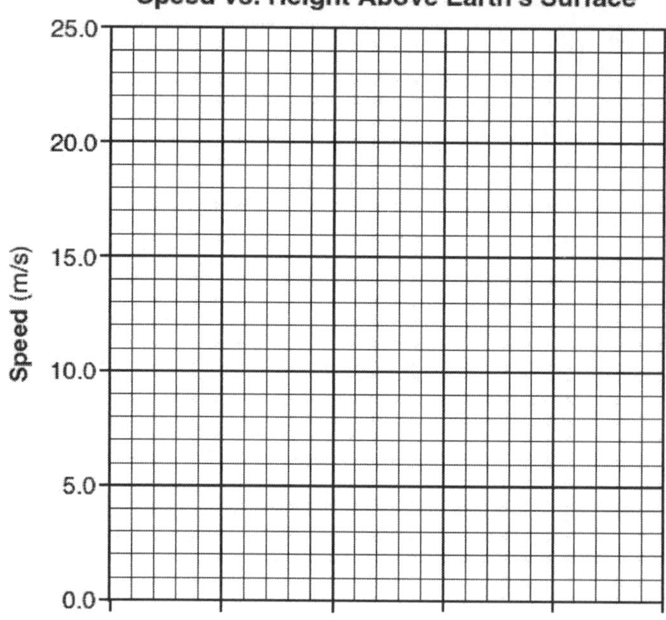

Height Above Earth's Surface (m)

5. A rock is dropped from a bridge. What happens to the magnitude of the acceleration and the speed of the rock as it falls?
   1. Both acceleration and speed increase.
   2. Both acceleration and speed remain the same.
   3. Acceleration increases and speed decreases.
   4. Acceleration remains the same and speed increases.

6. A rock falls from rest a vertical distance of 0.72 meter to the surface of a planet in 0.63 second. The magnitude of the acceleration due to gravity on the planet is
   1. 1.1 m/s^2
   2. 2.3 m/s^2
   3. 3.6 m/s^2
   4. 9.8 m/s^2

7. A ball is thrown straight downward with a speed of 0.50 meter per second from a height of 4.0 meters. What is the speed of the ball 0.70 second after it is released? [Neglect friction.]
   1. 0.50 m/s
   2. 7.4 m/s
   3. 9.8 m/s
   4. 15 m/s

8. An astronaut drops a hammer from 2.0 meters above the surface of the moon. If the acceleration due to gravity on the moon is 1.62 m/s^2, how long will it take for the hammer to fall to the Moon's surface?
   1. 0.62 s
   2. 1.2 s
   3. 1.6 s
   4. 2.5 s

Name: ______________________ Period: __________

Kinematics-Free Fall

9. An object is dropped from rest and falls freely 20 meters to Earth. When is the speed of the object 9.8 meters per second?
   1. during the entire first second of its fall
   2. at the end of its first second of fall
   3. during its entire time of fall
   4. after it had fallen 9.8 meters

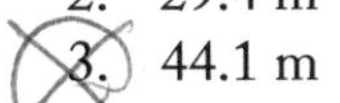

10. A ball is thrown vertically upward with an initial velocity of 29.4 meters per second. What is the maximum height reached by the ball? [Neglect friction.]
   1. 14.7 m
   2. 29.4 m
   3. 44.1 m
   4. 88.1 m

11. A soccer ball kicked on a level field has an initial vertical velocity component of 15 meters per second. Assuming the ball lands at the same height from which it was kicked, what is the total time the ball is in the air? [Neglect friction.]
   1. 0.654 s
   2. 1.53 s
   3. 3.06 s
   4. 6.12 s

12. What is the speed of a 2.5-kilogram mass after it has fallen freely from rest through a distance of 12 meters?
   1. 4.8 m/s
   2. 15 m/s
   3. 30 m/s
   4. 43 m/s

13. A baseball dropped from the roof of a tall building takes 3.1 seconds to hit the ground. How tall is the building? [Neglect friction.]
   1. 15 m
   2. 30 m
   3. 47 m
   4. 94 m

14. A ball dropped from rest falls freely until it hits the ground with a speed of 20 m/s. The time during which the ball is in free fall is approximately
   1. 1 s
   2. 2 s
   3. 0.5 s
   4. 10 s

15. A 25-newton weight falls freely from rest from the roof of a building. What is the total distance the weight falls in the first 1.0 second?
   1. 19.6 m
   2. 9.8 m
   3. 4.9 m
   4. 2.5 m

16. A basketball player jumped straight up to grab a rebound. If she was in the air for 0.80 second, how high did she jump?
   1. 0.50 m
   2. 0.78 m
   3. 1.2 m
   4. 3.1 m

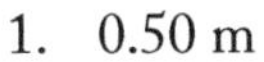

17. How far will a brick starting from rest fall freely in 3.0 seconds?
   1. 15 m
   2. 29 m
   3. 44 m
   4. 88 m

18. A ball thrown vertically upward reaches a maximum height of 30 meters above the surface of Earth. At its maximum height, the speed of the ball is
   1. 0.0 m/s
   2. 3.1 m/s
   3. 9.8 m/s
   4. 24 m/s

19. A rocket initially at rest on the ground lifts off vertically with a constant acceleration of 2.0×10^1 meters per second^2. How long will it take the rocket to reach an altitude of 9.0×10^3 meters?
   1. 3.0×10^1 s
   2. 4.3×10^1 s
   3. 4.5×10^2 s
   4. 9.0×10^2 s

20. Objects in free fall near the surface of Earth accelerate downward at 9.81 meters per second^2. Explain why a feather does not accelerate at this rate when dropped near the surface of Earth.

Name: ______________________________ Period: __________

Kinematics-Free Fall

21. A 1.0-kilogram ball is dropped from the roof of a building 40 meters tall. What is the approximate time of fall? [Neglect air resistance.]
    1. 2.9 s
    2. 2.0 s
    3. 4.1 s
    4. 8.2 s

22. A 0.25-kilogram baseball is thrown upward with a speed of 30 meters per second. Neglecting friction, the maximum height reached by the baseball is approximately
    1. 15 m
    2. 46 m
    3. 74 m
    4. 92 m

23. A rock falls from rest off a high cliff. How far has the rock fallen when its speed is 39.2 meters per second? [Neglect friction.]
    1. 19.6 m
    2. 44.1 m
    3. 78.3 m
    4. 123 m

24. An astronaut standing on a platform on the Moon drops a hammer. If the hammer falls 6.0 meters vertically in 2.7 seconds, what is its acceleration?
    1. 1.6 m/s^2
    2. 2.2 m/s^2
    3. 4.4 m/s^2
    4. 9.8 m/s^2

25. What is the time required for an object starting from rest to fall freely 500. meters near Earth's surface?
    1. 51.0 s
    2. 25.5 s
    3. 10.1 s
    4. 7.14 s

26. A 3.00-kilogram mass is thrown vertically upward with an initial speed of 9.80 meters per second. What is the maximum height this object will reach? [Neglect friction.]
    1. 1.00 m
    2. 4.90 m
    3. 9.80 m
    4. 19.6 m

27. Which diagram best represents the position of a ball, at equal time intervals, as it falls freely from rest near Earth's surface?

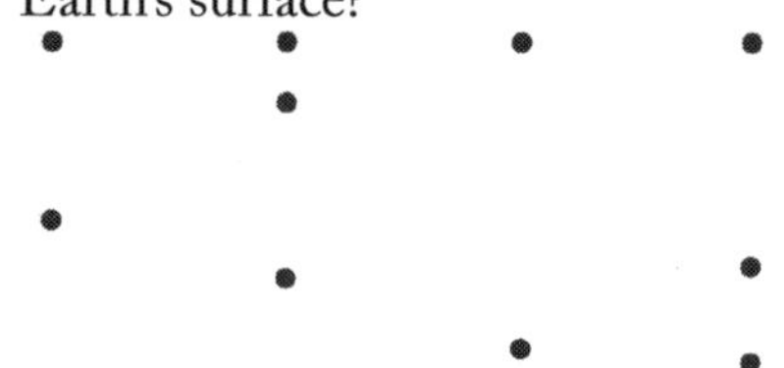

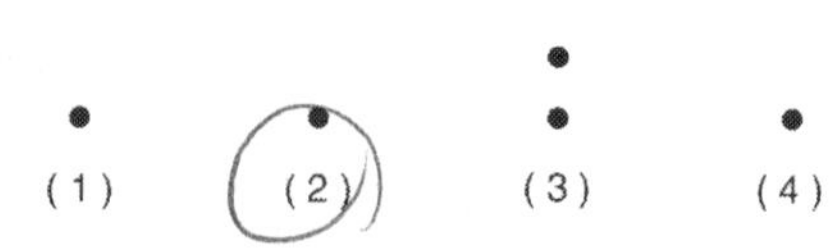

Name: ______________________________ Period: __________

Kinematics-Projectiles

1. A volleyball hit into the air has an initial speed of 10 meters per second. Which vector best represents the angle above the horizontal that the ball should be hit to remain in the air for the greatest amount of time?

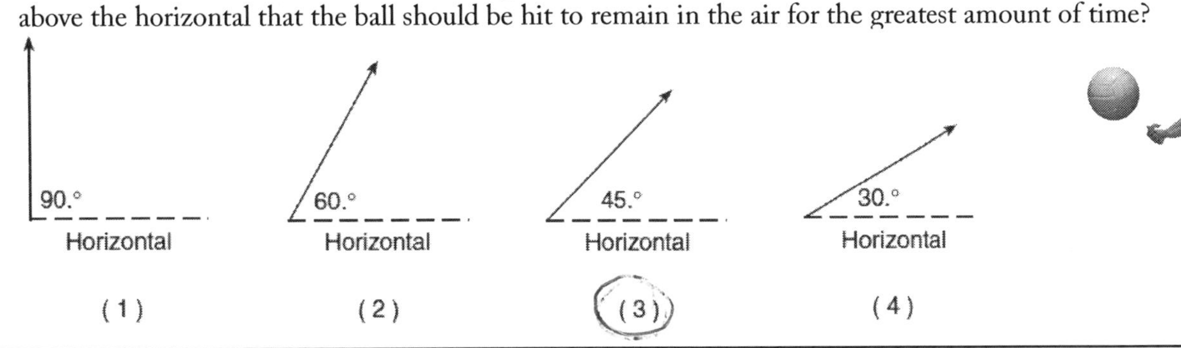

Base your answers to questions 2 through 4 on the information below.

A projectile is fired from the ground with an initial velocity of 250 meters per second at an angle of 60° above the horizontal.

2. On the diagram at right, use a protractor and ruler to draw a vector to represent the initial velocity of the projectile. Begin the vector at P, and use a scale of 1.0 centimeter = 50 meters per second.

3. Determine the horizontal component of the initial velocity.

4. Explain why the projectile has no acceleration in the horizontal direction. [Neglect friction.]

5. Two stones, A and B, are thrown horizontally from the top of a cliff. Stone A has an initial speed of 15 meters per second and stone B has an initial speed of 30 meters per second. Compared to the time it takes stone A to reach the ground, the time it takes stone B to reach the ground is
   1. the same
   2. twice as great
   3. half as great
   4. four times as great

6. The diagram below represents the path of a stunt car that is driven off a cliff, neglecting friction.

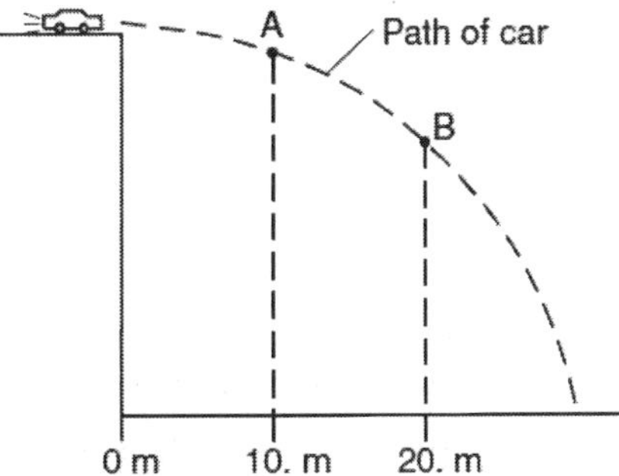

Compared to the horizontal component of the car's velocity at point A, the horizontal component of the car's velocity at point B is
   1. smaller
   2. greater
   3. the same

Name: ______________________________ Period: ____________

Kinematics-Projectiles

Base your answers to questions 7 through 9 on the information and diagram below.

An object was projected horizontally from a tall cliff. The diagram below represents the path of the object, neglecting friction.

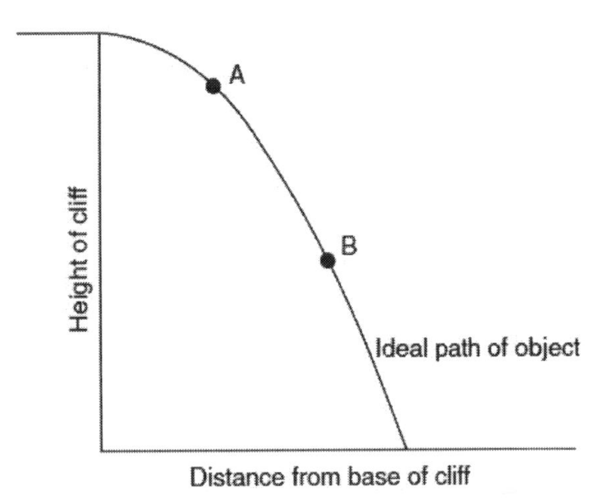

7. How does the magnitude of the horizontal component of the object's velocity at A compare with the magnitude of the horizontal component of the object's velocity at point B?

8. How does the magnitude of the vertical component of the object's velocity at point A compare with the magnitude of the vertical component of the object's velocity at point B?

9. On the diagram above, sketch a likely path of the horizontally projected object, assuming that it was subject to air resistance.

10. A golf ball is propelled with an initial velocity of 60 meters per second at 37° above the horizontal. The horizontal component of the golf ball's initial velocity is
    1. 30 m/s
    2. 36 m/s
    3. 40 m/s
    4. 48 m/s

11. A golf ball is hit with an initial velocity of 15 meters per second at an angle of 35° above the horizontal. What is the vertical component of the golf ball's initial velocity?
    1. 8.6 m/s
    2. 9.8 m/s
    3. 12 m/s
    4. 15 m/s

12. As shown in the diagram below, a student standing on the roof of a 50-meter-high building kicks a stone at a horizontal speed of 4 meters per second.

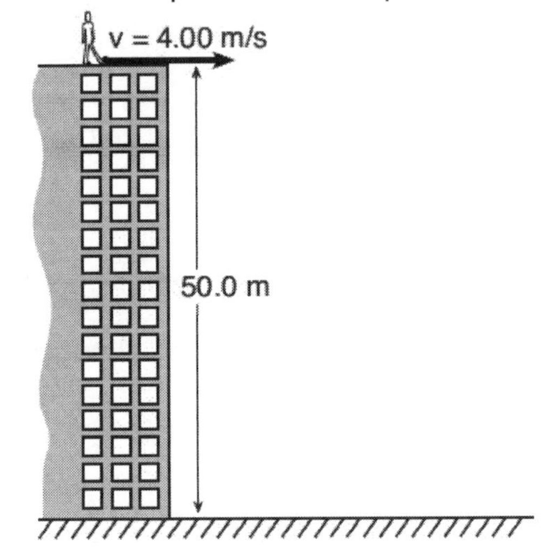

How much time is required for the stone to reach the level ground below? [Neglect friction.]
    1. 3.19 s
    2. 5.10 s
    3. 10.2 s
    4. 12.5 s

13. A machine launches a tennis ball at an angle of 25° above the horizontal at a speed of 14 m/s. The ball returns to level ground. Which combination of changes must produce an increase in time of flight of a second launch?
    1. decrease the launch angle and decrease the ball's initial speed
    2. decrease the launch angle and increase the ball's initial speed
    3. increase the launch angle and decrease the ball's initial speed
    4. increase the launch angle and increase the ball's initial speed

Name: ____________________ Period: __________

Kinematics-Projectiles

Base your answers to questions 14 through 16 on the information below.

A kicked soccer ball has an initial velocity of 25 meters per second at an angle of 40° above the horizontal, level ground. [Neglect friction.]

14. Calculate the magnitude of the vertical component of the ball's initial velocity. [Show all work, including the equation and substitution with units.

15. Calculate the maximum height the ball reaches above its initial position. [Show all work, including the equation and substitution with units.]

16. On the diagram, sketch the path of the ball's flight from its initial position at point P until it returns to level ground.

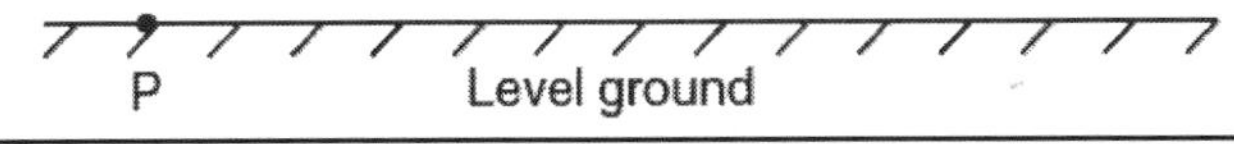

17. Two spheres, A and B, are simultaneously projected horizontally from the top of a tower. Sphere A has a horizontal speed of 40 m/s and sphere B has a horizontal speed of 20 m/s. Which statement best describes the time required for the spheres to reach the ground and the horizontal distance they travel?
    1. Both spheres hit the ground at the same time and at the same distance from the base of the tower.
    2. Both spheres hit the ground at the same time, but sphere A lands twice as far as sphere B from the base of the tower.
    3. Both spheres hit the ground at the same time, but sphere B lands twice as far as sphere A from the base of the tower.
    4. Sphere A hits the ground before sphere B, and sphere A lands twice as far as sphere B from the base of the tower.

Base your answers to questions 18 and 19 on the information below.

An outfielder throws a baseball to second base at a speed of 19.6 m/s and an angle of 30° above the horizontal.

18. Which pair represents the initial horizontal velocity (v_x) and initial vertical velocity (v_y) of the baseball?
    1. v_x = 17.0 m/s and v_y = 9.80 m/s
    2. v_x = 9.80 m/s and v_y = 17.0 m/s
    3. v_x = 19.4 m/s and v_y = 5.90 m/s
    4. v_x = 19.6 m/s and v_y = 19.6 m/s

19. If the ball was caught at the same height from which it was thrown, calculate the amount of time the ball was in the air, showing all work including the equation and substitution with units.

Name: ______________________________ Period: __________

Kinematics-Projectiles

20. Four projectiles, A, B, C, and D, were launched from, and returned to, level ground. The data table below shows the initial horizontal speed, initial vertical speed, and time of flight for each projectile.

Data Table

Projectile	Initial Horizontal Speed (m/s)	Initial Vertical Speed (m/s)	Time of Flight (s)
A	40.0	29.4	6.00
B	60.0	19.6	4.00
C	50.0	24.5	5.00
D	80.0	19.6	4.00

Which projectile traveled the greatest horizontal distance? [Neglect friction.]
1. A
2. B
3. C
4. D

21. Four identical projectiles are launched with the same initial speed, v, but at various angles above the level ground. Which diagram represents the initial velocity of the projectile that will have the largest total horizontal displacement? [Neglect air resistance.]

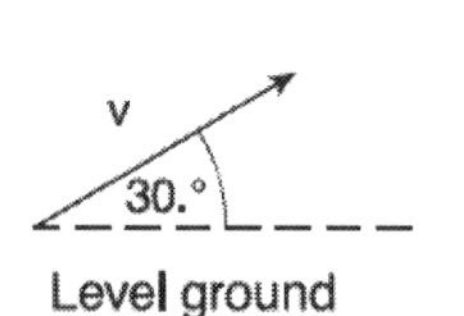

(1)

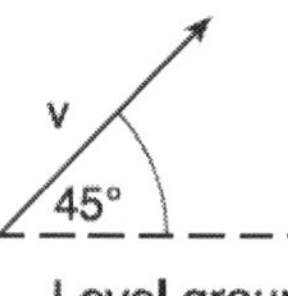

(2)

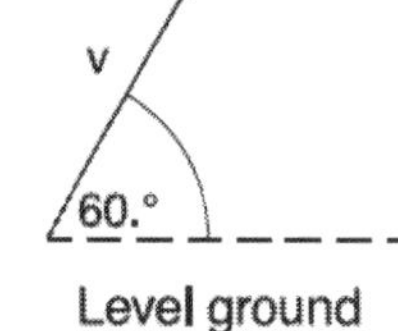

(3)

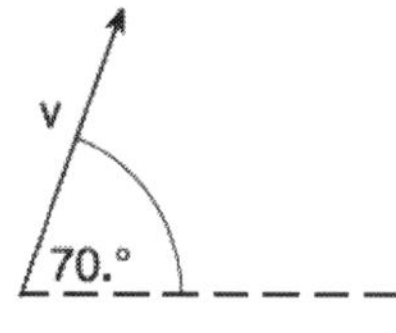

(4)

Base your answers to questions 22 and 23 on the information and diagram below.

A child kicks a ball with an initial velocity of 8.5 meters per second at an angle of 35° with the horizontal, as shown. The ball has an initial vertical velocity of 4.9 meters per second and a total time of flight of 1.0 second. [Neglect air resistance.]

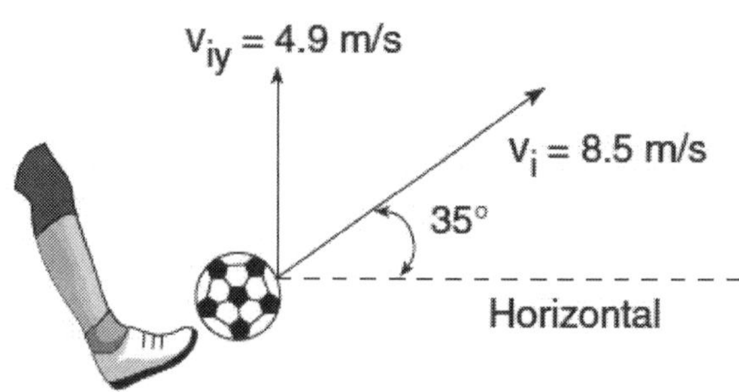

22. What is the horizontal component of the ball's initial velocity?

23. Find the maximum height reached by the ball.

Name:________________________________ Period: ____________

Kinematics-Projectiles

Base your answers to questions 24 through 26 on the information below.

The path of a stunt car driven horizontally off a cliff is represented in the diagram at right. After leaving the cliff, the car falls freely to point A in 0.50 second and to point B in 1.00 second.

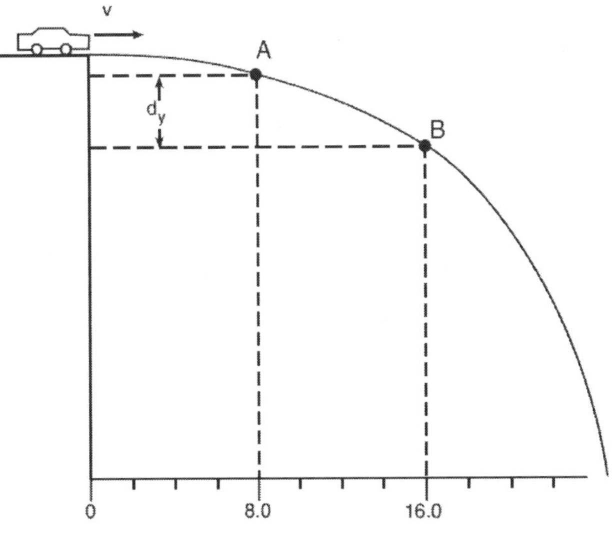

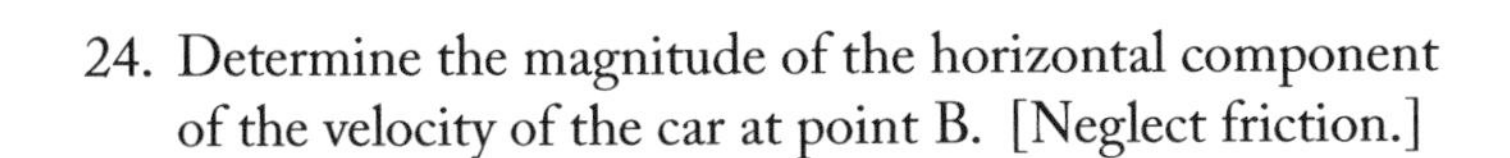

24. Determine the magnitude of the horizontal component of the velocity of the car at point B. [Neglect friction.]

25. Determine the magnitude of the vertical velocity of the car at point A.

26. Calculate the magnitude of the vertical displacement, d_y, of the car from point A to point B. [Neglect friction. Show all work, including the equation and substitution with units.]

Base your answers to questions 27 through 29 on the information and diagram below.

A projectile is launched into the air with an initial speed of v_i at a launch angle of 30° above the horizontal. The projectile lands on the ground 2.0 seconds later.

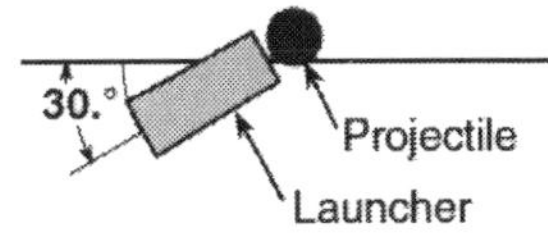

27. On the diagram, sketch the ideal path of the projectile.

28. How does the maximum altitude of the projectile change as the launch angle is increased from 30° to 45° above the horizontal? [Assume the same initial speed, v_i.]

29. How does the total horizontal distance traveled by the projectile change as the launch angle is increased from 30° to 45° above the horizontal? [Assume the same initial speed, v_i.]

Name:______________________________ Period: __________

Kinematics-Projectiles

Base your answers to questions 30 through 32 on the information and graph below.

A machine fired several projectiles at the same angle, θ, above the horizontal. Each projectile was fired with a different initial velocity, v_i. The graph below represents the relationship between the magnitude of the initial vertical velocity, v_{iy}, and the magnitude of the corresponding initial velocity, v_i, of these projectiles.

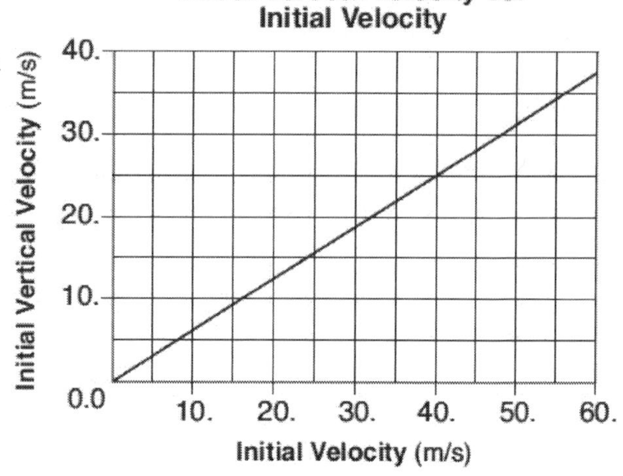

30. Determine the magnitude of the initial vertical velocity of the projectile, v_{iy}, when the magnitude of its initial velocity, v_i, was 40 meters per second.

31. Determine the angle, θ, above the horizontal at which the projectiles were fired.

32. Calculate the magnitude of the initial horizontal velocity of the projectile, v_{ix}, when the magnitude of its initial velocity, v_i, was 40 meters per second. [Show all work, including the equation and substitution with units.]

33. A plane flying horizontally above Earth's surface at 100 meters per second drops a crate. The crate strikes the ground 30 seconds later. What is the magnitude of the horizontal component of the crate's velocity just before it strikes the ground? [Neglect friction.]
    1. 0 m/s
    2. 100 m/s
    3. 294 m/s
    4. 394 m/s

34. The diagram below represents the path of an object after it was thrown.

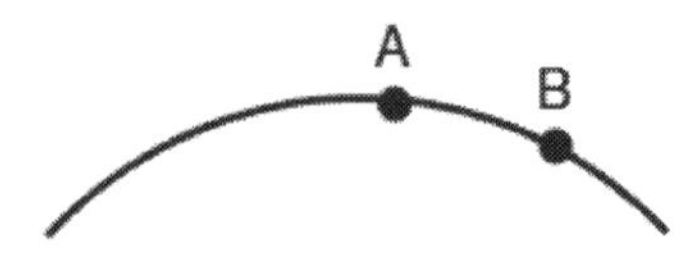

What happens to the object's acceleration as it travels from A to B?
1. It decreases.
2. It increases.
3. It remains the same.

35. A 0.2-kilogram red ball is thrown horizontally at a speed of 4 meters per second from a height of 3 meters. A 0.4-kilogram green ball is thrown horizontally from the same height at a speed of 8 meters per second. Compared to the time it takes the red ball to reach the ground, the time it takes the green ball to reach the ground is
    1. one-half as great
    2. twice as great
    3. the same
    4. four times as great

36. A ball is thrown at an angle of 38° to the horizontal. What happens to the magnitude of the ball's vertical acceleration during the total time interval that the ball is in the air?
    1. It decreases, then increases.
    2. It decreases, then remains the same.
    3. It increases, then decreases.
    4. It remains the same.

Name: ______________________________ Period: __________

Kinematics-Projectiles

Base your answers to questions 37 through 39 on the information and diagram below.

A projectile is launched horizontally at a speed of 30 meters per second from a platform located a vertical distance h above the ground. The projectile strikes the ground after time t at horizontal distance d from the base of the platform. [Neglect friction.]

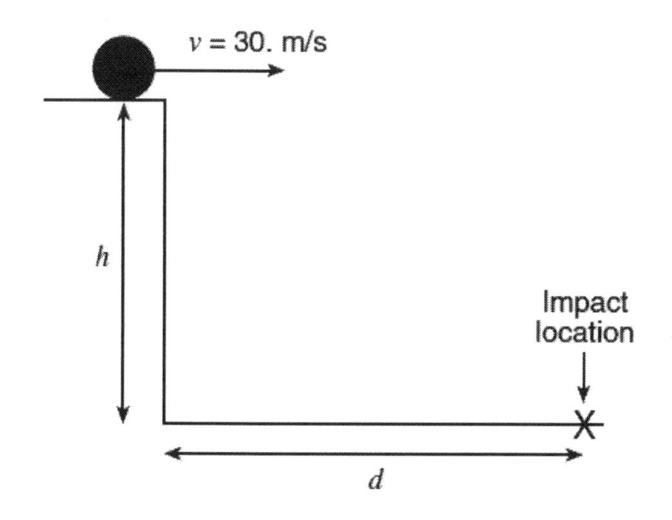

37. On the diagram above, sketch the theoretical path of the projectile.

38. Calculate the horizontal distance, d, if the projectile's time of flight is 2.5 seconds. [Show all work, including the equation and substitution with units.]

39. Express the projectile's total time of flight, t, in terms of the vertical distance, h, and the acceleration due to gravity, g. [Write an appropriate equation and solve it for t.]

40. A ball is thrown horizontally at a speed of 24 meters per second from the top of a cliff. If the ball hits the ground 4.0 seconds later, approximately how high is the cliff?
    1. 6.0 m
    2. 39 m
    3. 78 m
    4. 96 m

41. A golf ball is given an initial speed of 20 meters per second and returns to level ground. Which launch angle above level ground results in the ball traveling the greatest horizontal distance? [Neglect friction.]
    1. 60°
    2. 45°
    3. 30°
    4. 15°

42. A golf ball is hit at an angle of 45° above the horizontal. What is the acceleration of the golf ball at the highest point in its trajectory? [Neglect friction.]
    1. 9.8 m/s^2 upward
    2. 9.8 m/s^2 downward
    3. 6.9 m/s^2 horizontal
    4. 0.0 m/s^2

43. A soccer player kicks a ball with an initial velocity of 10 meters per second at an angle of 30° above the horizontal. The magnitude of the horizontal component of the ball's initial velocity is
    1. 5.0 m/s
    2. 8.7 m/s
    3. 9.8 m/s
    4. 10 m/s

44. A vector makes an angle, θ, with the horizontal. The horizontal and vertical components of the vector will be equal in magnitude if angle θ is
    1. 30°
    2. 45°
    3. 60°
    4. 90°

Name: ______________________ Period: __________

Kinematics-Projectiles

45. The diagram below represents a setup for demonstrating motion.

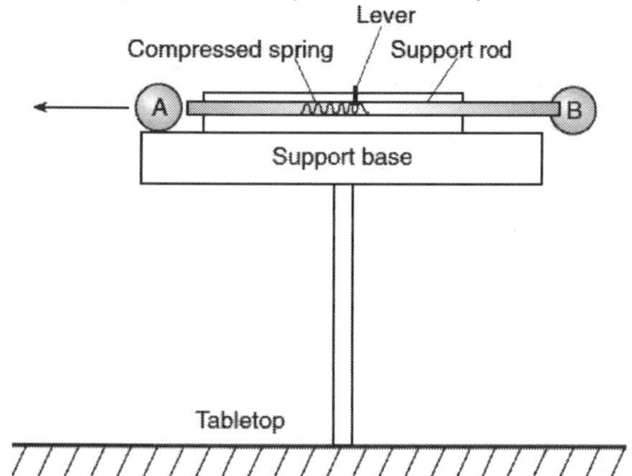

When the lever is released, the support rod withdraws from ball B, allowing it to fall. At the same instant, the rod contacts ball A, propelling it horizontally to the left. Which statement describes the motion that is observed after the lever is released and the balls fall? [Neglect friction.]
1. Ball A travels at constant velocity.
2. Ball A hits the tabletop at the same time as ball B.
3. Ball B hits the tabletop before ball A.
4. Ball B travels with an increasing acceleration.

46. A projectile launched at an angle of 45° above the horizontal travels through the air. Compared to the projectile's theoretical path with no air friction, the actual trajectory of the projectile with air friction is
1. lower and shorter
2. lower and longer
3. higher and shorter
4. higher and longer

47. A projectile is fired with an initial velocity of 120 meters per second at an angle, θ, above the horizontal. If the projectile's initial horizontal speed is 55 meters per second, then angle θ measures approximately
1. 13°
2. 27°
3. 63°
4. 75°

48. A baseball is thrown at an angle of 40° above the horizontal. The horizontal component of the baseball's initial velocity is 12 meters per second. What is the magnitude of the ball's initial velocity?
1. 7.71 m/s
2. 9.20 m/s
3. 15.7 m/s
4. 18.7 m/s

49. A toy rocket is launched twice into the air from level ground and returns to level ground. The rocket is first launched with initial speed v at an angle of 45° above the horizontal. It is launched the second time with the same initial speed, but with the launch angle increased to 60° above the horizontal. Describe how both the total horizontal distance the rocket travels and the time in the air are affected by the increase in launch angle. [Neglect friction.]

50. A projectile is launched at an angle above the ground. The horizontal component of the projectile's velocity, v_x, is initially 40 meters per second. The vertical component of the projectile's velocity, v_y, is initially 30 meters per second. What are the components of the projectile's velocity after 2.0 seconds of flight? [Neglect friction.]
1. v_x=40 m/s and v_y=10 m/s
2. v_x=40 m/s and v_y=30 m/s
3. v_x=20 m/s and v_y=10 m/s
4. v_x=20 m/s and v_y=30 m/s

Name: ______________________________ Period: ____________

Kinematics-Projectiles

51. A ball is thrown with an initial speed of 10 meters per second. At what angle above the horizontal should the ball be thrown to reach the greatest height?
    1. 0°
    2. 30°
    3. 45°
    4. 90°

52. The components of a 15-meters-per-second velocity at an angle of 60° above the horizontal are
    1. 7.5 m/s vertical and 13 m/s horizontal
    2. 13 m/s vertical and 7.5 m/s horizontal
    3. 6.0 m/s vertical and 9.0 m/s horizontal
    4. 9.0 m/s vertical and 6.0 m/s horizontal

53. Which combination of initial horizontal velocity, (v_H) and initial vertical velocity, (v_V) results in the greatest horizontal range for a projectile over level ground? [Neglect friction.]

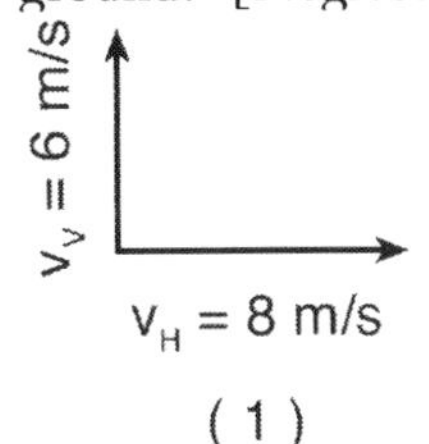

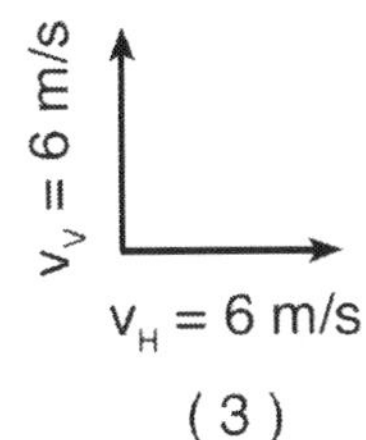

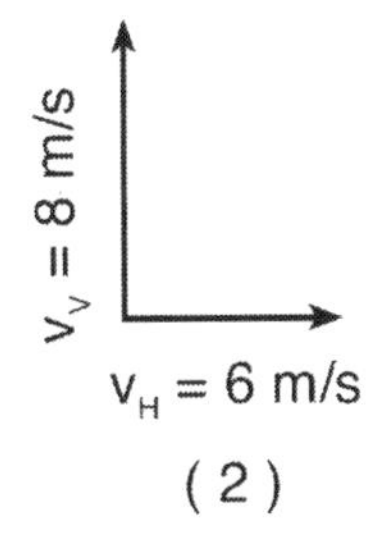

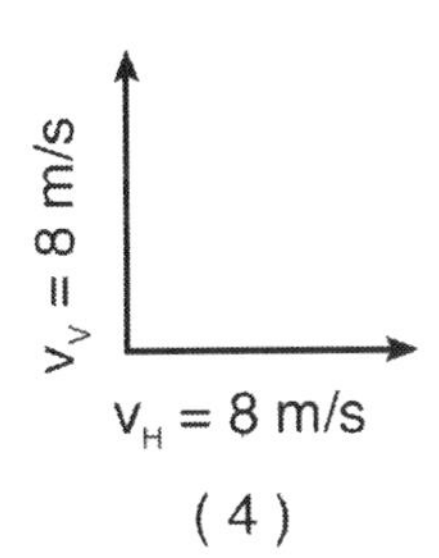

54. Without air resistance, a kicked ball would reach a maximum height of 6.7 meters and land 38 meters away. With air resistance, the ball would travel
    1. 6.7 m vertically and more than 38 m horizontally
    2. 38 m horizontally and less than 6.7 m vertically
    3. more than 6.7 m vertically and less than 38 m horizontally
    4. less than 38 m horizontally and less than 6.7 m vertically

Base your answers to questions 55 and 56 on the information below and on your knowledge of physics.

A football is thrown at an angle of 30° above the horizontal. The magnitude of the horizontal component of the ball's initial velocity is 13.0 meters per second. The magnitude of the vertical component of the ball's initial velocity is 7.5 meters per second. [Neglect friction.]

55. On the axes below, draw a graph representing the relationship between the horizontal displacement of the football and the time the football is in the air.

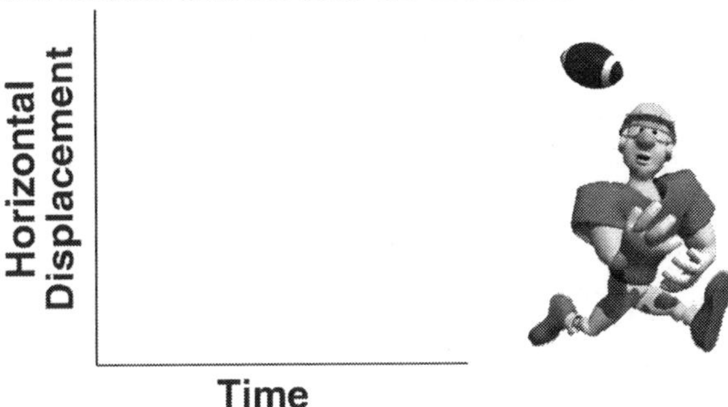

56. The football is caught at the same height from which it is thrown. Calculate the total time the football was in the air. [Show all work, including the equation and substitution with units.]

Name: ______________________________ Period: __________

Dynamics-Newton's 1st Law

1. As shown in the diagram, an open box and its contents have a combined mass of 5.0 kilograms. A horizontal force of 15 newtons is required to push the box at a constant speed of 1.5 meters per second across a level surface.

v = 1.5 m/s
Box
F = 15 N
5.0 kg
Level surface

The inertia of the box and its contents increases if there is an increase in the
   1. speed of the box
   2. mass of the contents of the box
   3. magnitude of the horizontal force applied to the box
   4. coefficient of kinetic friction between the box and the level surface

2. Which unit is equivalent to a newton per kilogram?
   1. m/s^2
   2. W/m
   3. J·s
   4. kg·m/s

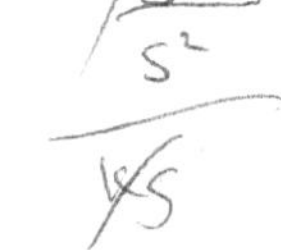

3. Which object has the most inertia?
   1. A 0.001-kilogram bumblebee traveling at 2 meters per second
   2. A 0.1-kilogram baseball traveling at 20 meters per second
   3. A 5-kilogram bowling ball traveling at 3 meters per second
   4. A 10-kilogram sled at rest

4. If the sum of all the forces acting on a moving object is zero, the object will
   1. slow down and stop
   2. change the direction of its motion
   3. accelerate uniformly
   4. continue moving with constant velocity

5. The mass of a high school football player is approximately
   1. 10^0 kg
   2. 10^1 kg
   3. 10^2 kg
   4. 10^3 kg

6. Which object has the greatest inertia?
   1. A 5-kg mass moving at 10 m/s
   2. A 10-kg mass moving at 1 m/s
   3. A 15-kg mass moving at 10 m/s
   4. A 20-kg mass moving at 1 m/s

7. The data table below lists the mass and speed of four different objects

Data Table

Object	Mass (kg)	Speed (m/s)
A	4.0	6.0
B	6.0	5.0
C	8.0	3.0
D	16.0	1.5

Which object has the greatest inertia?
   1. A
   2. B
   3. C
   4. D

8. A 0.50-kilogram cart is rolling at a speed of 0.40 meter per second. If the speed of the cart is doubled, the inertia of the cart is
   1. halved
   2. doubled
   3. quadrupled
   4. unchanged

9. Which person has the greatest inertia?
   1. A 110-kg wrestler resting on a mat
   2. A 90-kg man walking at 2 m/s
   3. A 70-kg long-distance runner traveling 5 m/s
   4. A 50-kg girl sprinting at 10 m/s

10. Which object has the greatest inertia?
   1. a falling leaf
   2. a softball in flight
   3. a seated high school student
   4. a rising helium-filled toy balloon

Name: ______________________ Period: __________

Dynamics-Newton's 1st Law

11. A lab cart is loaded with different masses and moved at various velocities. Which diagram shows the cart-mass system with the greatest inertia?

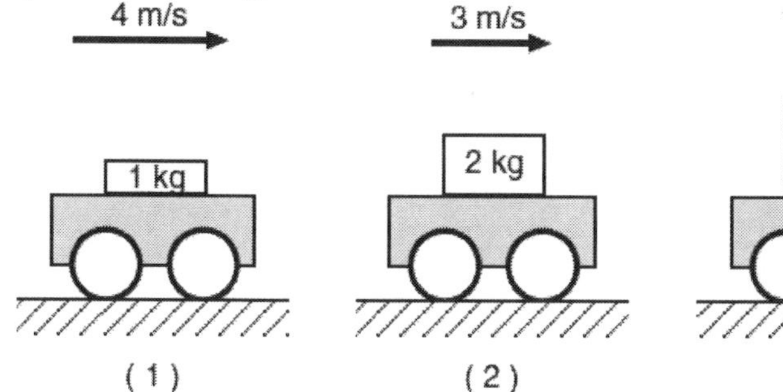

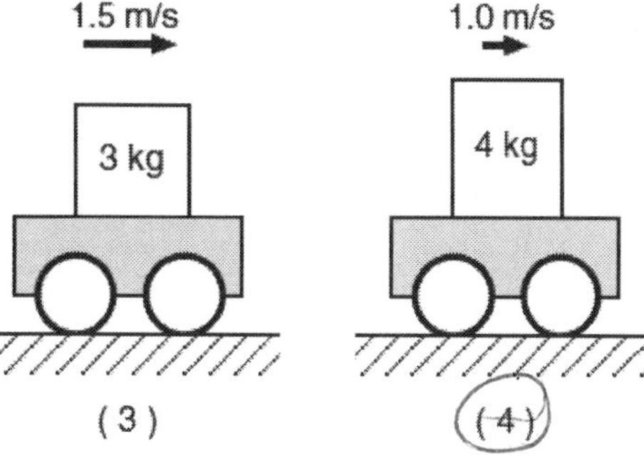

12. Which object has the greatest inertia?
    1. A 5-kg object moving at 5 m/s
    2. A 10-kg object moving at 3 m/s
    3. A 15-kg object moving at 1 m/s
    4. A 20-kg object at rest

13. A force of 1 newton is equivalent to 1
    1. $kg \cdot m/s^2$
    2. $kg \cdot m/s$
    3. $kg \cdot m^2/s^2$
    4. $kg^2 \cdot m^2/s^2$

14. Which object has the greatest inertia?
    1. a 1-kg object moving at 15 m/s
    2. a 5-kg object at rest
    3. a 10-kg object moving at 2 m/s
    4. a 15-kg object at rest

15. Which cart has the greatest inertia?
    1. a 1-kg cart traveling at 4 m/s
    2. a 2-kg cart traveling at 3 m/s
    3. a 3-kg cart traveling at 2 m/s
    4. a 4-kg cart traveling at 1 m/s

16. Which object has the greatest inertia?
    1. a 15-kg mass traveling at 5 m/s
    2. a 10-kg mass traveling at 10 m/s
    3. a 10-kg mass traveling at 5 m/s
    4. a 5-kg mass traveling at 15 m/s

17. Which object has the greatest inertia?
    1. a 0.010-kg bullet traveling at 90 m/s
    2. a 30-kg child traveling at 10 m/s on her bike
    3. a 490-kg elephant walking with a speed of 1 m/s
    4. a 1500-kg car at rest in a parking lot

18. A 15-kilogram cart is at rest on a horizontal surface. A 5-kilogram box is placed in the cart. Compared to the mass and inertia of the cart, the cart-box system has
    1. more mass and more inertia
    2. more mass and the same inertia
    3. the same mass and more inertia
    4. less mass and more inertia

19. A different force is applied to each of four different blocks on a frictionless, horizontal surface. In which diagram does the block have the greatest inertia 2.0 seconds after starting from rest?

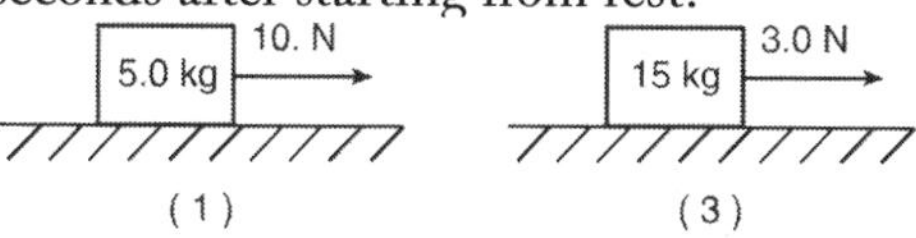

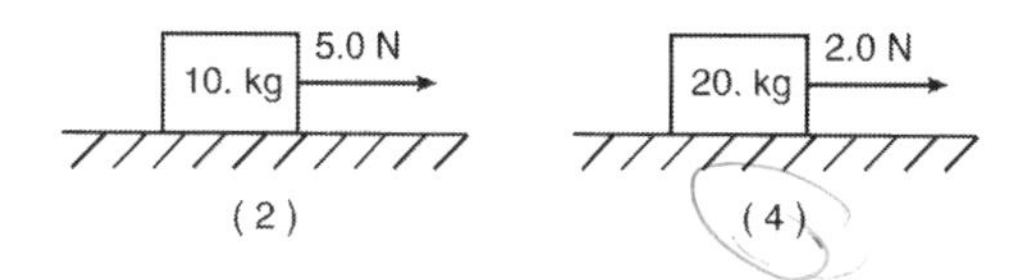

Name:_______________ Period: _______________

Dynamics-Newton's 2nd Law

1. A constant unbalanced force is applied to an object for a period of time. Which graph best represents the acceleration of the object as a function of elapsed time?

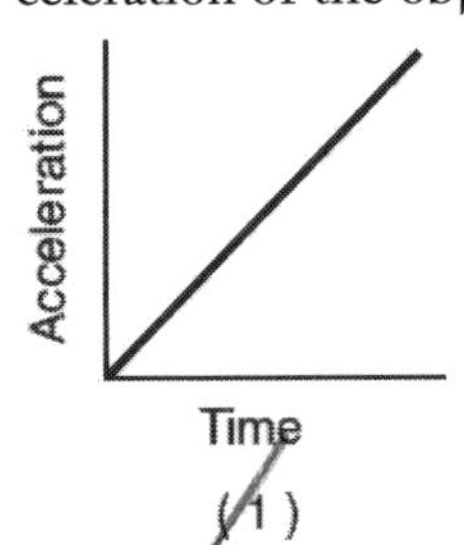

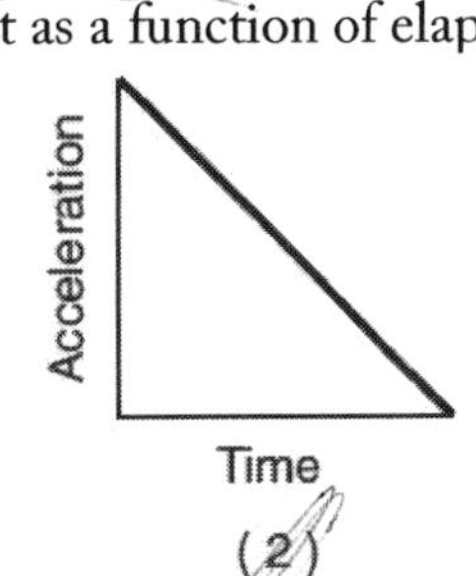

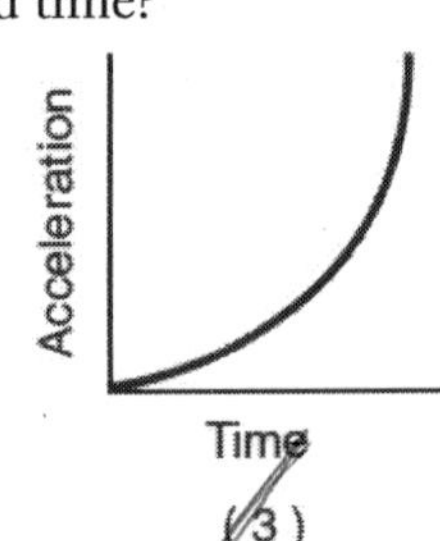

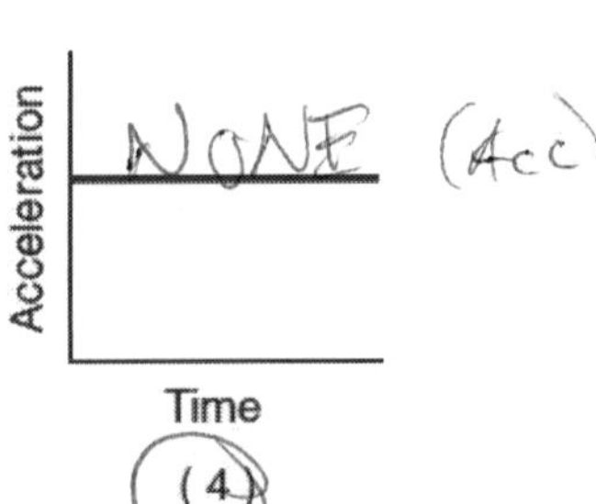

2. The diagram below shows a horizontal 12-newton force being applied to two blocks, A and B, initially at rest on a horizontal, frictionless surface. Block A has a mass of 1 kilogram and block B has a mass of 2 kilograms.

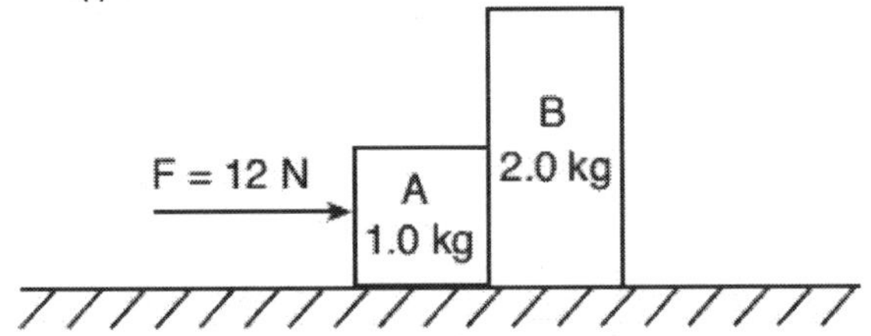

Frictionless surface

The magnitude of the acceleration of block B is
1. 6.0 m/s^2
2. 2.0 m/s^2
3. 3.0 m/s^2
4. 4.0 m/s^2

3. Which body is in equilibrium?
1. a satellite moving around Earth in a circular orbit
2. a cart rolling down a frictionless incline
3. an apple falling freely toward the surface of Earth
4. a block sliding at constant velocity across a tabletop

4. The weight of a typical high school physics student is closest to
1. 1500 N
2. 600 N
3. 120 N
4. 60 N

Base your answers to questions 5 and 6 on the diagram below, which shows a 1-newton metal disk resting on an index card that is balanced on top of a glass.

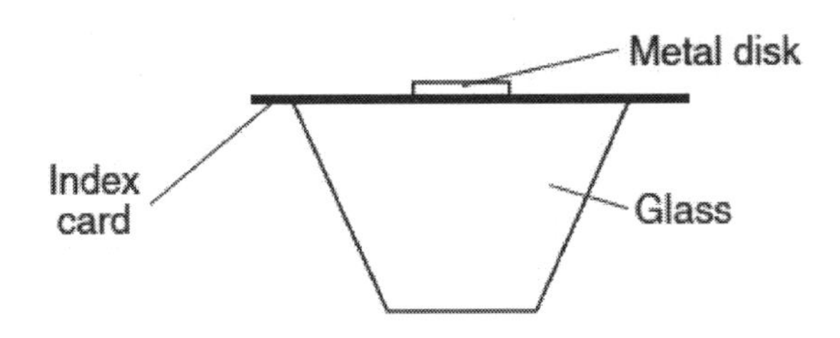

5. What is the net force acting on the disk?
1. 1.0 N
2. 2.0 N
3. 0 N
4. 9.8 N

6. When the index card is quickly pulled away from the glass in a horizontal direction, the disk falls straight down into the glass. This action is a result of the disk's
1. inertia
2. charge
3. shape
4. temperature

7. A student is standing in an elevator that is accelerating downward. The force that the student exerts on the floor of the elevator must be
1. less than the weight of the student when at rest
2. greater than the weight of the student when at rest
3. less than the force of the floor on the student
4. greater than the force of the floor on the student

Name: ________________________ Period: ____________

Dynamics-Newton's 2nd Law

8. The diagram below represents two concurrent forces.

Which vector below represents the force that will produce equilibrium with these two forces?

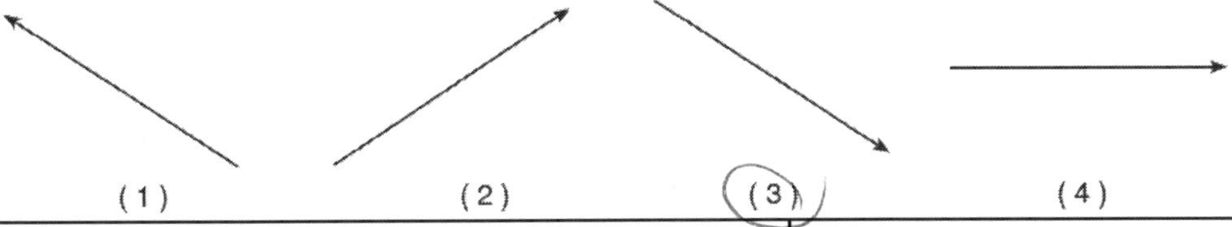

9. In the diagram below, a 20-newton force due north and a 20-newton force due east act concurrently on an object, as shown in the diagram below.

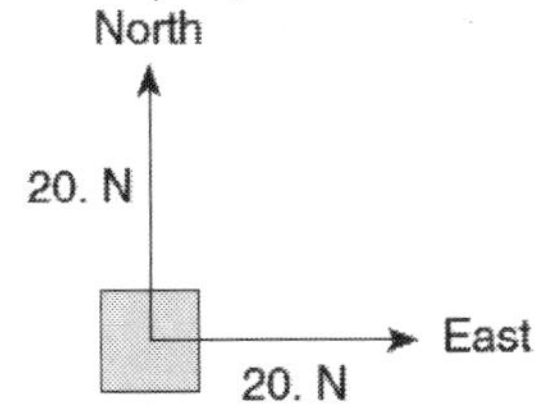

The additional force necessary to bring the object into a state of equilibrium is
1. 20 N, northeast
2. 20 N, southwest
3. 28 N, northeast
4. 28 N, southwest

10. A man standing on a scale in an elevator notices that the scale reads 30 newtons greater than his normal weight. Which type of movement of the elevator could cause this greater-than-normal reading?
1. accelerating upward
2. accelerating downward
3. moving upward at constant speed
4. moving downward at constant speed

11. Two forces, F_1 and F_2, are applied to a block on a frictionless, horizontal surface as shown below.

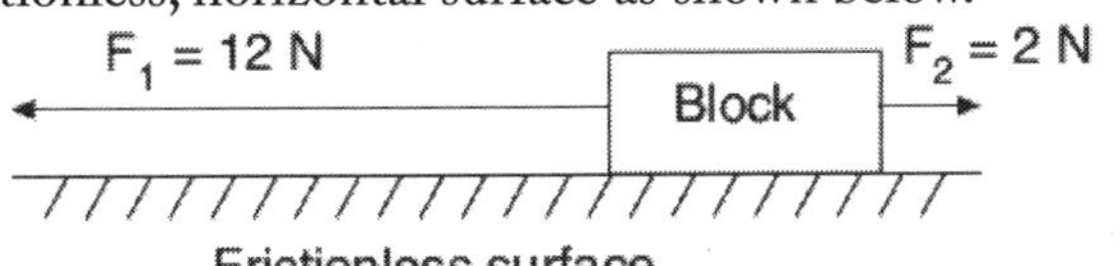

If the magnitude of the block's acceleration is 2 m/s², what is the mass of the block?
1. 1 kg
2. 5 kg
3. 6 kg
4. 7 kg

Base your answers to questions 12 and 13 on the information and diagram below.

A soccer ball is kicked from point P_i at an angle above a horizontal field. The ball follows an ideal path before landing on the field at point P_f.

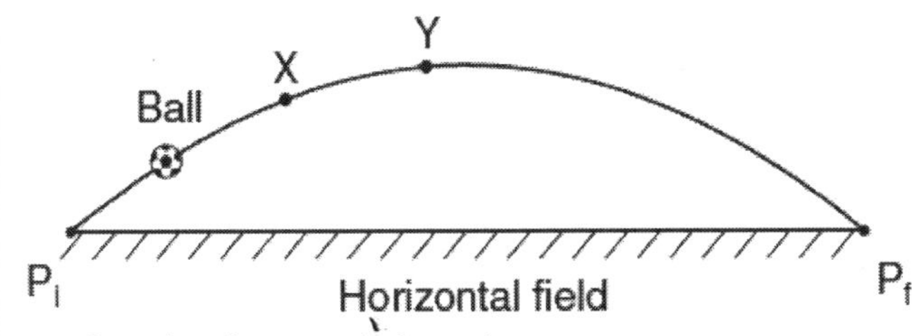

12. On the diagram below, draw an arrow to represent the direction of the net force on the ball when it is at position X. Label the arrow F_{net}. [Neglect friction.]

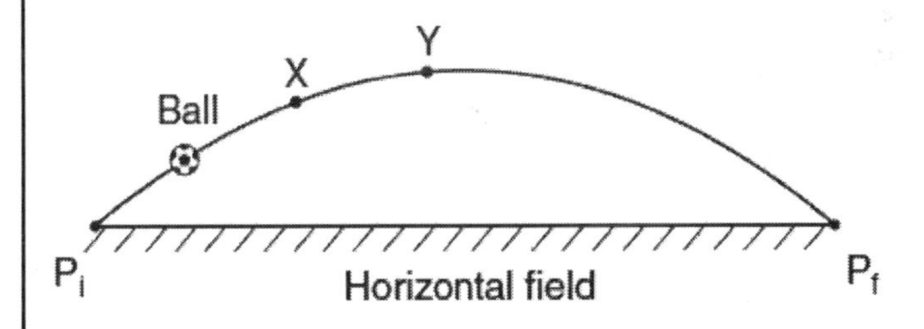

13. On the diagram above, draw an arrow to represent the direction of the acceleration of the ball at position Y. Label the arrow *a*. [Neglect friction.]

14. A 5-newton force could have perpendicular components of
1. 1 N and 4 N
2. 2 N and 3 N
3. 3 N and 4 N
4. 5 N and 5 N

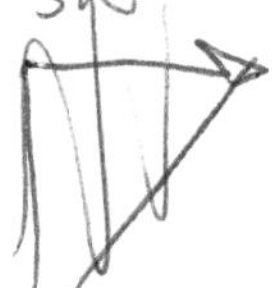

Name: ______________________ Period: __________

Dynamics-Newton's 2nd Law

15. Which graph best represents the motion of an object in equilibrium?

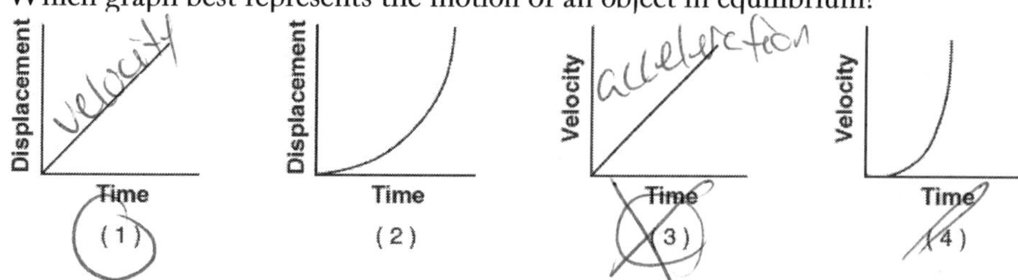

16. Which diagram represents a box in equilibrium?

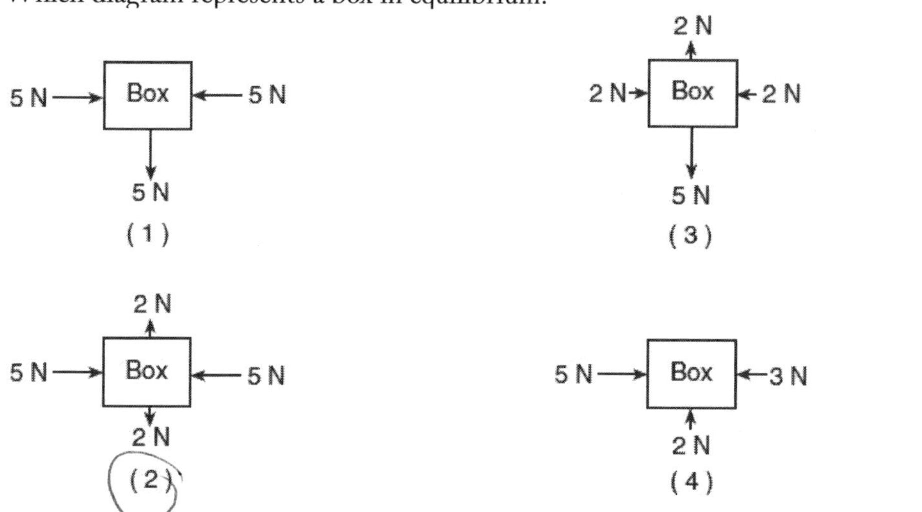

17. The diagram below shows a 5-kilogram block at rest on a horizontal, frictionless table.

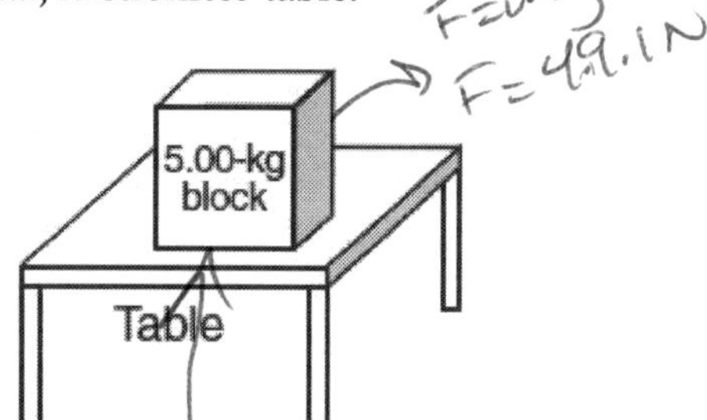

Which diagram best represents the force exerted on the block by the table?

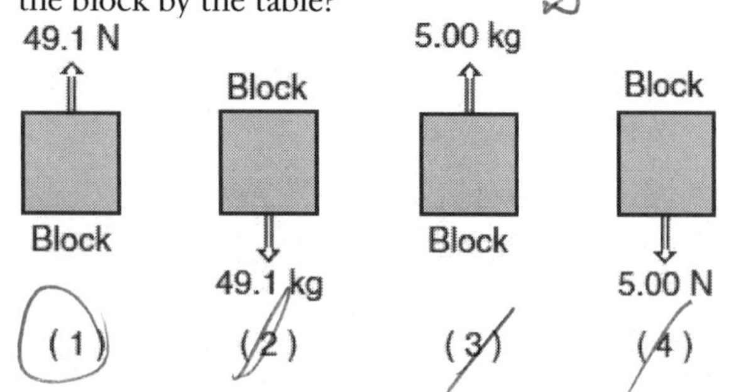

Base your answers to questions 18 and 19 on the information below.

The instant before a batter hits a 0.14-kilogram baseball, the velocity of the ball is 45 meters per second west. The instant after the batter hits the ball, the ball's velocity is 35 meters per second east. The bat and ball are in contact for 1.0×10^{-2} second.

18. Determine the magnitude and direction of the average acceleration of the baseball while it is in contact with the bat.

19. Calculate the magnitude of the average force the bat exerts on the ball while they are in contact. [Show all work, including the equation and substitution with units.]

Name: ______________________________ Period: __________

Dynamics-Newton's 2nd Law

20. The vector diagram below represents two forces, F_1 and F_2, simultaneously acting on an object.

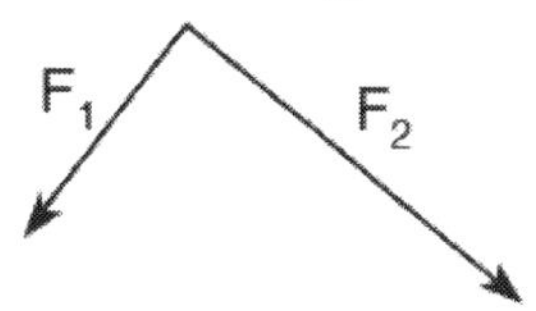

Which vector best represents the resultant of the two forces?

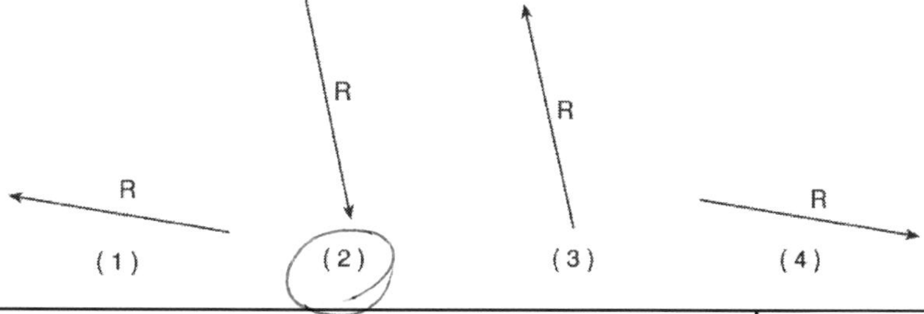

21. Two forces act concurrently on an object on a horizontal, frictionless surface, as shown in the diagram below.

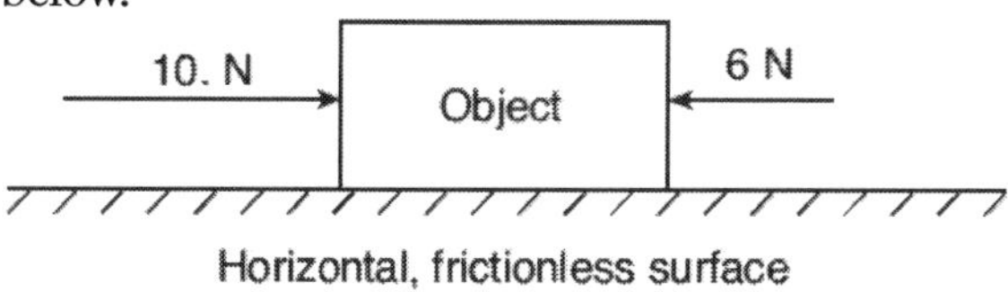

What additional force, when applied to the object, will establish equilibrium?
1. 16 N toward the right
2. 16 N toward the left
3. 4 N toward the right
4. 4 N toward the left

22. A 3-newton force and a 4-newton force are acting concurrently on a point. Which force could *not* produce equilibrium with these two forces?
1. 1 N
2. 7 N
3. 9 N
4. 4 N

23. The diagram shows a worker using a rope to pull a cart.

The worker's pull on the handle of the cart can best be described as a force having
1. magnitude, only
2. direction, only
3. both magnitude and direction
4. neither magnitude nor direction

Base your answers to questions 24 through 26 on the information and diagram below.

Force A with a magnitude of 5.6 newtons and force B with a magnitude of 9.4 newtons act concurrently on point P.

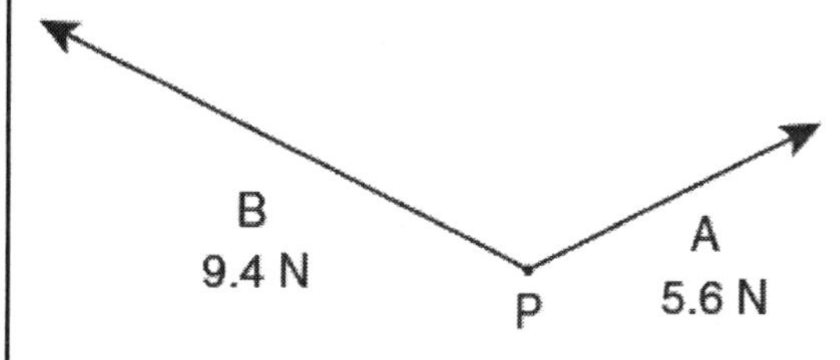

24. Determine the scale used in the diagram.

1 cm = ______ N

25. On the diagram, use a ruler and protractor to construct a vector representing the resultant of forces A and B.

26. Determine the magnitude of the resultant force.

Name:______________________________ Period: ____________

Dynamics-Newton's 2nd Law

27. Two 30-newton forces act concurrently on an object. In which diagram would the forces produce a resultant with a magnitude of 30 newtons?

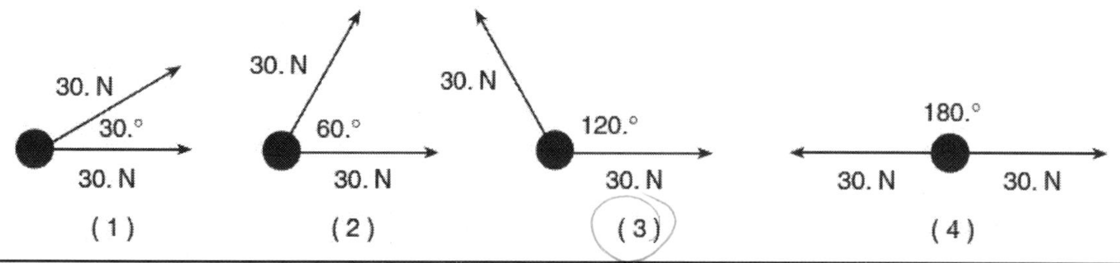

Base your answers to questions 28 through 30 on the information and diagram below.

In the scaled diagram, two forces, F_1 and F_2, act on a 4.0-kilogram block at point P. Force F_1 has a magnitude of 12 newtons, and is directed toward the right.

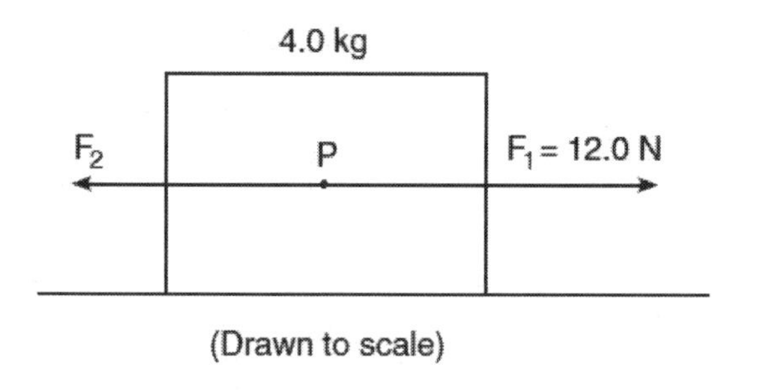

28. Using a ruler and the scaled diagram, determine the magnitude of F_2 in newtons.

29. Determine the magnitude of the net force acting on the block.

30. Calculate the magnitude of the acceleration of the block. [Show all work, including the equation and substitution with units.]

31. The diagram below shows a force of magnitude F applied to a mass at an angle θ relative to a horizontal frictionless surface.

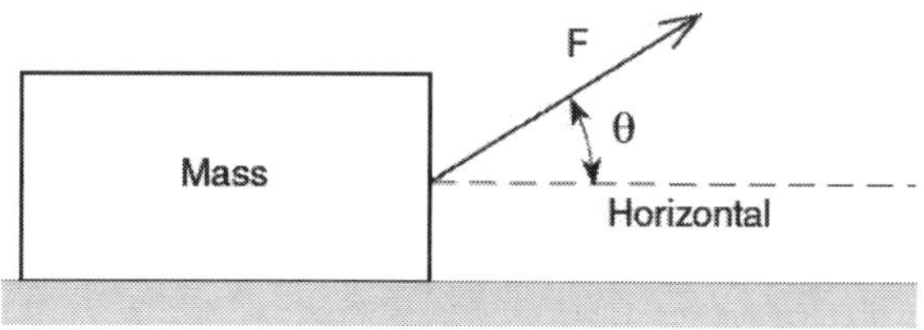

As angle θ is increased, the horizontal acceleration of the mass
1. decreases
2. increases
3. remains the same

32. Forces A and B have a resultant R. Force A and resultant R are represented in the diagram below.

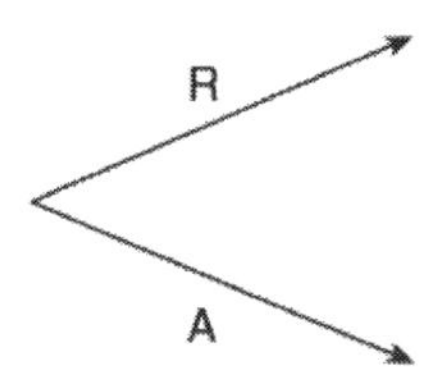

Which vector best represents force B?

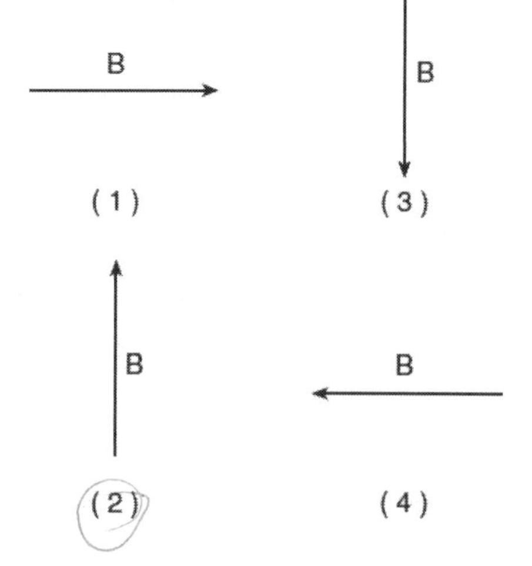

Name:________________________________ Period: ____________

Dynamics-Newton's 2nd Law

Base your answers to questions 33 and 34 on the information below.

A soccer player accelerates a 0.50-kilogram soccer ball by kicking it with a net force of 5 newtons.

33. Calculate the magnitude of the acceleration of the ball. [Show all work, including the equation and substitution with units.]

34. What is the magnitude of the force of the soccer ball on the player's foot?

35. The vector diagram below represents the horizontal component, F_H, and the vertical component, F_V, of a 24-newton force acting at 35° above the horizontal.

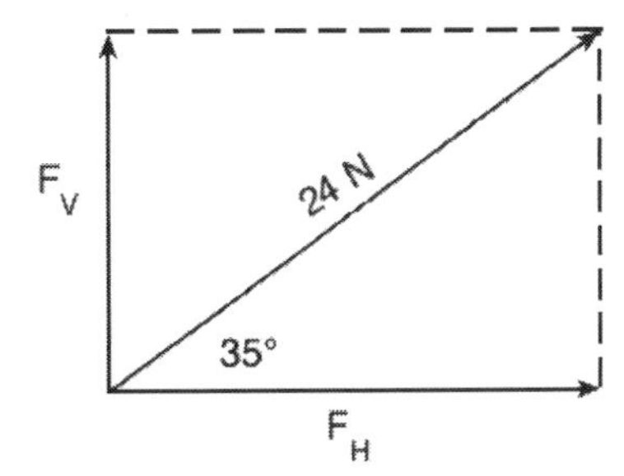

What are the magnitudes of the horizontal and vertical components?
1. $F_H = 3.5$ N and $F_V = 4.9$ N
2. $F_H = 4.9$ N and $F_V = 3.5$ N
3. $F_H = 14$ N and $F_V = 20$ N
4. $F_H = 20$ N and $F_V = 14$ N

36. Two forces act concurrently on an object. Their resultant force has the largest magnitude when the angle between the forces is
1. 0°
2. 30°
3. 90°
4. 180°

37. A 0.50-kilogram frog is at rest on the bank surrounding a pond of water. As the frog leaps from the bank, the magnitude of the acceleration of the frog is 3.0 meters per second². Calculate the magnitude of the net force exerted on the frog as it leaps. [Show all work, including the equation and substitution with units.]

38. Which graph best represents the motion of an object that is *not* in equilibrium as it travels along a straight line?

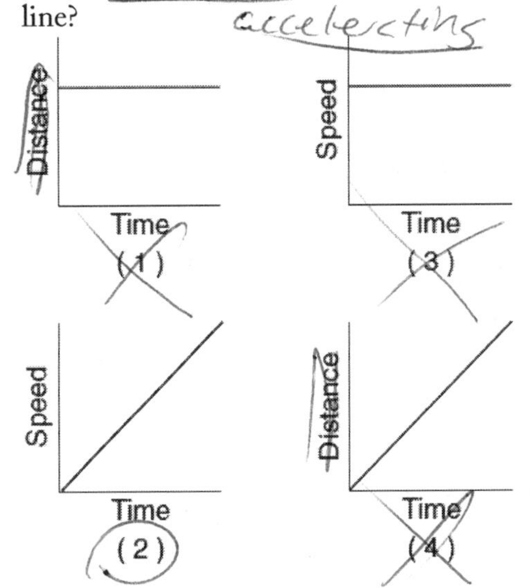

39. The diagram below represents a 5-newton force and a 12-newton force acting on point P.

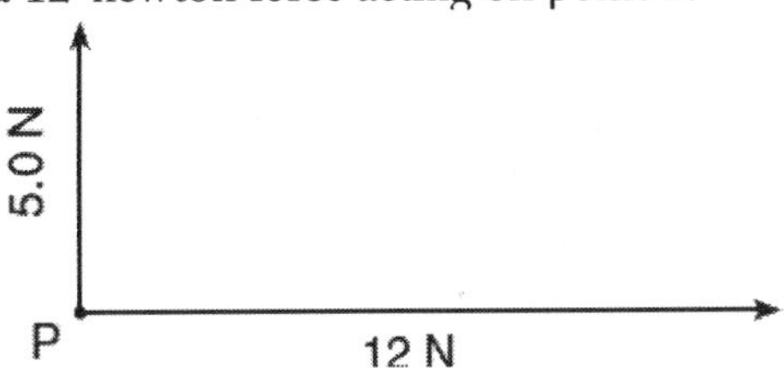

The resultant of the two forces has a magnitude of
1. 5 N
2. 7 N
3. 12 N
4. 13 N

Name:________________________________ Period: ____________

Dynamics-Newton's 2nd Law

40. Which pair of forces acting concurrently on an object will produce the resultant of greatest magnitude?

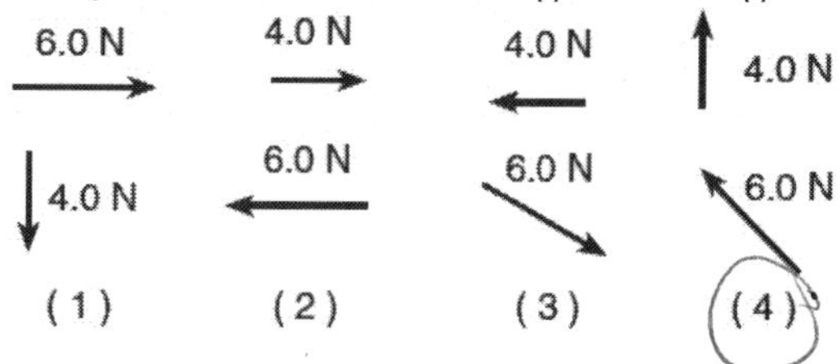

41. A 5-newton force and a 7-newton force act concurrently on a point. As the angle between the forces is increased from 0° to 180°, the magnitude of the resultant of the two forces changes from
    1. 0 N to 12 N
    2. 2 N to 12 N
    3. 12 N to 2 N
    4. 12 N to 0 N

42. A force of 25 newtons east and a force of 25 newtons west act concurrently on a 5-kilogram cart. What is the acceleration of the cart?
    1. 1.0 m/s^2 west
    2. 0.20 m/s^2 east
    3. 5.0 m/s^2 east
    4. 0 m/s^2

43. A high school physics student is sitting in a seat reading this question. The magnitude of the force with which the seat is pushing up on the student to support him is closest to
    1. 0 N
    2. 60 N
    3. 600 N
    4. 6,000 N

44. As the angle between two concurrent forces decreases, the magnitude of the force required to produce equilibrium
    1. decreases
    2. increases
    3. remains the same

45. A 60-kg skydiver is falling at a constant speed near the surface of Earth. The magnitude of the force of air friction acting on the skydiver is approximately
    1. 0 N
    2. 6 N
    3. 60 N
    4. 600 N

46. The weight of a chicken egg is most nearly equal to
    1. 10^{-3} N
    2. 10^{-2} N
    3. 10^{0} N
    4. 10^{2} N

47. A 1.5-kilogram lab cart is accelerated uniformly from rest to a speed of 2.0 meters per second in 0.50 second. What is the magnitude of the force producing this acceleration?
    1. 0.70 N
    2. 1.5 N
    3. 3.0 N
    4. 6.0 N

48. Which body is in equilibrium?
    1. a satellite orbiting Earth in a circular orbit
    2. a ball falling freely toward the surface of Earth
    3. a car moving with a constant speed along a straight, level road
    4. a projectile at the highest point in its trajectory

49. The diagram below represents a force vector, A, and a resultant vector, R.

Which force vector B below could be added to force vector A to produce resultant vector R?

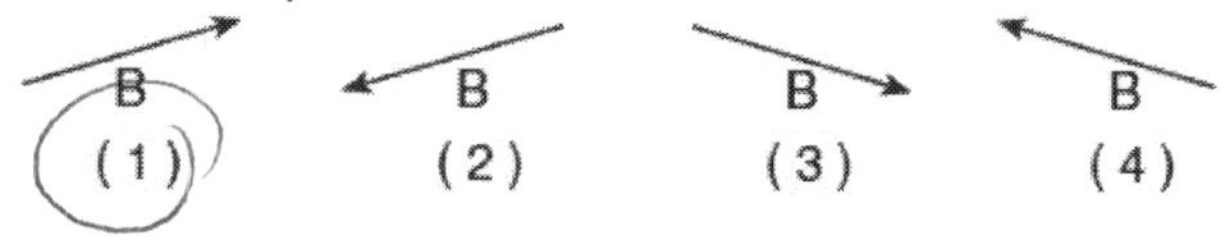

50. A 25-newton horizontal force northward and a 35-newton horizontal force southward act concurrently on a 15-kilogram object on a frictionless surface. What is the magnitude of the object's acceleration?
    1. 0.67 m/s^2
    2. 1.7 m/s^2
    3. 2.3 m/s^2
    4. 4.0 m/s^2

Name: ______________________________ Period: __________

Dynamics-Newton's 2nd Law

51. A woman is standing on a bathroom scale in an elevator car. If the scale reads a value greater than the weight of the woman at rest, the elevator car could be moving
    1. downward at constant speed
    2. upward at constant speed
    3. downward at increasing speed
    4. upward at increasing speed

52. A net force of 10 newtons accelerates an object at 5.0 meters per second^2. What net force would be required to accelerate the same object at 1.0 meter per second^2?
    1. 1.0 N
    2. 2.0 N
    3. 5.0 N
    4. 50 N

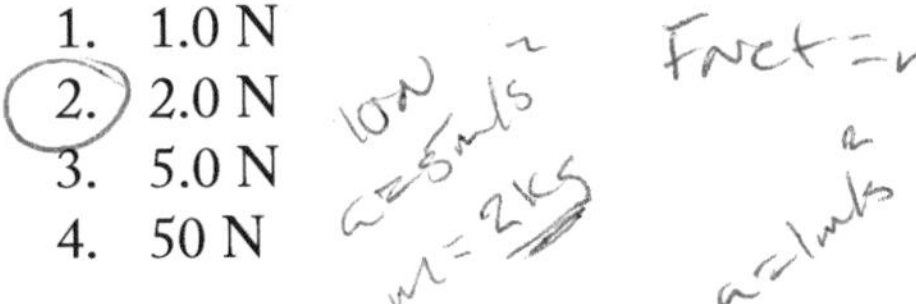

53. A 6.0-newton force and an 8.0-newton force act concurrently on a point. As the angle between these forces increases from 0° to 90°, the magnitude of their resultant
    1. decreases
    2. increases
    3. remains the same

54. Which situation describes an object that has no unbalanced force acting on it?
    1. an apple in free fall
    2. a satellite orbiting Earth
    3. a hockey puck moving at constant velocity across ice
    4. a laboratory cart moving down a frictionless 30° incline

55. Two 20-newton forces act concurrently on an object. What angle between these forces will produce a resultant force with the greatest magnitude?
    1. 0°
    2. 45°
    3. 90°
    4. 180°

56. Which situation represents a person in equilibrium?
    1. a child gaining speed while sliding down a slide
    2. a woman accelerating upward in an elevator
    3. a man standing still on a bathroom scale
    4. a teenager driving around a corner in his car

57. A number of 1-newton horizontal forces are exerted on a block on a frictionless, horizontal surface. Which top-view diagram shows the forces producing the greatest magnitude of acceleration of the block?

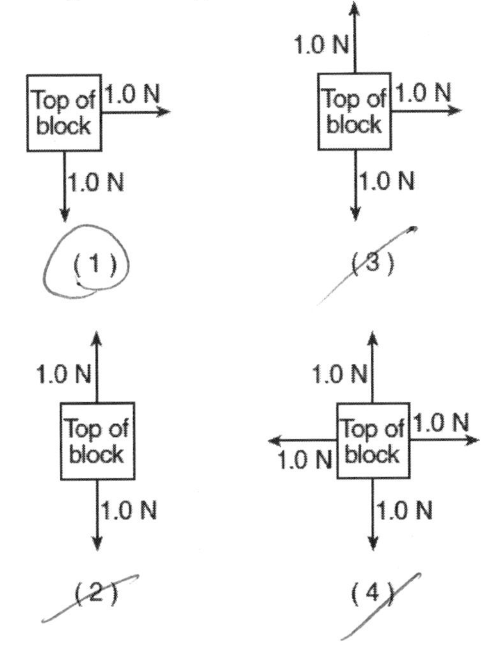

58. A rock is thrown straight up into the air. At the highest point of the rock's path, the magnitude of the net force acting on the rock is
    1. less than the magnitude of the rock's weight, but greater than zero
    2. greater than the magnitude of the rock's weight
    3. the same as the magnitude of the rock's weight
    4. zero

59. Four forces act concurrently on a block on a horizontal surface as shown in the diagram below.

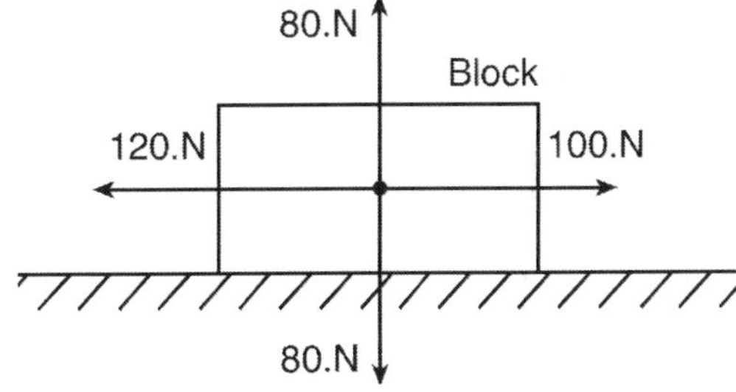

As a result of these forces, the block
1. moves at constant speed to the right
2. moves at a constant speed to the left
3. accelerates to the right
4. accelerates to the left

Name:____ Period: _______________

Dynamics-Newton's 2nd Law

Base your answers to questions 60 through 63 on the information below and diagram at right as well as your knowledge of physics.

Two forces, a 60-newton force east and an 80-newton force north, act concurrently on an object located at point P, as shown.

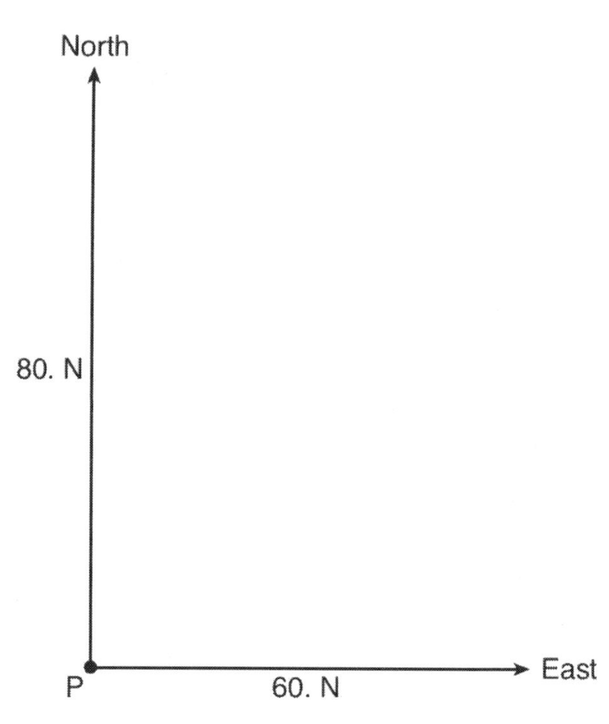

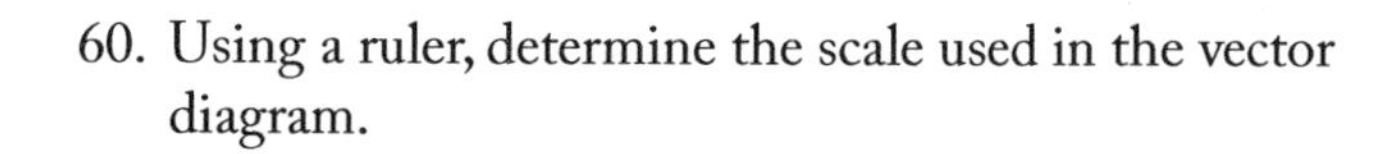

60. Using a ruler, determine the scale used in the vector diagram.

61. Draw the resultant force vector to scale on the diagram. Label the vector "R."

62. Determine the magnitude of the resultant force, R.

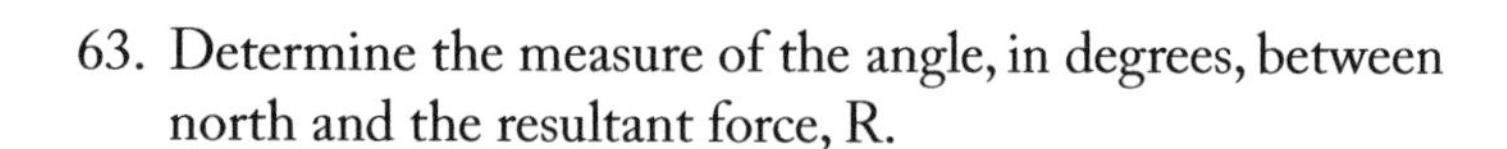

63. Determine the measure of the angle, in degrees, between north and the resultant force, R.

64. A 4.0-kilogram object is accelerated at 3.0 meters per second^2 north by an unbalanced force. The same unbalanced force acting on a 2.0-kilogram object will accelerate this object toward the north at
   1. 12 m/s^2
   2. 6.0 m/s^2
   3. 3.0 m/s^2
   4. 1.5 m/s^2

65. A 750-newton person stands in an elevator that is accelerating downward. The upward force of the elevator floor on the person must be
   1. equal to 0 N
   2. less than 750 N
   3. equal to 750 N
   4. greater than 750 N

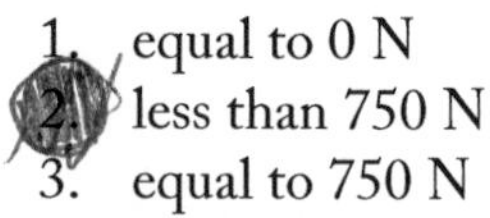

66. A 160-kilogram space vehicle is traveling along a straight line at a constant speed of 800 meters per second. The magnitude of the net force on the space vehicle is
   1. 0 N
   2. 1.60×10^2 N
   3. 8.00×10^2 N
   4. 1.28×10^5 N

67. A student throws a 5.0-newton ball straight up. What is the net force on the ball at its maximum height?
   1. 0.0 N
   2. 5.0 N, up
   3. 5.0 N, down
   4. 9.8 N, down

68. An object is in equilibrium. Which force vector diagram could represent the force(s) acting on the object?

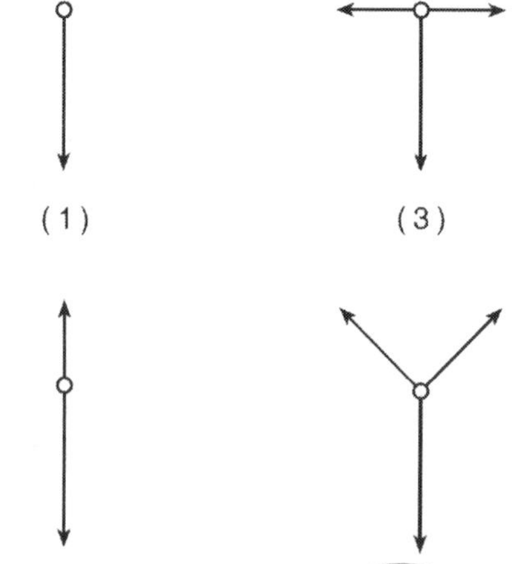

Name: ______________________________ Period: __________

Dynamics-Newton's 3rd Law

1. A student pulls a 60-newton sled with a force having a magnitude of 20 newtons. What is the magnitude of the force that the sled exerts on the student?

   1. 20 N
   2. 40 N
   3. 60 N
   4. 80 N

2. If a 65-kilogram astronaut exerts a force with a magnitude of 50 newtons on a satellite that she is repairing, the magnitude of the force that the satellite exerts on her is
   1. 0 N
   2. 50 N less than her weight
   3. 50 N more than her weight
   4. 50 N

3. A 400-newton girl standing on a dock exerts a force of 100 newtons on a 10,000-newton sailboat as she pushes it away from the dock. How much force does the sailboat exert on the girl?
   1. 25 N
   2. 100 N
   3. 400 N
   4. 10,000 N

4. A carpenter hits a nail with a hammer. Compared to the magnitude of the force the hammer exerts on the nail, the magnitude of the force the nail exerts on the hammer during contact is
   1. less
   2. greater
   3. the same

5. A woman is pushing a baby stroller. Compared to the magnitude of the force exerted on the stroller by the woman, the magnitude of the force exerted on the woman by the stroller is
   1. zero
   2. smaller, but greater than zero
   3. larger
   4. the same

6. When a child squeezes the nozzle of a garden hose, water shoots out of the hose toward the east. What is the compass direction of the force being exerted on the child by the nozzle?

7. A 100-kg boy and a 50-kg girl, each holding a spring scale, pull against each other as shown in the diagram below.

The graph below shows the relationship between the magnitude of the force that the boy applies on his spring scale and time.

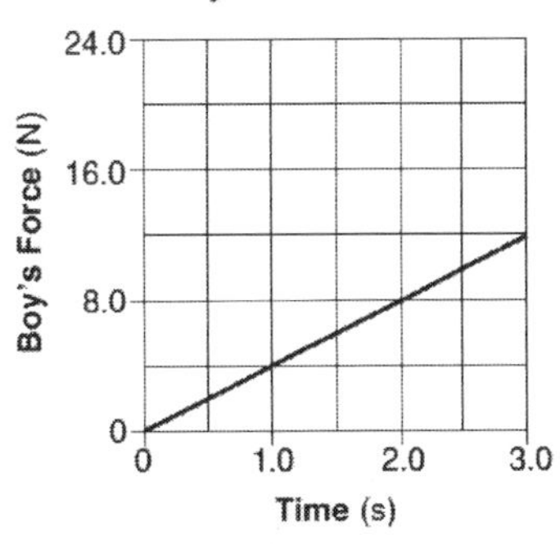

Which graph best represents the relationship between the magnitude of the force that the girl applies on her spring scale and time?

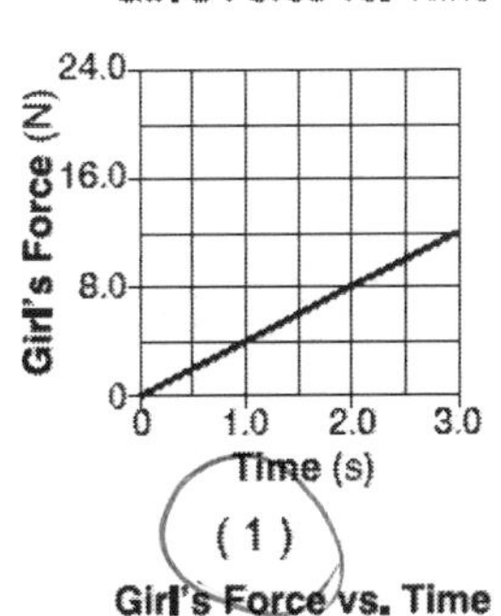

(1)

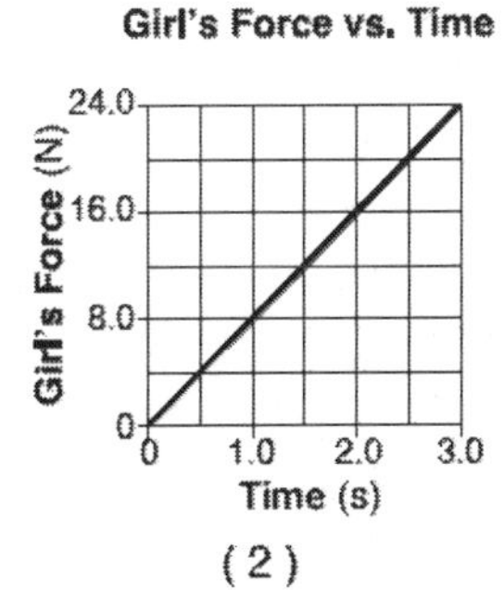

(2)

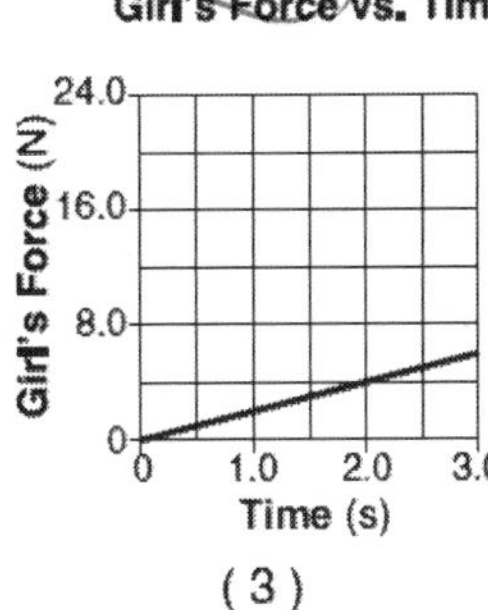

(3)

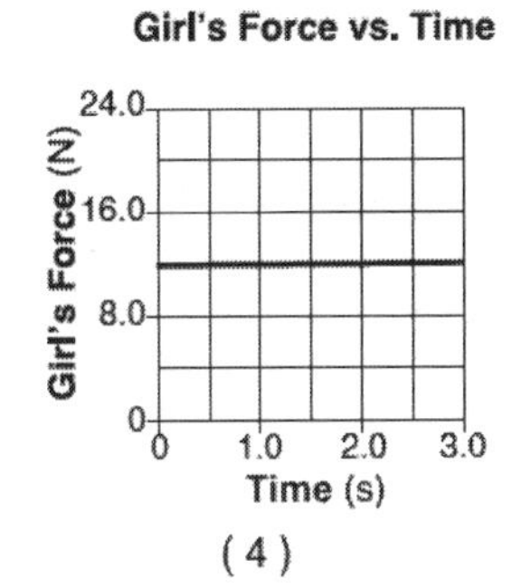

(4)

Name:____________________________________ Period: ______________

Dynamics-Newton's 3rd Law

8. The diagram below shows a compressed spring between two carts initially at rest on a horizontal, frictionless surface. Cart A has a mass of 2 kilograms and cart B has a mass of 1 kilogram. A string holds the carts together.

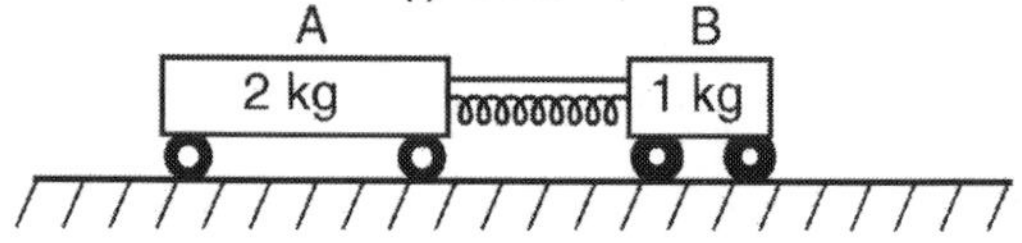

The string is cut and the carts move apart. Compared to the magnitude of the force the spring exerts on cart A, the magnitude of the force the spring exerts on cart B is

1. the same
2. half as great
3. twice as great
4. four times as great

9. A baseball bat exerts a force of magnitude F on a ball. If the mass of the bat is three times the mass of the ball, the magnitude of the force of the ball on the bat is

1. F
2. 2F
3. 3F
4. F/3

10. As a 5.0×10^2-newton basketball player jumps from the floor up toward the basket, the magnitude of the force of her feet on the floor is 1.0×10^3 newtons. As she jumps, the magnitude of the force of the floor on her feet is

1. 25 N
2. 100 N
3. 500 N
4. 1,000 N

Name: ______________________________ Period: __________

Dynamics-Friction

1. Which vector diagram best represents a cart slowing down as it travels to the right on a horizontal surface?

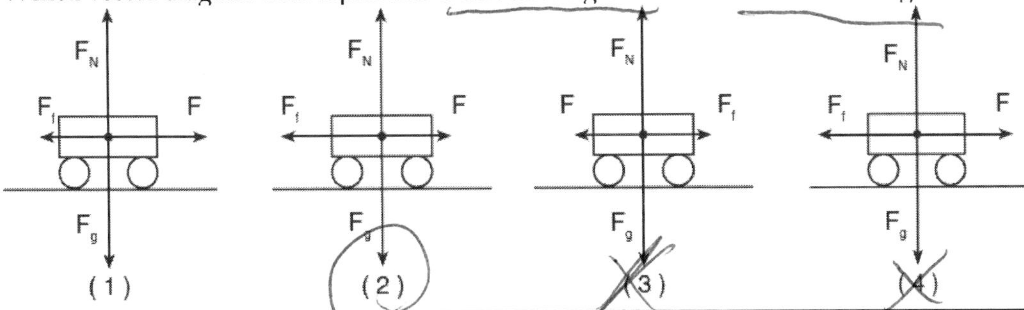

Base your answers to questions 2 and 3 on the information below.

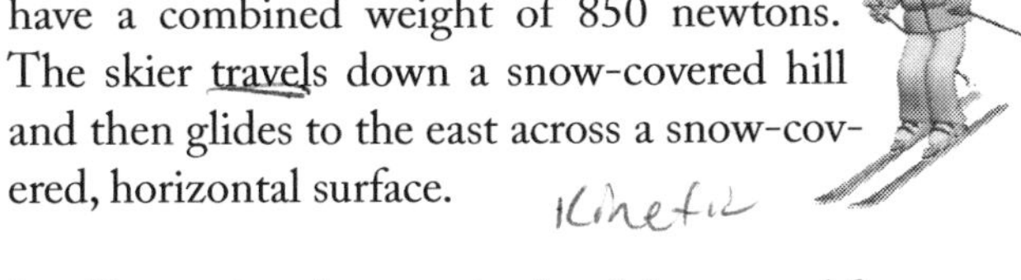

A student and the waxed skis she is wearing have a combined weight of 850 newtons. The skier travels down a snow-covered hill and then glides to the east across a snow-covered, horizontal surface.

2. Determine the magnitude of the normal force exerted by the snow on the skis as the skier glides across the horizontal surface.

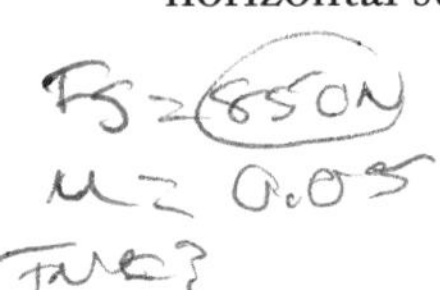

3. Calculate the magnitude of the force of friction acting on the skis as the skier glides across the snow-covered, horizontal surface. [Show all work, including the equation and substitution with units.]

4. The coefficient of kinetic friction between a 780-newton crate and a level warehouse floor is 0.200. Calculate the magnitude of the horizontal force required to move the crate across the floor at constant speed. [Show all work, including the equation and substitution with units.]

Base your answers to questions 5 through 8 on the information below.

An ice skater applies a horizontal force to a 20-kg block on frictionless, level ice, causing the block to accelerate uniformly at 1.4 m/s^2 to the right. After the skater stops pushing the block, it slides onto a region of ice that is covered with a thin layer of sand. The coefficient of kinetic friction between the block and the sand-covered ice is 0.28.

5. Calculate the magnitude of the force applied to the block by the skater.

6. On the diagram below, starting at point A, draw a vector to represent the force applied to the block by the skater. Begin the vector at point A and use a scale of 1 cm = 10 newtons.

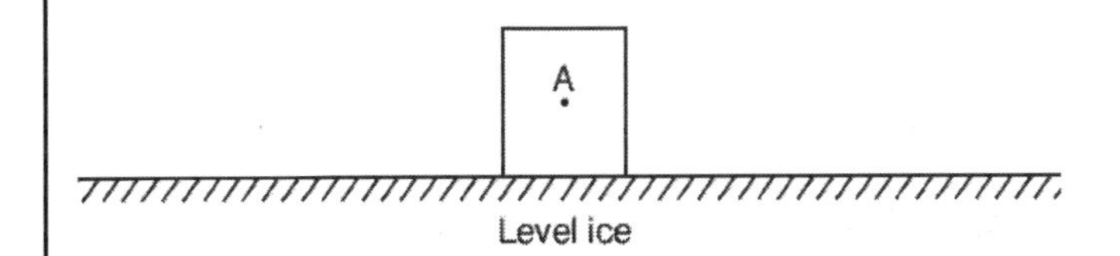

7. Determine the magnitude of the normal force acting on the block.

8. Calculate the magnitude of the force of friction acting on the block as it slides over the sand-covered ice. [Show all work, including the equation and substitution with units.]

Name:________________________________ Period: ____________

Dynamics-Friction

Base your answers to questions 9 through 13 on the information below.

A manufacturer's advertisement claims that their 1,250-kilogram (12,300-newton) sports car can accelerate on a level road from 0 to 60 miles per hour (0 to 26.8 meters per second) in 3.75 seconds.

9. Determine the acceleration, in meters per second^2, of the car according to the advertisement.

10. Calculate the net force required to give the car the acceleration claimed in the advertisement. [Show all work, including the equation and substitution with units.]

11. What is the normal force exerted by the road on the car?

12. The coefficient of friction between the car's tires and the road is 0.80. Calculate the maximum force of friction between the car's tires and the road. [Show all work, including the equation and substitution with units.]

13. Using the values for the forces you have calculated, explain whether or not the manufacturer's claim for the car's acceleration is possible.

Base your answers to questions 14 and 15 on the information and diagram below.

A force of 60 newtons is applied to a rope to pull a sled across a horizontal surface at a constant velocity. The rope is at an angle of 30 degrees above the horizontal.

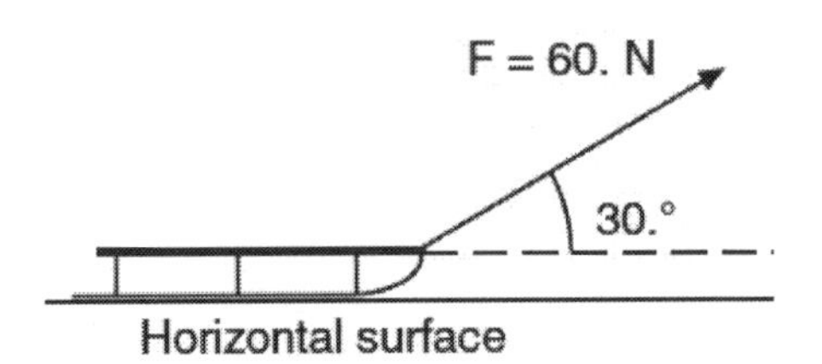

14. Calculate the magnitude of the component of the 60-newton force that is parallel to the horizontal surface. [Show all work, including the equation and substitution with units.]

15. Determine the magnitude of the frictional force acting on the sled.

16. A child pulls a wagon at a constant velocity along a level sidewalk. The child does this by applying a 22-newton force to the wagon handle, which is inclined at 35° to the sidewalk as shown below.

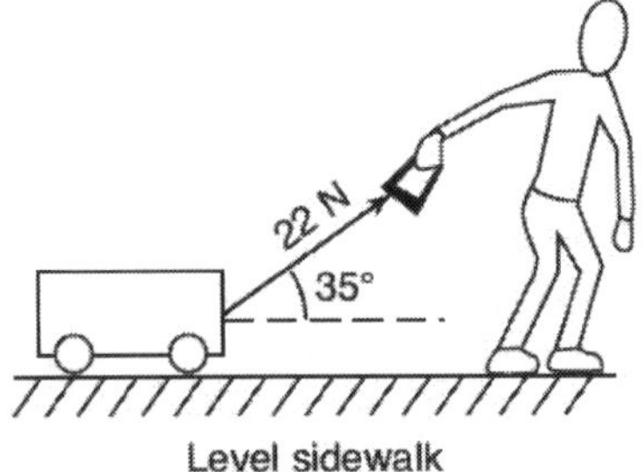

What is the magnitude of the force of friction on the wagon?
1. 11 N
2. 13 N
3. 18 N
4. 22 N

Name: ____________________ Period: __________

Dynamics-Friction

Base your answers to questions 17 through 21 on the information and diagram below.

A horizontal force of 8 newtons is used to pull a 20-newton wooden box moving toward the right along a horizontal, wood surface, as shown (not drawn to scale).

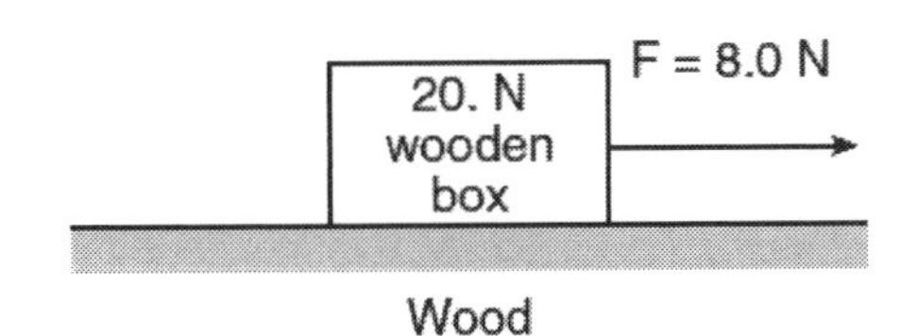

17. Starting at point P on the diagram below, use a metric ruler and a scale of 1 cm = 10 N to draw a vector representing the normal force acting on the box. Label the vector F_N.

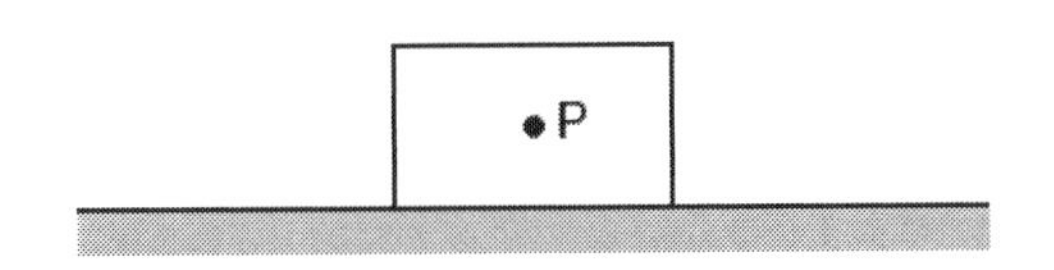

18. Calculate the magnitude of the frictional force acting on the box. [Show all work, including the equation and substitution with units.]

19. Determine the magnitude of the net force acting on the box.

20. Determine the mass of the box.

21. Calculate the magnitude of the acceleration of the box. [Show all work, including the equation and substitution with units.]

22. The diagram below shows a 4-kilogram object accelerating at 10 m/s^2 on a rough horizontal surface.

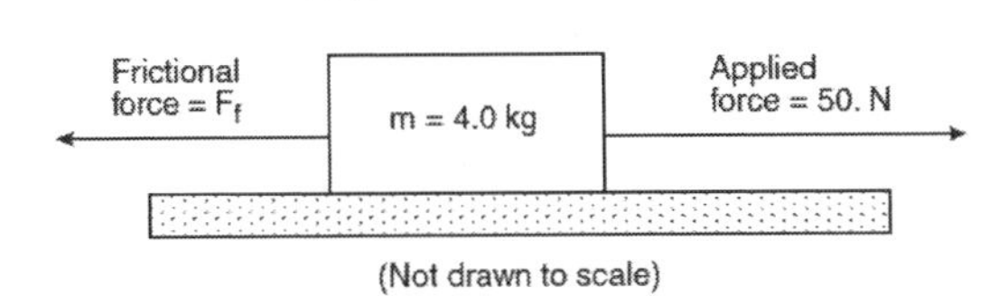

(Not drawn to scale)

What is the magnitude of the frictional force F_f acting on the object?
1. 5 N
2. 10 N
3. 20 N
4. 40 N

23. A car's performance is tested on various horizontal road surfaces. The brakes are applied, causing the rubber tires of the car to slide along the road without rolling. The tires encounter the greatest force of friction to stop the car on
1. dry concrete
2. dry asphalt
3. wet concrete
4. wet asphalt

24. When a 12-newton horizontal force is applied to a box on a horizontal tabletop, the box remains at rest. The force of static friction acting on the box is
1. 0 N
2. between 0 N and 12 N
3. 12 N
4. greater than 12 N

25. A box is pushed toward the right across a classroom floor. The force of friction on the box is directed toward the
1. left
2. right
3. ceiling
4. floor

26. A skier on waxed skis is pulled at constant speed across level snow by a horizontal force of 39 newtons. Calculate the normal force exerted on the skier.

Name:_______________________________ Period: ____________

Dynamics-Friction

Base your answers to questions 27 through 31 on the information below. Show all work, including the equation and substitution with units.

A force of 10 newtons toward the right is exerted on a wooden crate initially moving to the right on a horizontal wooden floor. The crate weighs 25 newtons.

27. Calculate the magnitude of the force of friction between the crate and the floor.

28. On the diagram below, draw and label all vertical forces acting on the crate.

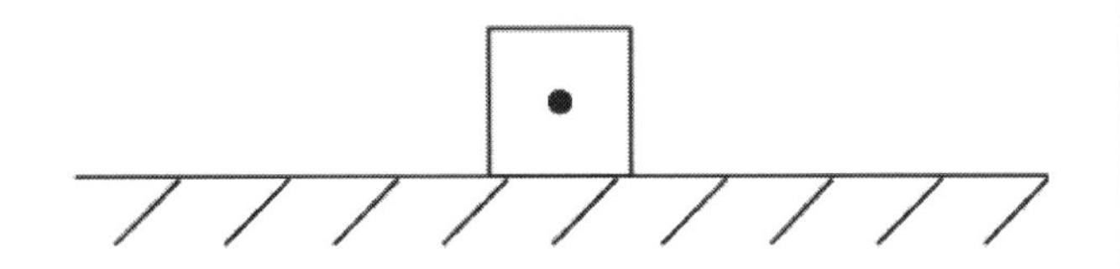

29. On the diagram above, draw and label all horizontal forces acting on the crate.

30. What is the magnitude of the net force acting on the crate?

31. Is the crate accelerating? Explain your answer.

Base your answers to questions 32 through 34 on the information and diagram below.

A 10-kg box, sliding to the right across a rough horizontal floor, accelerates at -2 m/s^2 due to the force of friction.

32. Calculate the magnitude of the net force acting on the box. [Show all work, including the equation and substitution with units.]

33. On the diagram below, draw a vector representing the net force acting on the box. Begin the vector at point P and use a scale of 1 cm = 10 newtons.

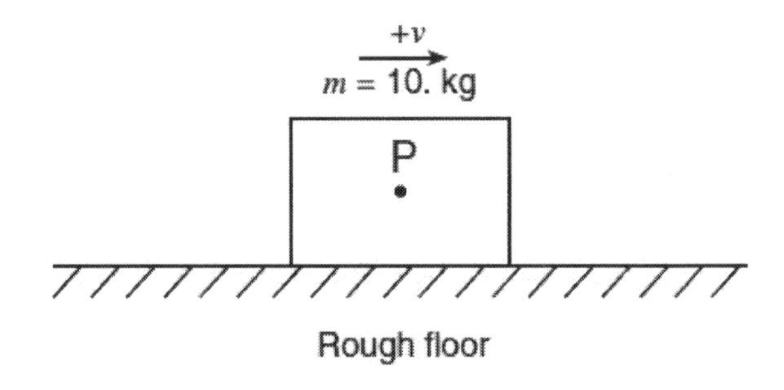

34. Calculate the coefficient of kinetic friction between the box and the floor. [Show all work, including the equation and substitution with units.]

35. A 10-kg rubber block is pulled horizontally at constant velocity across a sheet of ice. Calculate the magnitude of the force of friction acting on the block. [Show all work, including the equation and substitution with units.]

Name: ________________________ Period: ____________

Dynamics-Friction

36. What is the magnitude of the force needed to keep a 60-newton rubber block moving across level, dry asphalt in a straight line at a constant speed of 2 meters per second?
    1. 40 N
    2. 51 N
    3. 60 N
    4. 120 N

37. The force required to start an object sliding across a uniform horizontal surface is larger than the force required to keep the object sliding at a constant velocity. The magnitudes of the required forces are different in these situations because the force of kinetic friction
    1. is greater than the force of static friction
    2. is less than the force of static friction
    3. increases as the speed of the object relative to the surface increases
    4. decreases as the speed of the object relative to the surface increases

38. The diagram below shows a block sliding down a plane inclined at angle θ with the horizontal.

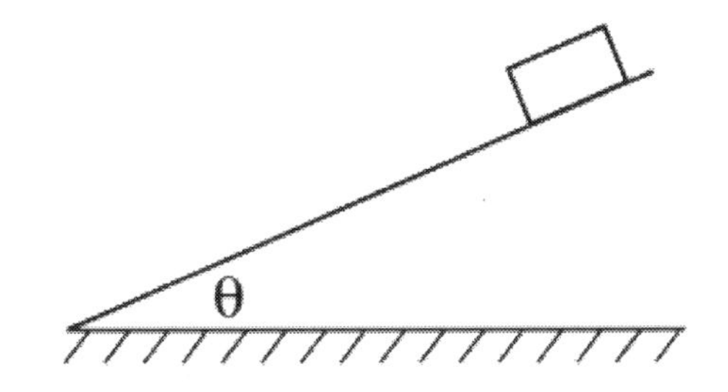

As angle θ is increased, the coefficient of kinetic friction between the bottom surface of the block and the surface of the incline will
    1. decrease
    2. increase
    3. remain the same

39. An airplane is moving with a constant velocity in level flight. Compare the magnitude of the forward force provided by the engines to the magnitude of the backward frictional drag force.

40. An 80-kilogram skier slides on waxed skis along a horizontal surface of snow at constant velocity while pushing with his poles. What is the horizontal component of the force pushing him forward?
    1. 0.05 N
    2. 0.4 N
    3. 40 N
    4. 4 N

41. Compared to the force needed to start sliding a crate across a rough level floor, the force needed to keep it sliding is
    1. less
    2. greater
    3. the same

42. A 1500-kilogram car accelerates at 5 meters per second^2 on a level, dry, asphalt road. Determine the magnitude of the net horizontal force acting on the car.

43. A 0.50-kilogram puck sliding on a horizontal shuffleboard court is slowed to rest by a frictional force of 1.2 newtons. What is the coefficient of kinetic friction between the puck and the surface of the shuffleboard court?
    1. 0.24
    2. 0.42
    3. 0.60
    4. 4.1

44. An 8.0-newton wooden block slides across a horizontal wooden floor at constant velocity. What is the magnitude of the force of kinetic friction between the block and the floor?
    1. 2.4 N
    2. 3.4 N
    3. 8.0 N
    4. 27 N

Name: ____________________ Period: __________

Dynamics-Friction

Base your answers to questions 45 through 47 on the information below and your knowledge of physics.

A horizontal 20-newton force is applied to a 5.0-kilogram box to push it across a rough, horizontal floor at a constant velocity of 3.0 meters per second to the right.

45. Determine the magnitude of the force of friction acting on the box.

46. Calculate the weight of the box. [Show all work, including the equation and substitution with units.]

47. Calculate the coefficient of kinetic friction between the box and the floor. [Show all work, including the equation and substitution with units.]

Base your answers to questions 48 through 51 on the information below and your knowledge of physics.

The diagram below represents a 4.0-newton force applied to a 0.200-kilogram copper block sliding to the right on a horizontal steel table.

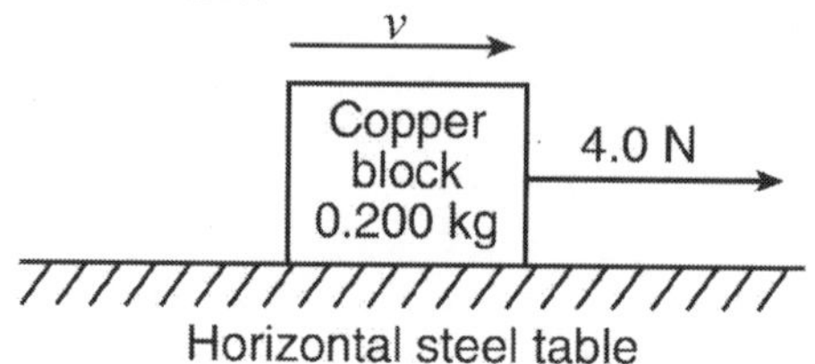

48. Determine the weight of the block.

49. Calculate the magnitude of the force of friction acting on the moving block. [Show all work, including the equation and substitution with units.]

50. Determine the magnitude of the net force acting on the moving block.

51. Describe what happens to the magnitude of the velocity of the block as the block slides across the table.

Name:________________________ Period: ____________

Dynamics-Ramps and Inclines

1. The diagram at right represents a block at rest on an incline.

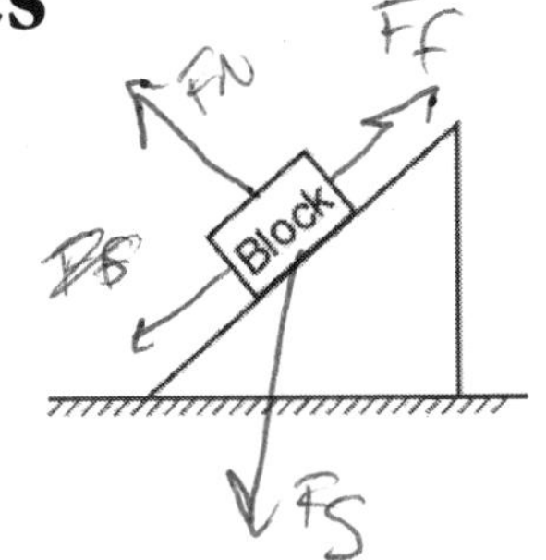

Which diagram below best represents the forces acting on the block? (F_f = frictional force, F_N = normal force, and F_w = weight.)

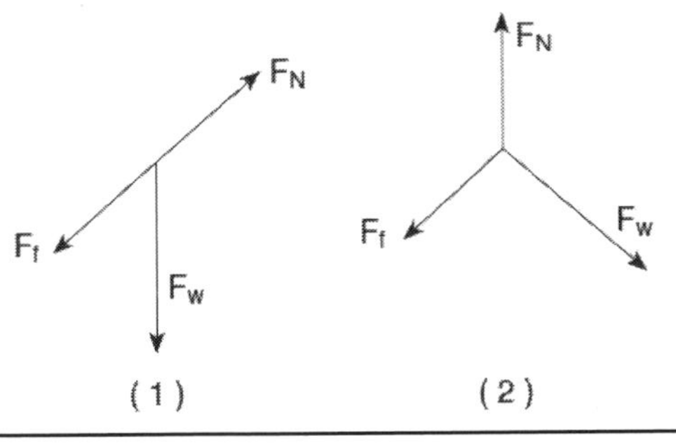

2. The diagram below shows a 50-kilogram crate on a frictionless plane at angle θ to the horizontal. The crate is pushed at constant speed up the incline from point A to point B by force F.

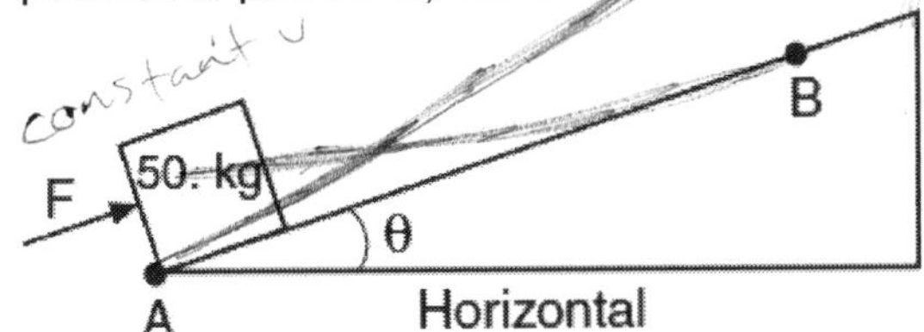

If angle θ were increased, what would be the effect on the magnitude of force F and the total work W done on the crate as it is moved from A to B?

1. W would remain the same and the magnitude of F would decrease.
2. W would remain the same and the magnitude of F would increase.
3. W would increase and the magnitude of F would decrease.
4. W would increase and the magnitude of F would increase.

3. In the diagram below, a 10-kilogram block is at rest on a plane inclined at 15° to the horizontal.

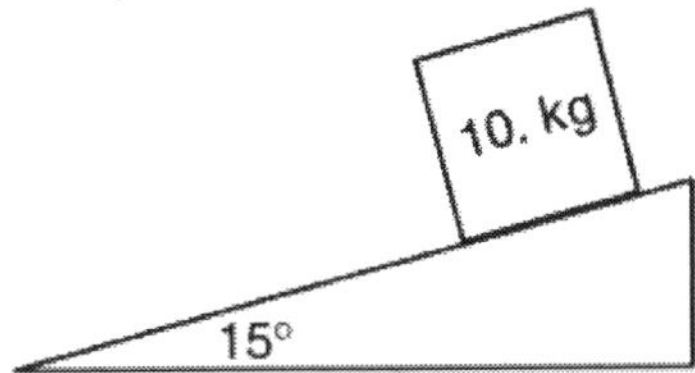

As the angle of the incline is increased to 30°, the mass of the block will

1. decrease
2. increase
3. remain the same

4. A block weighing 10 newtons is on a ramp inclined at 30° to the horizontal. A 3-newton force of friction, F_f, acts on the block as it is pulled up the ramp at constant velocity with force F, which is parallel to the ramp, as shown in the diagram below.

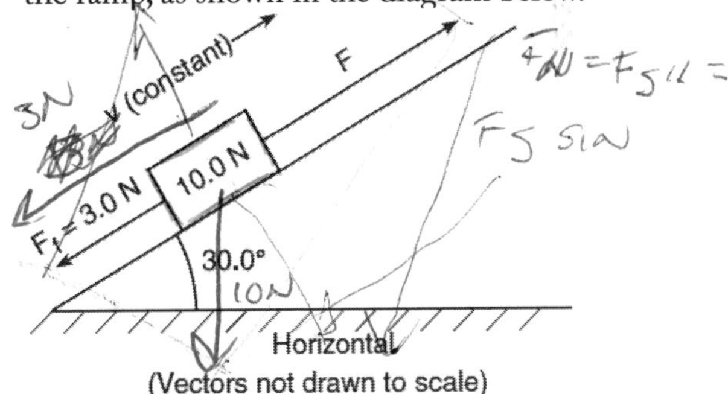

What is the magnitude of force F?

1. 7 N
2. 8 N
3. 10 N
4. 13 N

5. The diagram below shows a 1.0×10^5-newton truck at rest on a hill that makes an angle of 8.0° with the horizontal.

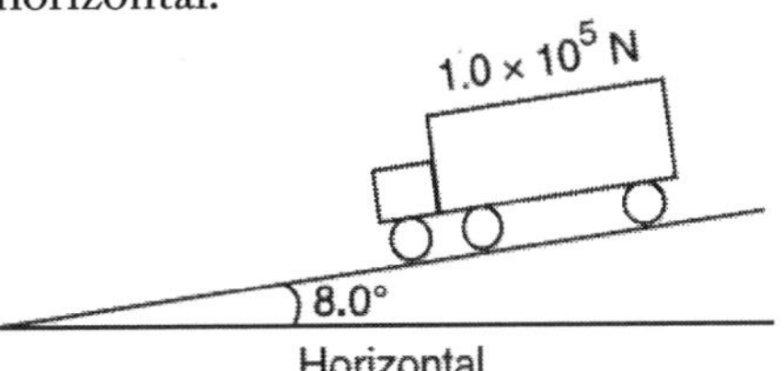

What is the component of the truck's weight parallel to the hill?

1. 1.4×10^3 N
2. 1.0×10^4 N
3. 1.4×10^4 N
4. 9.9×10^4 N

Name: ________________________________ Period: __________

Dynamics-Ramps and Inclines

6. The diagram below shows a sled and rider sliding down a snow-covered hill that makes an angle of 30° with the horizontal.

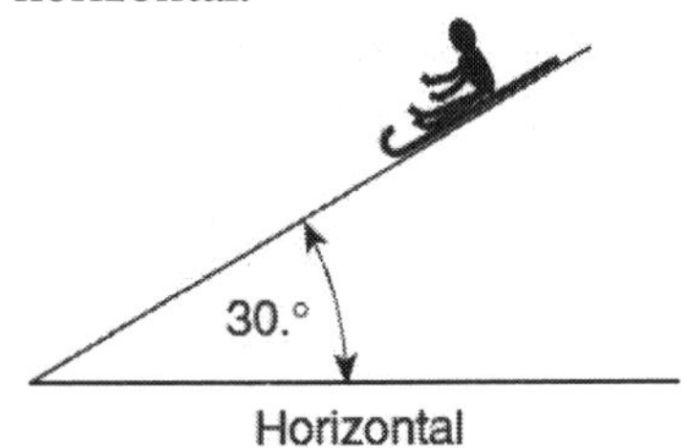

Which vector best represents the direction of the normal force, F_N, exerted by the hill on the sled?

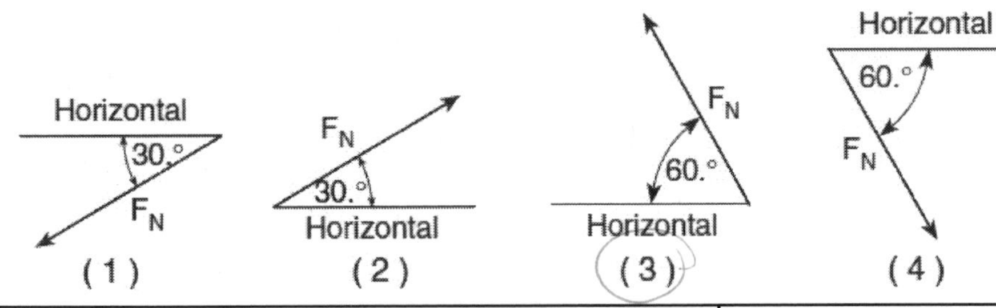

7. Three forces act on a box on an inclined plane as shown in the diagram below. [Vectors are not drawn to scale.]

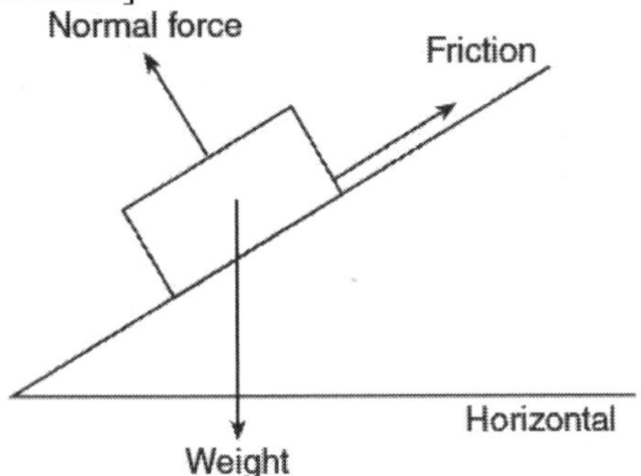

If the box is at rest, the net force acting on it is equal to
1. the weight
2. the normal force
3. friction
4. zero

8. An 8.0-newton block is accelerating down a frictionless ramp inclined at 15° to the horizontal, as shown in the diagram below.

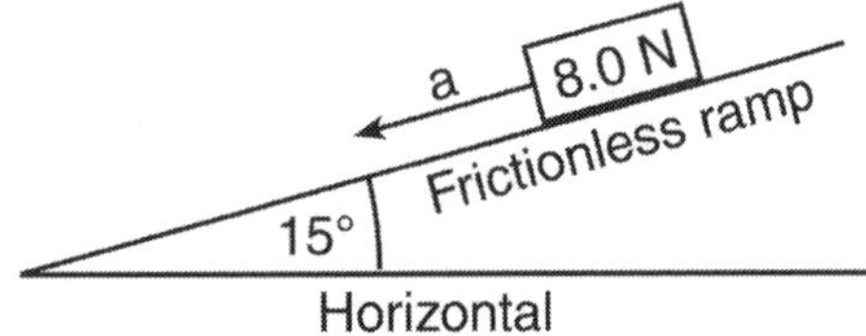

What is the magnitude of the net force causing the block's acceleration?
1. 0 N
2. 2.1 N
3. 7.7 N
4. 8.0 N

Name:________________________________ Period: ____________

UCM-Circular Motion

Base your answers to questions 1 and 2 on the information and diagram below.

The diagram shows the top view of a 65-kilogram student at point A on an amusement park ride. The ride spins the student in a horizontal circle of radius 2.5 meters, at a constant speed of 8.6 meters per second. The floor is lowered and the student remains against the wall without falling to the floor.

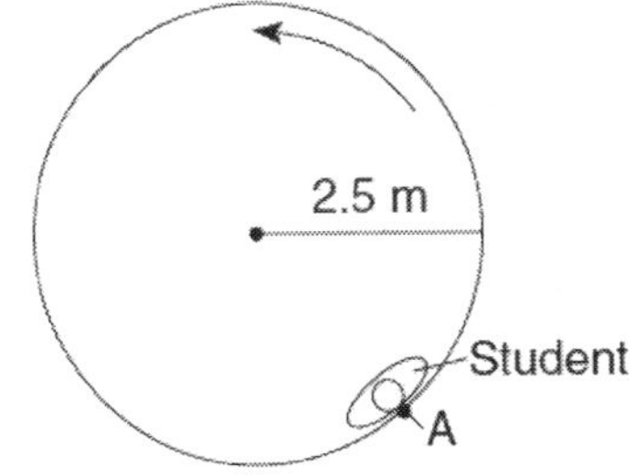

1. Which vector best represents the direction of the centripetal acceleration of the student at point A?

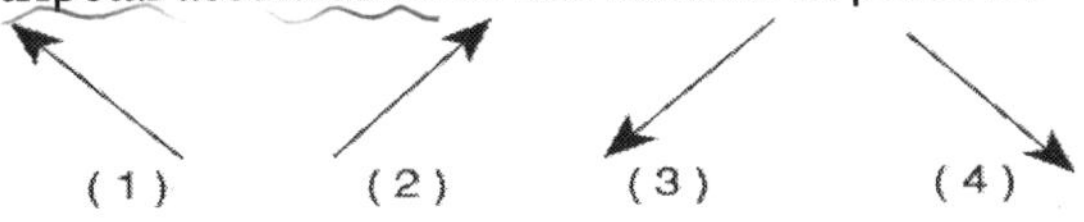

2. The magnitude of the centripetal force acting on the student at point A is approximately
   1. 1.2×10^4 N
   2. 1.9×10^3 N
   3. 2.2×10^2 N
   4. 3.0×10^1 N

3. The magnitude of the centripetal force acting on an object traveling in a horizontal, circular path will *decrease* if the
   1. radius of the path is increased
   2. mass of the object is increased
   3. direction of motion of the object is reversed
   4. speed of the object is increased

4. Centripetal force F_C acts on a car going around a curve. If the speed of the car were twice as great, the magnitude of the centripetal force necessary to keep the car moving in the same path would be
   1. F_C
   2. $2F_C$
   3. $F_C/2$
   4. $4F_C$

5. A car travels at constant speed around a section of horizontal, circular track. On the diagram below, draw an arrow at point P to represent the direction of the centripetal acceleration of the car when it is at point P.

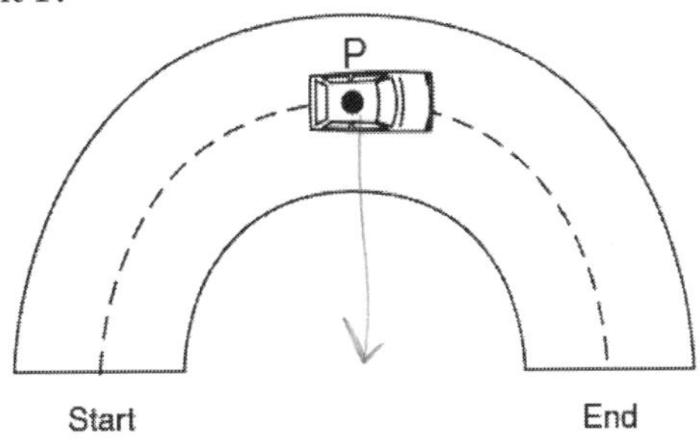

6. A child is riding on a merry-go-round. As the speed of the merry-go-round is doubled, the magnitude of the centripetal force acting on the child
   1. remains the same
   2. is doubled
   3. is halved
   4. is quadrupled

7. A ball attached to a string is moved at constant speed in a horizontal circular path. A target is located near the path of the ball as shown in the diagram.

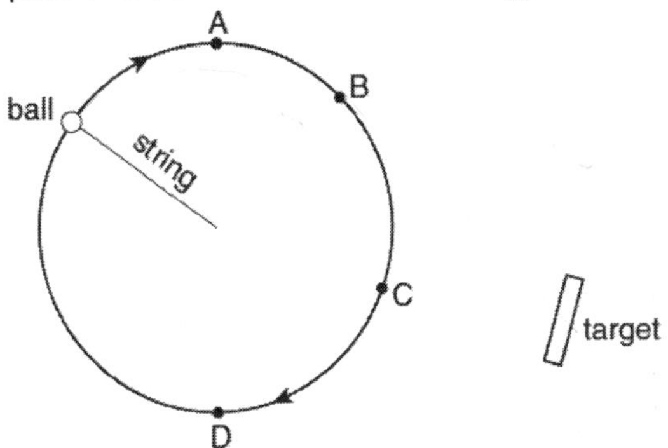

At which point along the ball's path should the string be released, if the ball is to hit the target?
   1. A
   2. B
   3. C
   4. D

8. Which unit is equivalent to meters per second?
   1. Hz·s
   2. Hz·m
   3. s/Hz
   4. m/Hz

Name:________________________ Period: ____________

UCM-Circular Motion

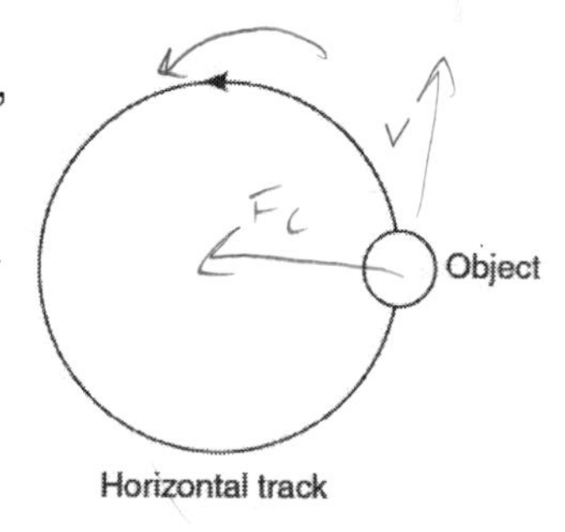

9. The diagram at right shows an object moving counterclockwise around a horizontal, circular track.

   Which diagram represents the direction of both the object's velocity and the centripetal force acting on the object when it is in the position shown?

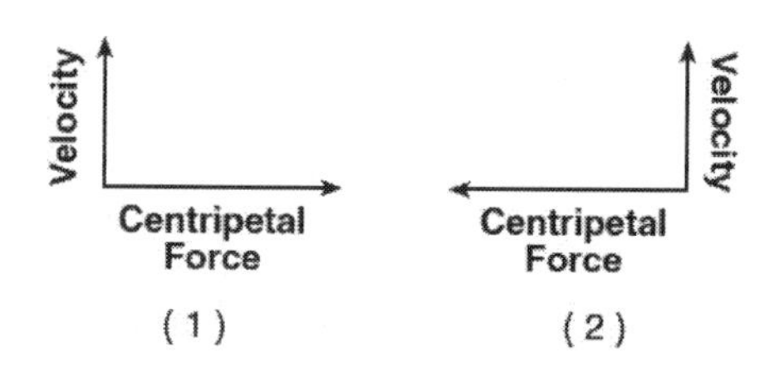

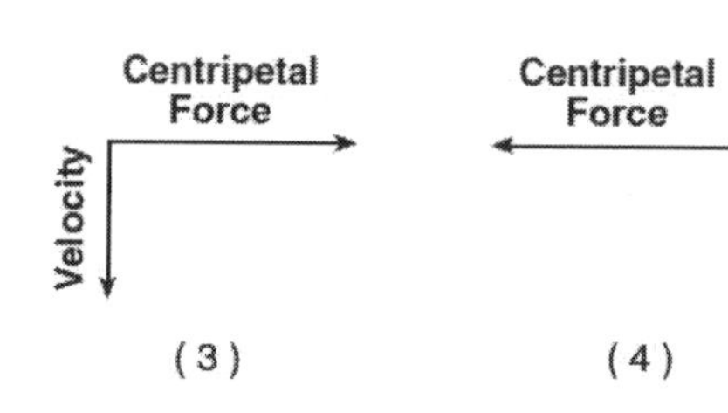

---

10. Which graph best represents the relationship between the magnitude of the centripetal acceleration and the speed of an object moving in a circle of constant radius?

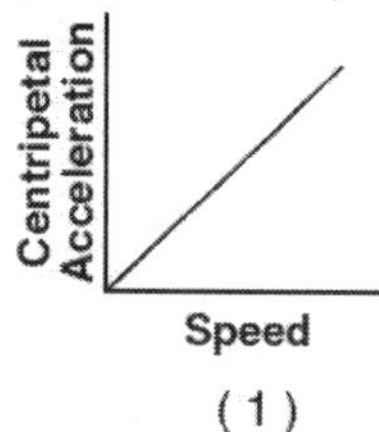

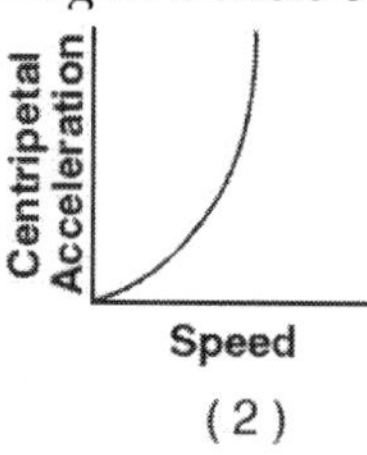

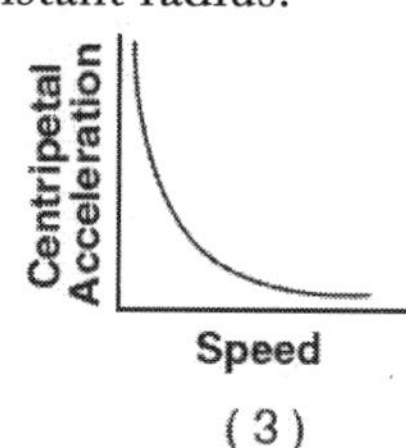

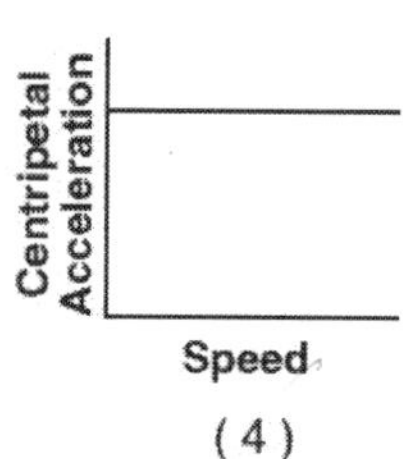

Base your answers to questions 11 through 13 on the information and data table below.

In an experiment, a student measured the length and period of a simple pendulum. The data table lists the length (l) of the pendulum in meters and the square of the period (T^2) of the pendulum in seconds^2.

Length (ℓ) (meters)	Square of Period (T^2) (seconds^2)
0.100	0.410
0.300	1.18
0.500	1.91
0.700	2.87
0.900	3.60

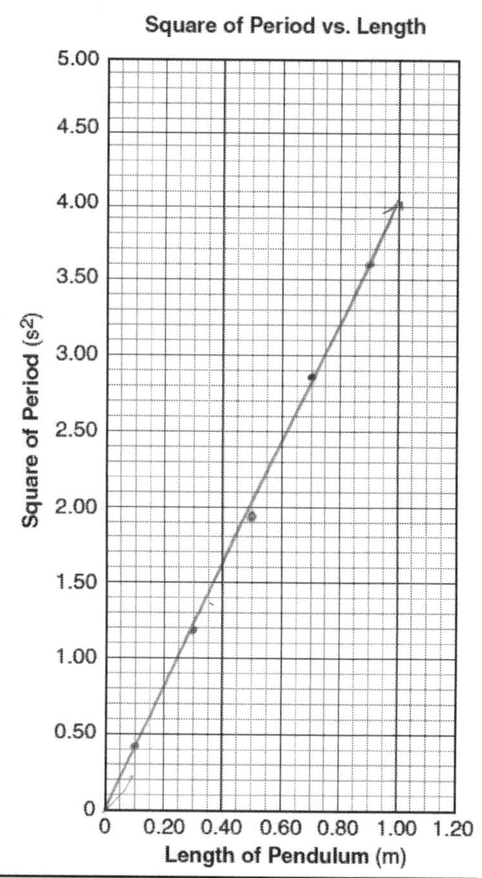

11. Using the information in the data table, construct a graph on the grid provided by plotting the data points for the square of period versus length, and then drawing the best-fit straight line.

12. Using your graph, determine the time in seconds it would take this pendulum to make one complete swing if it were 0.200 meter long.

13. The period of a pendulum is related to its length by the formula: $T^2 = \left(\frac{4\pi^2}{g}\right) \bullet l$ If g represents the acceleration due to gravity, explain how the graph you have drawn could be used to calculate the value of g.

Name: ______________________________ Period: ____________

UCM-Circular Motion

14. A 1.0×10^3-kilogram car travels at a constant speed of 20 meters per second around a horizontal circular track. Which diagram correctly represents the direction of the car's velocity (v) and the direction of the centripetal force (F_C) acting on the car at one particular moment?

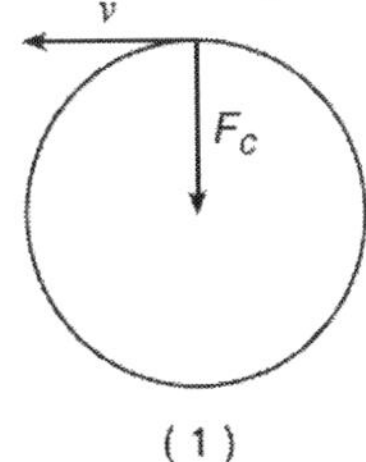

(1)

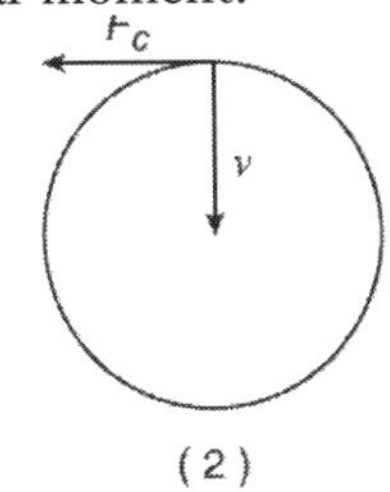

(2)

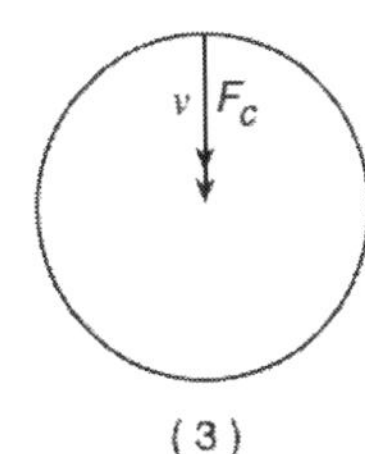

(3)

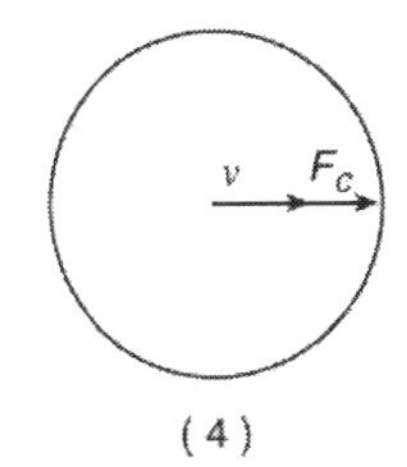

(4)

15. A baby and a stroller have a total mass of 20 kilograms. A force of 36 newtons keeps the stroller moving in a circular path with a radius of 5.0 meters. Calculate the speed at which the stroller moves around the curve. [Show all work, including the equation and substitution with units.]

16. The diagram below shows a 5.0-kilogram bucket of water being swung in a horizontal circle of 0.70-meter radius at a constant speed of 2.0 meters per second.

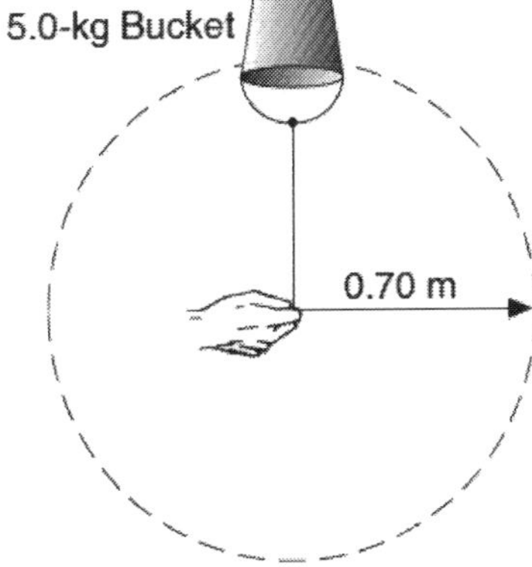

The magnitude of the centripetal force on the bucket of water is approximately
1. 5.7 N
2. 14 N
3. 29 N
4. 200 N

17. In the diagram below, S is a point on a car tire rotating at a constant rate.

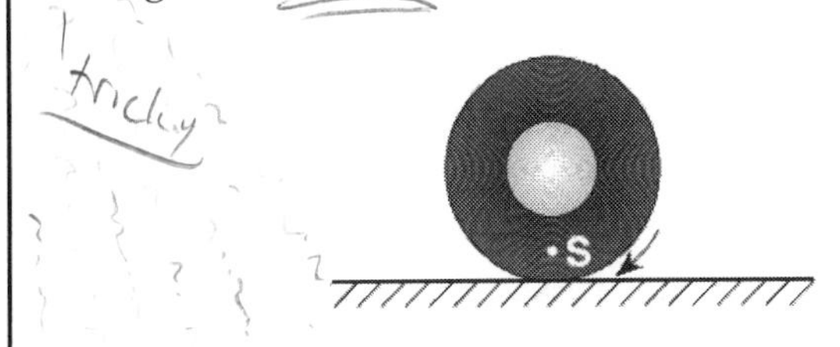

Which graph best represents the magnitude of the centripetal acceleration of point S as a function of time?

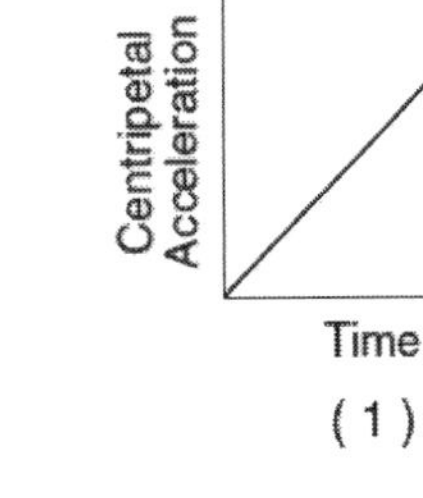

(1)

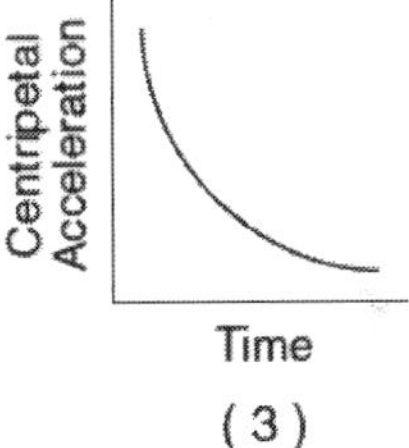

(3)

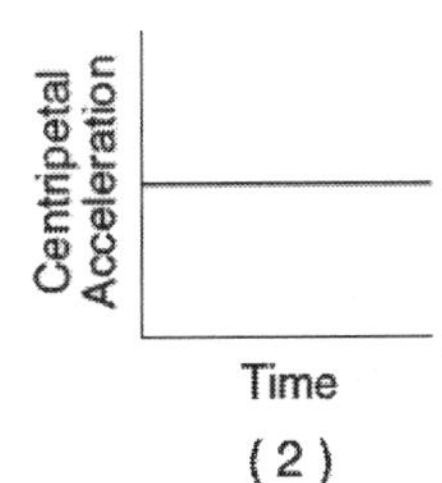

(2)

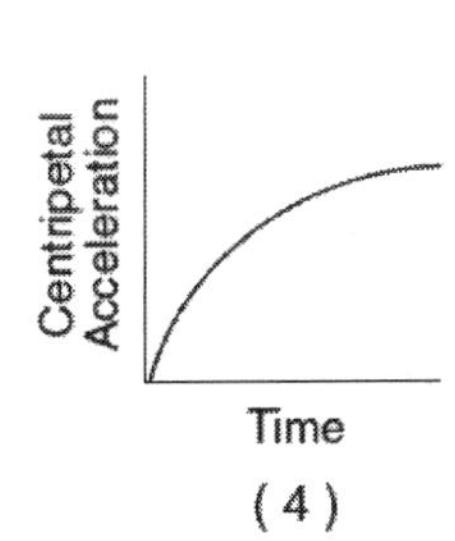

(4)

18. A 0.50-kilogram object moves in a horizontal circular path with a radius of 0.25 meter at a constant speed of 4.0 meters per second. What is the magnitude of the object's acceleration?
1. 8.0 m/s^2
2. 16 m/s^2
3. 32 m/s^2
4. 64 m/s^2

Name:_______________________________ Period: ___________

UCM-Circular Motion

Base your answers to questions 19 and 20 on the information below.

A go-cart travels around a flat, horizontal, circular track with a radius of 25 meters. The mass of the go-cart with the rider is 200 kilograms. The magnitude of the maximum centripetal force exerted by the track on the go-cart is 1200 newtons.

19. What is the maximum speed the 200-kilogram go-cart can travel without sliding off the track?
    1. 8.0 m/s
    2. 12 m/s
    3. 150 m/s
    4. 170 m/s

20. Which change would increase the maximum speed at which the go-cart could travel without sliding off this track?
    1. Decrease the coefficient of friction between the go-cart and the track.
    2. Decrease the radius of the track.
    3. Increase the radius of the track.
    4. Increase the mass of the go-cart.

21. A car moves with a constant speed in a clockwise direction around a circular path of radius r, as represented in the diagram below.

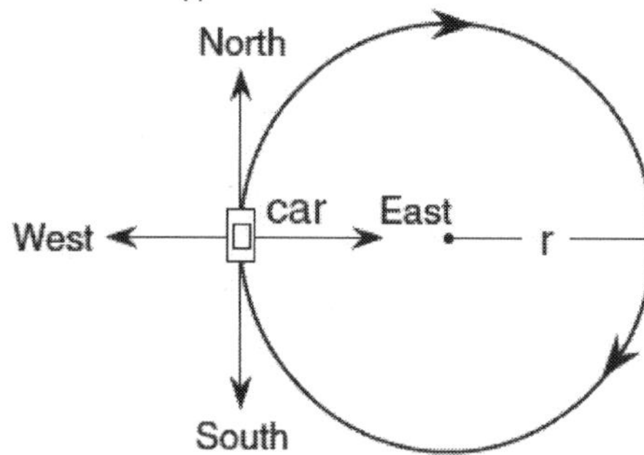

When the car is in the position shown, its acceleration is directed toward the
1. north
2. west
3. south
4. east

Base your answers to questions 22 and 23 on the information below.

In an experiment, a 0.028-kilogram rubber stopper is attached to one end of a string. A student whirls the stopper overhead in a horizontal circle with a radius of 1.0 meter. The stopper completes 10 revolutions in 10 seconds.

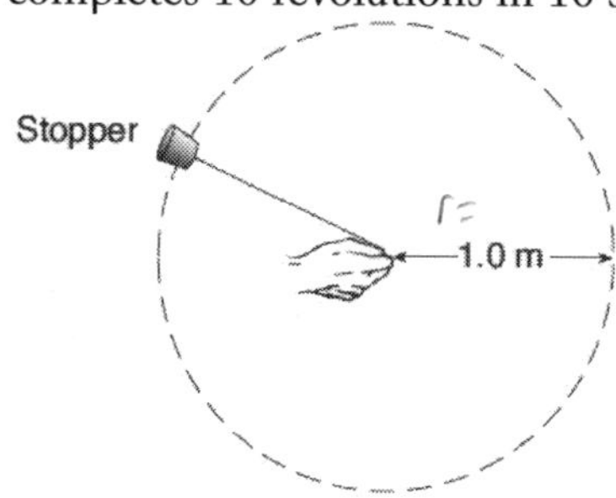

(Not drawn to scale)

22. Determine the speed of the whirling stopper.

23. Calculate the magnitude of the centripetal force on the whirling stopper. [Show all work, including the equation and substitution with units.]

24. The diagram below represents a mass, m, being swung clockwise at constant speed in a horizontal circle.

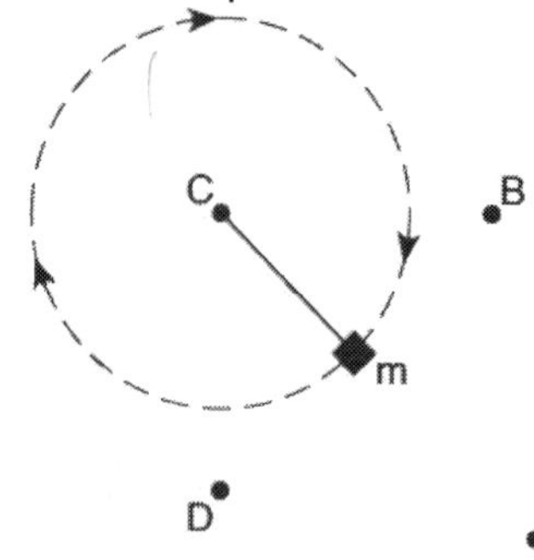

At the instant shown, the centripetal force acting on mass m is directed toward point
1. A
2. B
3. C
4. D

Name: ______________________ Period: __________

UCM-Circular Motion

Base your answers to questions 25 and 26 on the information below.

A 2.0×10^3-kilogram car travels at a constant speed of 12 meters per second around a circular curve of radius 30 meters.

25. What is the magnitude of the centripetal acceleration of the car as it goes around the curve?
    1. 0.40 m/s^2
    2. 4.8 m/s^2
    3. 800 m/s^2
    4. 9,600 m/s^2

26. As the car goes around the curve, the centripetal force is directed
    1. toward the center of the circular curve
    2. away from the center of the circular curve
    3. tangent to the curve in the direction of motion
    4. tangent to the curve opposite the direction of motion

27. A car round a horizontal curve of constant radius at a constant speed. Which diagram best represents the directions of both the car's velocity, v, and acceleration, a?

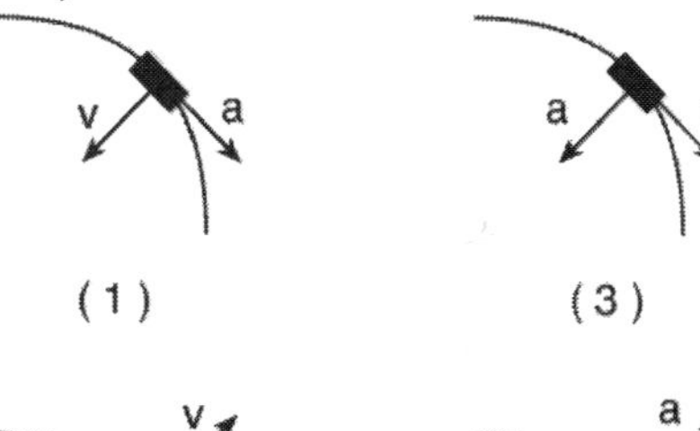

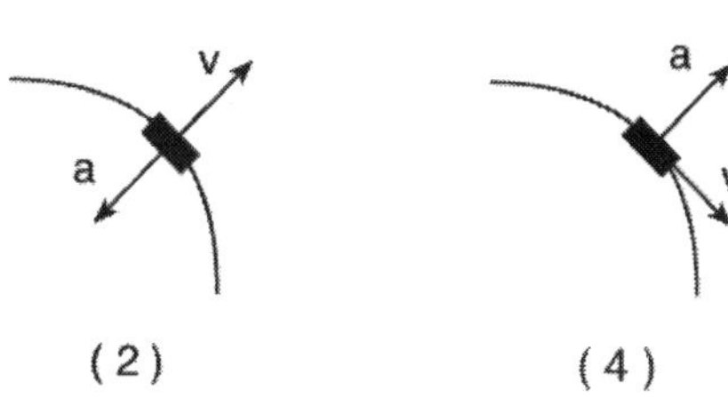

28. A 1750-kilogram car travels at a constant speed of 15 meters per second around a horizontal circular track with a radius of 45 meters. The magnitude of the centripetal force acting on the car is
    1. 5 N
    2. 583 N
    3. 8750 N
    4. 3.94×10^5 N

Base your answers to questions 29 through 31 on the information below.

The combined mass of a race car and its driver is 600 kilograms. Traveling at constant speed, the car completes one lap around a circular track of radius 160 meters in 36 seconds.

29. Calculate the speed of the car. [Show all work, including the equation and substitution with units.]

30. On the diagram below, draw an arrow to represent the direction of the net force acting on the car when it is in position A.

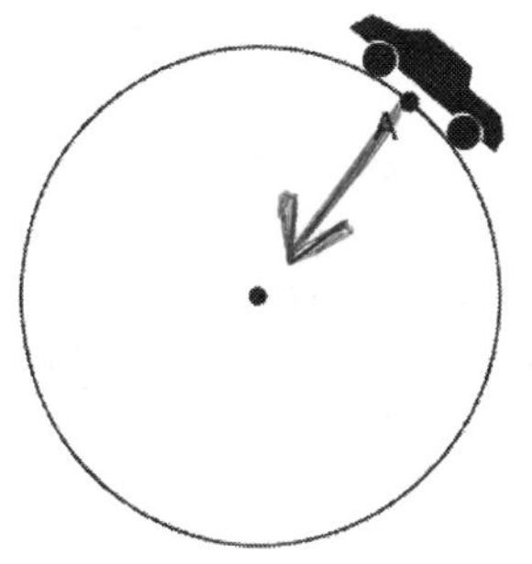

31. Calculate the magnitude of the centripetal acceleration of the car. [Show all work, including the equation and substitution with units.]

32. A ball of mass M at the end of a string is swung in a horizontal circular path of radius R at constant speed V. Which combination of changes would require the greatest increase in the centripetal force acting on the ball?
    1. doubling V and doubling R
    2. doubling V and halving R
    3. halving V and doubling R
    4. halving V and halving R

Name: ______________________________ Period: __________

UCM-Circular Motion

Base your answers to questions 33 through 36 on the information and table below.

Length (meters)	Period (seconds)
0.05	0.30
0.20	0.90
0.40	1.30
0.60	1.60
0.80	1.80
1.00	2.00

In a laboratory exercise, a student kept the mass and amplitude of swing of a simple pendulum constant. The length of the pendulum was increased and the period of the pendulum was measured. The student recorded the data in the table. You are to construct a graph on the grid provided following the directions below.

Period vs. Length of Pendulum

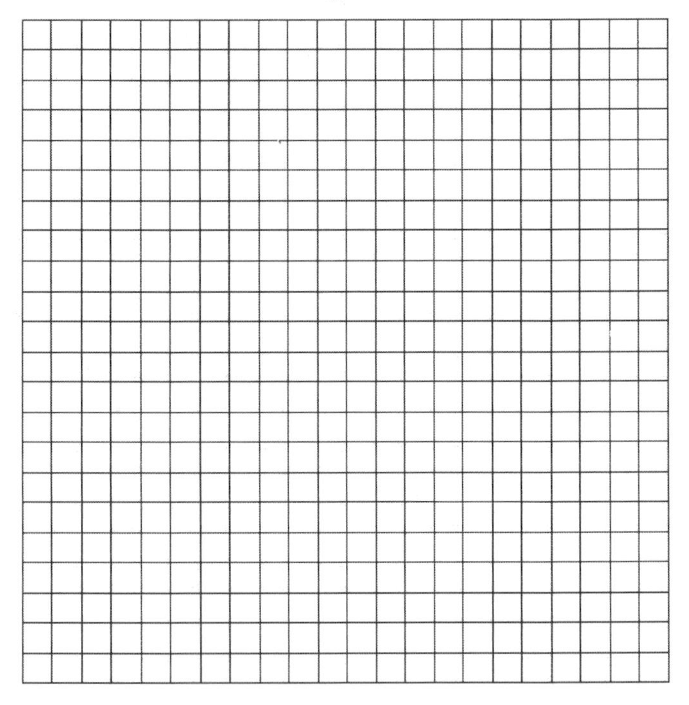

33. Label each axis with the appropriate physical quantity and unit, and mark an appropriate scale on each axis.

34. Plot the data points for period versus pendulum length.

35. Draw the best-fit line or curve for the data graphed.

36. Using your graph, determine the period of a pendulum whose length is 0.25 meter.

37. In the diagram below, a cart travels clockwise at constant speed in a horizontal circle.

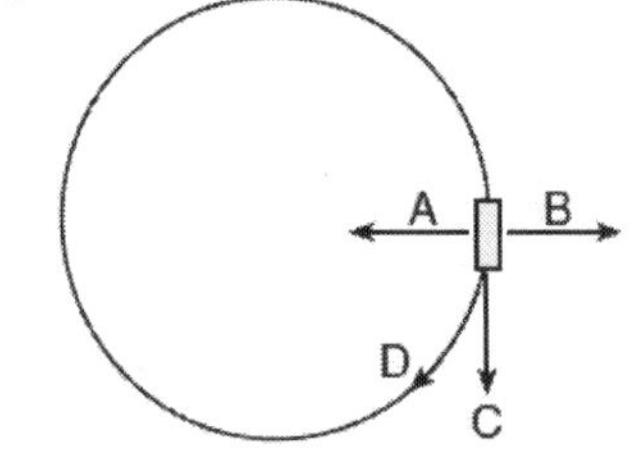

At the position shown in the diagram, which arrow indicates the direction of the centripetal acceleration of the cart?

1. A
2. B
3. C
4. D

38. The centripetal force acting on the space shuttle as it orbits Earth is equal to the shuttle's

1. inertia
2. momentum
3. velocity
4. weight

Name: ______________________ Period: __________

UCM-Circular Motion

Base your answers to questions 39 through 42 on the information and diagram below.

In an experiment, a rubber stopper is attached to one end of a string that is passed through a plastic tube before weights are attached to the other end. The stopper is whirled in a horizontal circular path at constant speed.

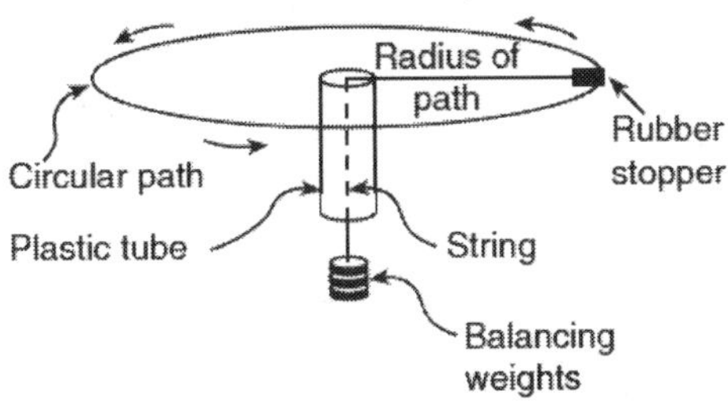

39. On the diagram of the top view (below), draw the path of the rubber stopper if the string breaks at the position shown.

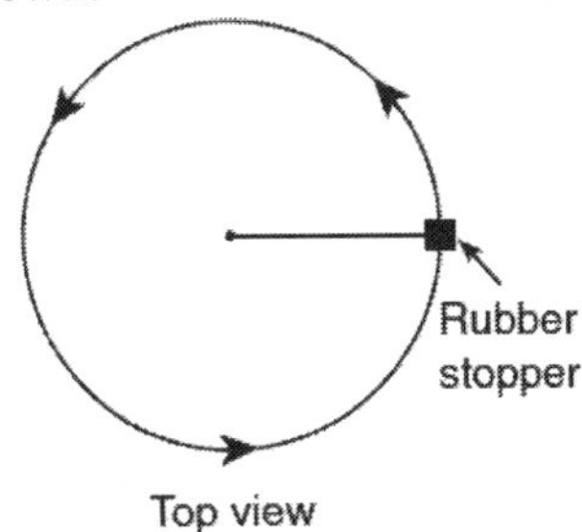

40. Describe what would happen to the radius of the circle if the student whirls the stopper at a greater speed without changing the balancing weights.

41. List three measurements that must be taken to show that the magnitude of the centripetal force is equal to the balancing weights.

42. The rubber stopper is now whirled in a vertical circle at the same speed. On the diagram, draw and label vectors to indicate the direction of the weight (F_g) and the direction of the centripetal force (F_C) at the position shown.

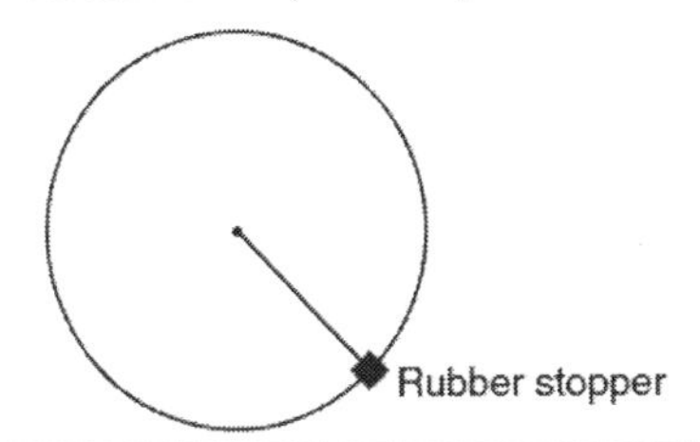

43. An unbalanced force of 40 newtons keeps a 5.0-kilogram object traveling in a circle of radius 2.0 meters. What is the speed of the object?
   1. 8.0 m/s
   2. 2.0 m/s
   3. 16 m/s
   4. 4.0 m/s

44. A student on an amusement park ride moves in a circular path with a radius of 3.5 meters once every 8.9 seconds. The student moves at an average speed of
   1. 0.39 m/s
   2. 1.2 m/s
   3. 2.5 m/s
   4. 4.3 m/s

45. A stone on the end of a string is whirled clockwise at constant speed in a horizontal circle as shown in the diagram. Which pair of arrows best represents the directions of the stone's velocity, v, and acceleration, a, at the position shown?

Top view
String
Stone

v a
(1)

v a
(3)

v a
(2)

v a
(4)

Name:______________________ Period: ____________

UCM-Circular Motion

Base your answers to questions 46 and 47 on the information below.

A 28-gram rubber stopper is attached to a string and whirled clockwise in a horizontal circle with a radius of 0.80 meter. The diagram in your answer booklet represents the motion of the rubber stopper. The stopper maintains a constant speed of 2.5 meters per second.

46. Calculate the magnitude of the centripetal acceleration of the stopper. [Show all work, including the equation and substitution with units.]

47. On the diagram below, draw an arrow showing the direction of the centripetal force acting on the stopper when it is at the position shown.

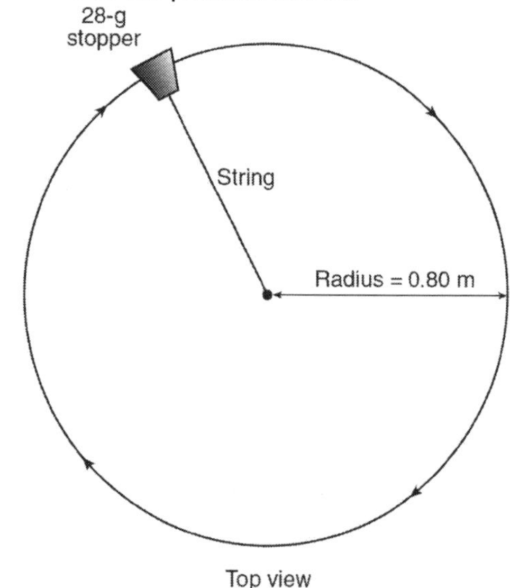

Top view

48. A 1.0×10^3-kilogram car travels at a constant speed of 20 meters per second around a horizontal circular track. The diameter of the track is 1.0×10^2 meters. The magnitude of the car's centripetal acceleration is
 1. $0.20\ m/s^2$
 2. $2.0\ m/s^2$
 3. $8.0\ m/s^2$
 4. $4.0\ m/s^2$

Base your answers to questions 49 and 50 on the information and diagram below.

A 1.5×10^3-kg car is driven at a constant speed of 12 meters per second counterclockwise around a horizontal circular track having a radius of 50 meters, as represented below.

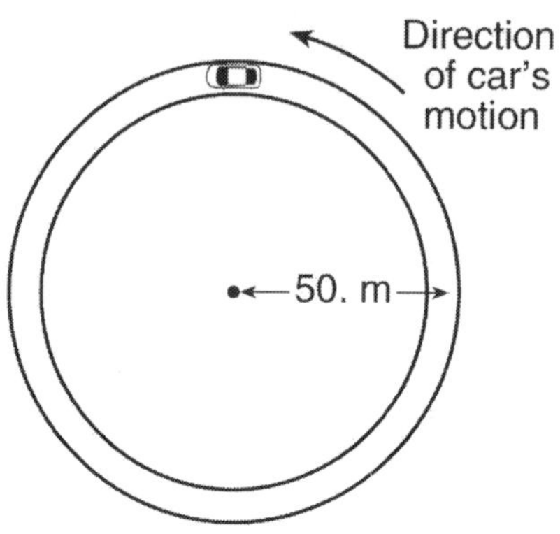

Track, as Viewed from Above

49. On the diagram above, draw a vector to indicate the direction of the velocity of the car when it is at the position shown. Start the arrow on the car.

50. Calculate the magnitude of the centripetal acceleration of the car. [Show all work, including the equation and substitution with units.

51. A body, B, is moving at constant speed in a horizontal circular path around point P. Which diagram shows the direction of the velocity (v) and the direction of the centripetal force (F_c) acting on the body?

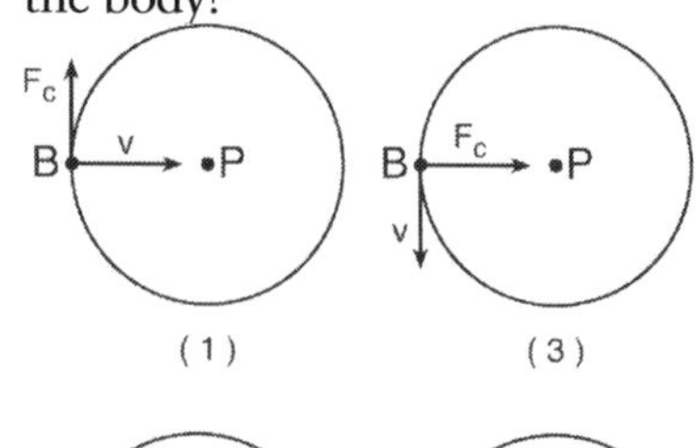

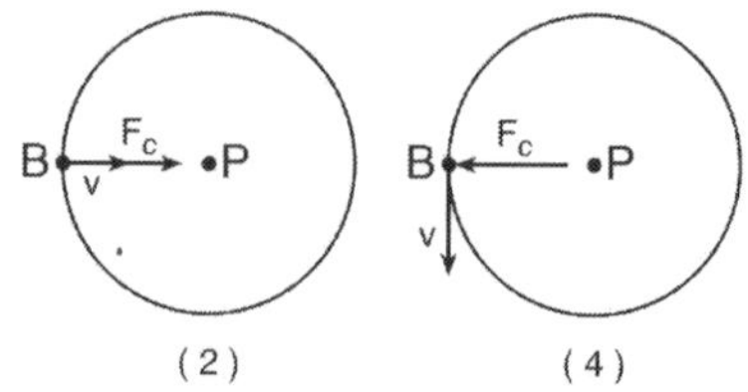

Name: ______________________ Period: ____________

UCM-Circular Motion

52. A boy pushes his sister on a swing. What is the frequency of oscillation of his sister on the swing if the boy counts 90 complete swings in 300 seconds?
    1. 0.30 Hz
    2. 2.0 Hz
    3. 1.5 Hz
    4. 18 Hz

Name:_______________________________ Period: __________

UCM-Gravity

1. A space probe is launched into space from Earth's surface. Which graph represents the relationship between the magnitude of the gravitational force exerted on Earth by the space probe and the distance between the space probe and the center of Earth?

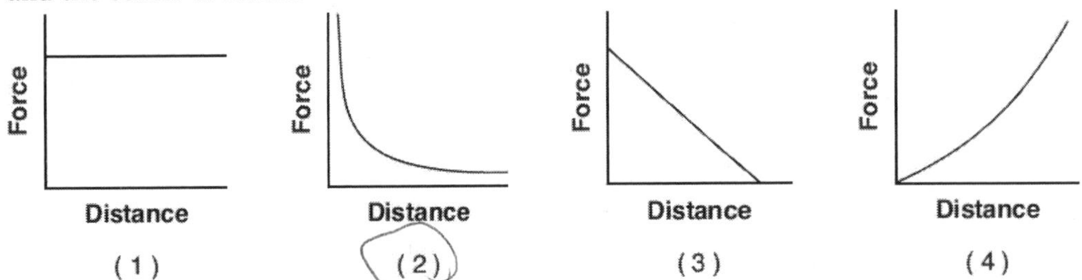

2. The diagram shows two bowling balls, A and B, each having a mass of 7 kilograms, placed 2 meters apart.

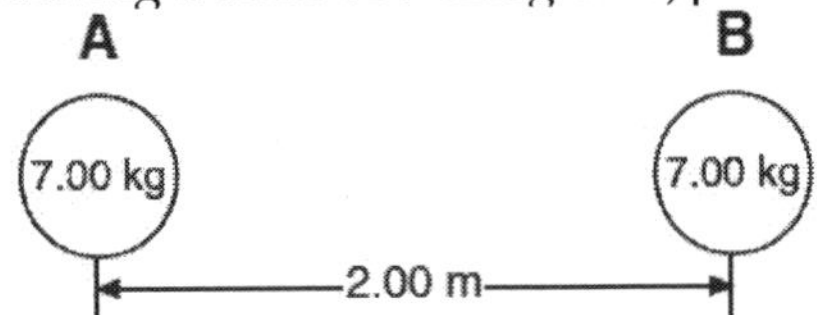

What is the magnitude of the gravitational force exerted by ball A on ball B?
1. 8.17×10^{-9} N
2. 1.63×10^{-9} N
3. 8.17×10^{-10} N
4. 1.17×10^{-10} N

3. A 60-kg physics student would weigh 1560 N on the surface of planet X. What is the magnitude of the acceleration due to gravity on the surface of planet X?
1. 0.038 m/s^2
2. 6.1 m/s^2
3. 9.8 m/s^2
4. 26 m/s^2

4. Earth's mass is approximately 81 times the mass of the Moon. If Earth exerts a gravitational force of magnitude F on the Moon, the magnitude of the gravitational force of the Moon on Earth is
1. F
2. F/81
3. 9F
4. 81F

5. An object weighs 100 newtons on Earth's surface. When it is moved to a point one Earth radius above Earth's surface, it will weigh
1. 25 N
2. 50 N
3. 100 N
4. 400 N

6. A container of rocks with a mass of 65 kilograms is brought back from the Moon's surface where the acceleration due to gravity is 1.62 meters per second^2. What is the weight of the container of rocks on Earth's surface?
1. 638 N
2. 394 N
3. 105 N
4. 65 N

7. The graph below represents the relationship between gravitational force and mass for objects near the surface of Earth.

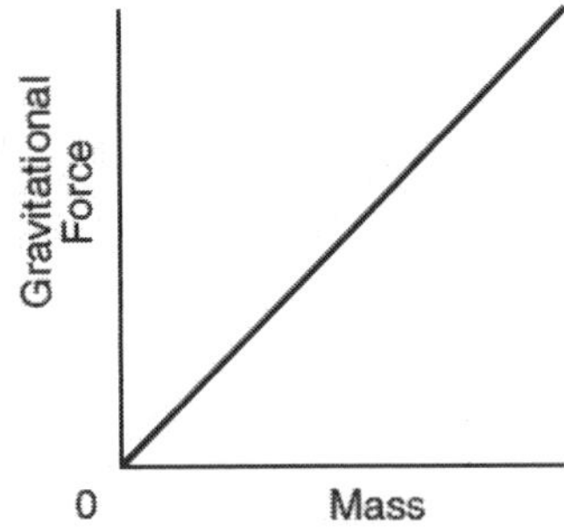

The slope of the graph represents the
1. acceleration due to gravity
2. universal gravitational constant
3. momentum of objects
4. weight of objects

8. A person weighing 785 newtons on the surface of Earth would weigh 298 newtons on the surface of Mars. What is the magnitude of the gravitational field strength on the surface of Mars?
1. 2.63 N/kg
2. 3.72 N/kg
3. 6.09 N/kg
4. 9.81 N/kg

Name:________________________________ Period: ______________

UCM-Gravity

Base your answers to questions 9 through 11 on the passage and data table below.

The net force on a planet is due primarily to the other planets and the Sun. By taking into account all the forces acting on a planet, investigators calculated the orbit of each planet.

A small discrepancy between the calculated orbit and the observed orbit of the planet Uranus was noted. It appeared that the sum of the forces on Uranus did not equal its mass times its acceleration, unless there was another force on the planet that was not included in the calculation. Assuming that this force was exerted by an unobserved planet, two scientists working independently calculated where this unknown planet must be in order to account for the discrepancy. Astronomers pointed their telescopes in the predicted direction and found the planet we now call Neptune.

Data Table

Mass of the Sun	1.99×10^{30} kg
Mass of Uranus	8.73×10^{25} kg
Mass of Neptune	1.03×10^{26} kg
Mean distance of Uranus to the Sun	2.87×10^{12} m
Mean distance of Neptune to the Sun	4.50×10^{12} m

9. What fundamental force is the author referring to in this passage as a force between planets?

10. The diagram at right represents Neptune, Uranus, and the Sun in a straight line. Neptune is 1.63×10^{12} meters from Uranus.

    Calculate the magnitude of the interplanetary force of attraction between Uranus and Neptune at this point. [Show all work, including the equation and substitution with units.]

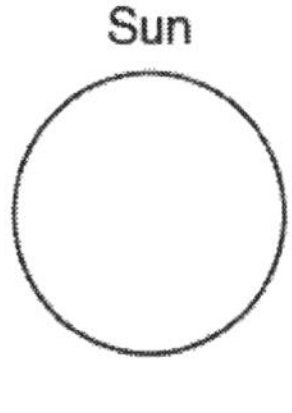

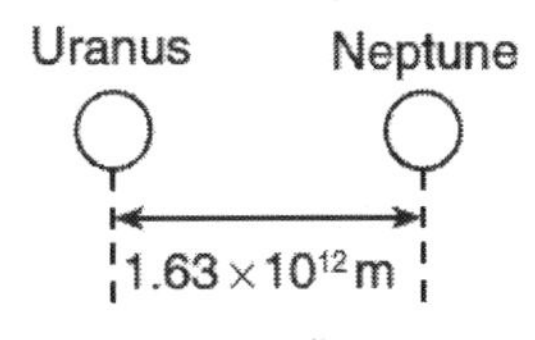

(Not drawn to scale)

11. The magnitude of the force the Sun exerts on Uranus is 1.41×10^{21} newtons. Explain how it is possible for the Sun to exert a greater force on Uranus than Neptune exerts on Uranus.

12. When Earth and the Moon are separated by a distance of 3.84×10^8 meters, the magnitude of the gravitational force of attraction between them is 2.0×10^{20} newtons. What would be the magnitude of this gravitational force of attraction if Earth and the Moon were separated by a distance of 1.92×10^8 meters?
    1. 5.0×10^{19} N
    2. 2.0×10^{20} N
    3. 4.0×10^{20} N
    4. 8.0×10^{20} N

13. An astronaut weighs 8.00×10^2 newtons on the surface of Earth. What is the weight of the astronaut 6.37×10^6 meters above the surface of Earth?
    1. 0.00 N
    2. 2.00×10^2 N
    3. 1.60×10^3 N
    4. 3.20×10^3 N

Name:______________________________ Period: ____________

UCM-Gravity

Base your answers to questions 14 and 15 on the information and table below.

The weight of an object was determined at five different distances from the center of Earth. The results are shown in the table below. Position A represents results for the object at the surface of Earth.

Position	Distance from Earth's Center (m)	Weight (N)
A	6.37×10^6	1.0×10^3
B	1.27×10^7	2.5×10^2
C	1.91×10^7	1.1×10^2
D	2.55×10^7	6.3×10^1
E	3.19×10^7	4.0×10^1

14. The approximate mass of the object is
    1. 0.01 kg
    2. 10 kg
    3. 100 kg
    4. 1,000 kg

15. At what distance from the center of Earth is the weight of the object approximately 28 newtons?
    1. 3.5×10^7 m
    2. 3.8×10^7 m
    3. 4.1×10^7 m
    4. 4.5×10^7 m

16. Gravitational forces differ from electrostatic forces in that gravitational forces are
    1. attractive, only
    2. repulsive, only
    3. neither attractive nor repulsive
    4. both attractive and repulsive

17. The gravitational force of attraction between Earth and the Sun is 3.52×10^{22} newtons. Calculate the mass of the Sun. [Show all work, including the equation and substitution with units.]

18. A 5.0-kilogram sphere, starting from rest, falls freely 22 meters in 3.0 seconds near the surface of a planet. Compared to the acceleration due to gravity near Earth's surface, the acceleration due to gravity near the surface of the planet is approximately
    1. the same
    2. twice as great
    3. one-half as great
    4. four times as great

19. Which diagram best represents the gravitational field lines surrounding Earth?

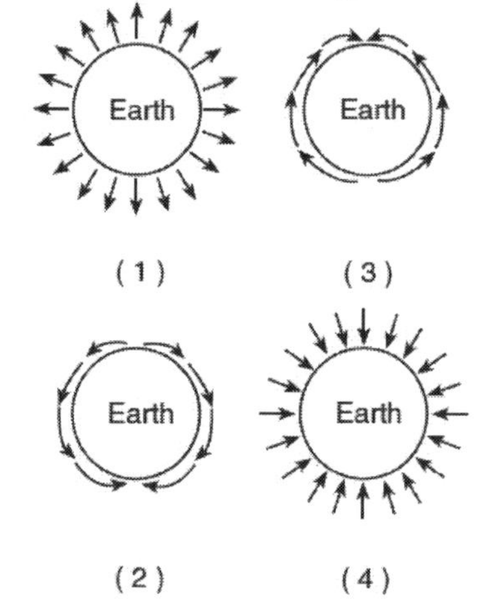

20. The diagram below represents two satellites of equal mass, A and B, in circular orbits around a planet.

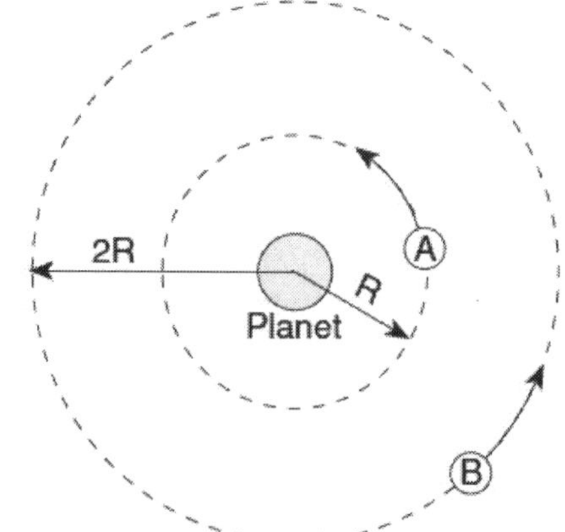

Compared to the magnitude of the gravitational force of attraction between satellite A and the planet, the magnitude of the gravitational force of attraction between satellite B and the planet is
    1. half as great
    2. twice as great
    3. one-fourth as great
    4. four times as great

Name: ______________________________ Period: ____________

UCM-Gravity

Base your answers to questions 21 and 22 on the information below. [Show all work, including the equation and substitution with units.]

Io (pronounced "EYE oh") is one of Jupiter's moons discovered by Galileo. Io is slightly larger than Earth's Moon.

The mass of Io is 8.93×10^{22} kilograms and the mass of Jupiter is 1.90×10^{27} kilograms. The distance between the centers of Io and Jupiter is 4.22×10^{8} meters.

21. Calculate the magnitude of the gravitational force of attraction that Jupiter exerts on Io.

22. Calculate the magnitude of the acceleration of Io due to the gravitational force exerted by Jupiter.

23. Which diagram best represents the gravitational forces, F_g, between a satellite, S, and Earth?

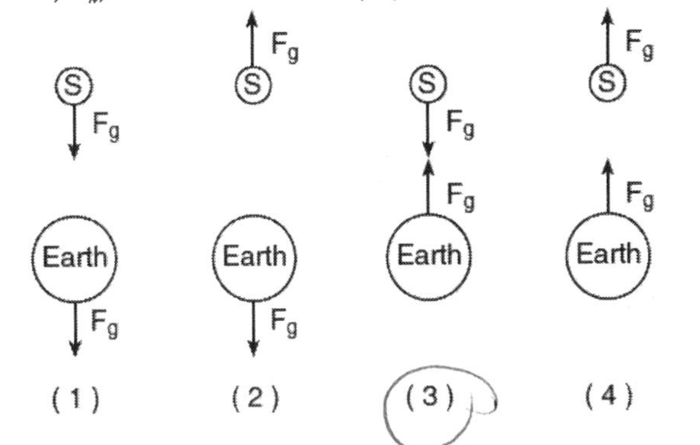

24. Two physics students have been selected by NASA to accompany astronauts on a future mission to the Moon. The students are to design and carry out a simple experiment to measure the acceleration due to gravity on the surface of the Moon.

    Describe an experiment that the students could conduct to measure the acceleration due to gravity on the Moon. Your description must include:
    - the equipment needed
    - what quantities would be measured using the equipment
    - what procedure the students should follow in conducting their experiment
    - what equations and/or calculations the students would need to do to arrive at a value for the acceleration due to gravity on the Moon.

25. As a meteor moves from a distance of 16 Earth radii to a distance of 2 Earth radii from the center of Earth, the magnitude of the gravitational force between the meteor and Earth becomes
    1. 1/8 as great
    2. 8 times as great
    3. 64 times as great
    4. 4 times as great

Name:______________________________ Period: ____________

UCM-Gravity

26. A 25-kilogram space probe fell freely with an acceleration of 2 meters per second^2 just before it landed on a distant planet. What is the weight of the space probe on that planet?
    1. 12.5 N
    2. 25 N
    3. 50 N
    4. 250 N

27. The acceleration due to gravity on the surface of planet X is 19.6 meters per second^2. If an object on the surface of this planet weighs 980 newtons, the mass of the object is
    1. 50 kg
    2. 100 kg
    3. 490 N
    4. 908 N

28. What is the acceleration due to gravity at a location where a 15-kilogram mass weighs 45 newtons?
    1. 675 m/s^2
    2. 9.81 m/s^2
    3. 3.00 m/s^2
    4. 0.333 m/s^2

29. As an astronaut travels from the surface of Earth to a position that is four times as far away from the center of Earth, the astronaut's
    1. mass decreases
    2. mass remains the same
    3. weight increases
    4. weight remains the same

30. A satellite weighs 200 newtons on the surface of Earth. What is its weight at a distance of one Earth radius above the surface of Earth?
    1. 50 N
    2. 100 N
    3. 400 N
    4. 800 N

31. A 2.00-kilogram object weighs 19.6 newtons on Earth. If the acceleration due to gravity on Mars is 3.71 meters per second^2, what is the object's mass on Mars?
    1. 2.64 kg
    2. 2.00 kg
    3. 19.6 N
    4. 7.42 N

32. A 1200-kilogram space vehicle travels at 4.8 meters per second along the level surface of Mars. If the magnitude of the gravitational field strength on the surface of Mars is 3.7 newtons per kilogram, the magnitude of the normal force acting on the vehicle is
    1. 320 N
    2. 930 N
    3. 4400 N
    4. 5800 N

33. What is the weight of a 2.00-kilogram object on the surface of Earth?
    1. 4.91 N
    2. 2.00 N
    3. 9.81 N
    4. 19.6 N

34. A 2.0-kilogram object is falling freely near Earth's surface. What is the magnitude of the gravitational force that Earth exerts on the object?
    1. 20 N
    2. 2.0 N
    3. 0.20 N
    4. 0.0 N

35. Calculate the magnitude of the centripetal force acting on Earth as it orbits the Sun, assuming a circular orbit and an orbital speed of 3.00×10^4 meters per second. [Show all work, including the equation and substitution with units.]

36. On a small planet, an astronaut uses a vertical force of 175 newtons to lift an 87.5-kilogram boulder at constant velocity to a height of 0.350 meter above the planet's surface. What is the magnitude of the gravitational field strength on the surface of the planet?
    1. 0.500 N/kg
    2. 2.00 N/kg
    3. 9.81 N/kg
    4. 61.3 N/kg

37. Calculate the magnitude of the average gravitational force between Earth and Moon. [Show all work, including the equation and substitution with units.]

Name: ______________________ Period: __________

UCM-Gravity

38. Which graph represents the relationship between the magnitude of the gravitational force exerted by Earth on a spacecraft and the distance between the center of the spacecraft and center of Earth? [Assume constant mass for the spacecraft.]

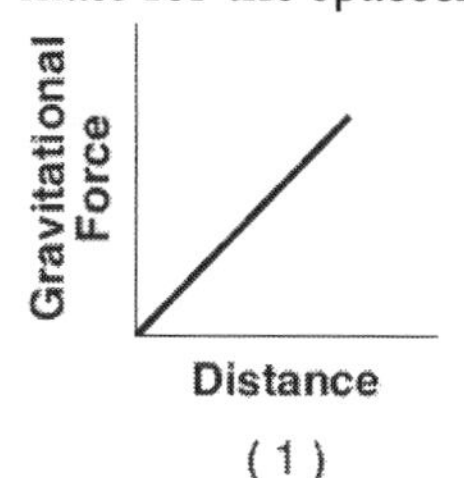

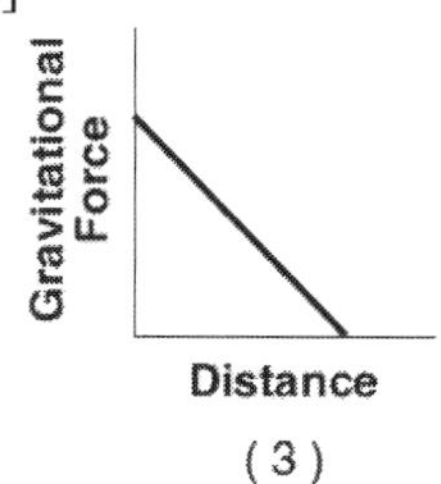

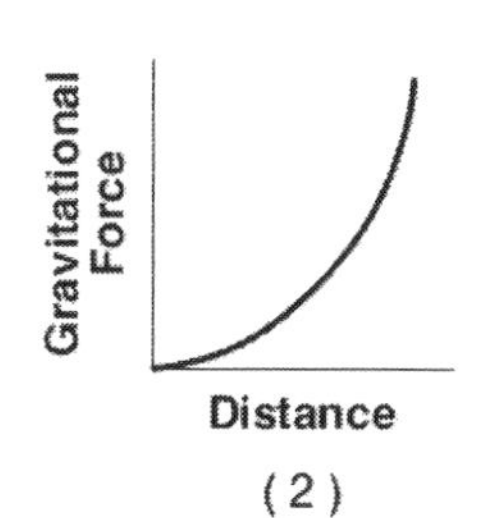

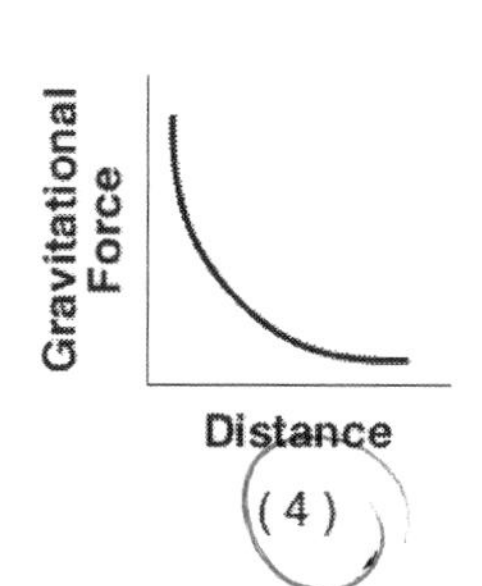

39. In which diagram do the field lines best represent the gravitational field around Earth?

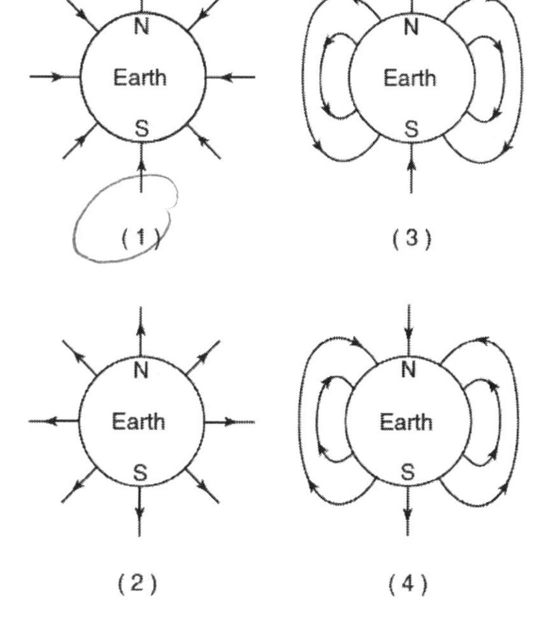

40. At a certain location, a gravitational force with a magnitude of 350 newtons acts on a 70-kilogram astronaut. What is the magnitude of the gravitational field strength at this location?
    1. 0.20 kg/N
    2. 5.0 N/kg
    3. 9.8 m/s²
    4. 25,000 N•kg

41. Which graph represents the relationship between the magnitude of the gravitational force, F_g, between two masses and the distance, r, between the centers of the masses?

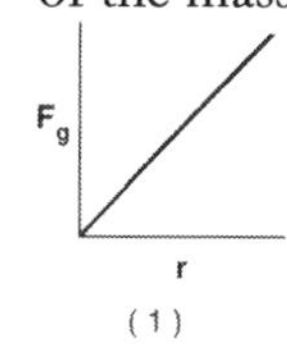

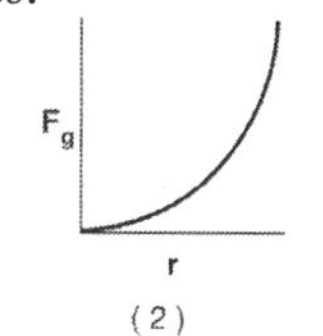

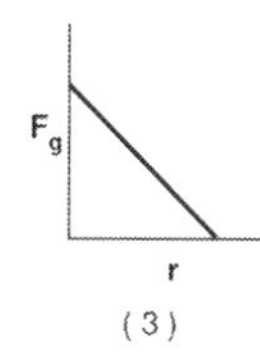

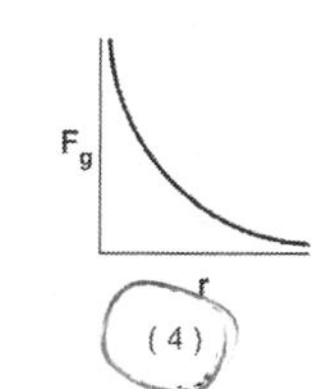

42. A 2.0-kilogram mass is located 3.0 meters above the surface of Earth. What is the magnitude of Earth's gravitational field strength at this location?
    1. 4.9 N/kg
    2. 2.0 N/kg
    3. 9.8 N/kg
    4. 20 N/kg

Name: ______________________________ Period: __________

UCM-Gravity

Base your answers to questions 43 through 45 on the information below and on your knowledge of physics. [Show all work, including the equation and substitution with units.]

Pluto orbits the Sun at an average distance of 5.91×10^{12} meters. Pluto's diameter is 2.30×10^{6} meters and its mass is 1.31×10^{22} kilograms.

Charon orbits Pluto with their centers separated by a distance of 1.96×10^{7} meters. Charon has a diameter of 1.21×10^{6} meters and a mass of 1.55×10^{21} kilograms.

43. Calculate the magnitude of the gravitational force of attraction that Pluto exerts on Charon.

44. Calculate the magnitude of the acceleration of Charon toward Pluto.

45. State the reason why the magnitude of the Sun's gravitational force on Pluto is greater than the magnitude of the Sun's gravitational force on Charon.

46. The Hubble telescope's orbit is 5.6×10^{5} meters above Earth's surface. The telescope has a mass of 1.1×10^{4} kilograms. Earth exerts a gravitational force of 9.1×10^{4} newtons on the telescope. The magnitude of Earth's gravitational field strength at this location is
    1. 1.5×10^{-20} N/kg
    2. 0.12 N/kg
    3. 8.3 N/kg
    4. 9.8 N/kg

Name: ______________________________ Period: ____________

Momentum-Impulse

1. A 1,200-kilogram car traveling at 10 meters per second hits a tree and is brought to rest in 0.10 second. What is the magnitude of the average force acting on the car to bring it to rest?
   1. 1.2×10^2 N
   2. 1.2×10^3 N
   3. 1.2×10^4 N
   4. 1.2×10^5 N

   m=1200kg
   v=10m/s
   t=0.1sec
   F=?

2. A 50-kilogram student threw a 0.40-kilogram ball with a speed of 20 meters per second. What was the magnitude of the impulse that the student exerted on the ball?
   1. 8.0 N·s
   2. 78 N·s
   3. 4.0×10^2 N·s
   4. 1.0×10^3 N·s

3. In the diagram below, a 60-kilogram rollerskater exerts a 10-newton force on a 30-kilogram rollerskater for 0.20 second.

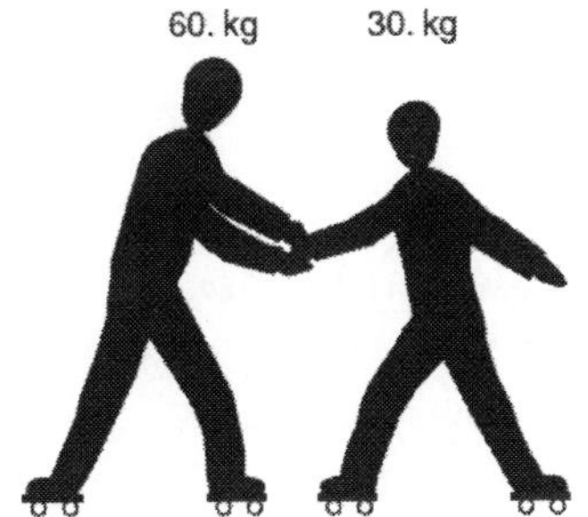

   What is the magnitude of the impulse applied to the 30-kilogram rollerskater?
   1. 50 N·s
   2. 2.0 N·s
   3. 6.0 N·s
   4. 12 N·s

4. Two carts are pushed apart by an expanding spring, as shown in the diagram below.

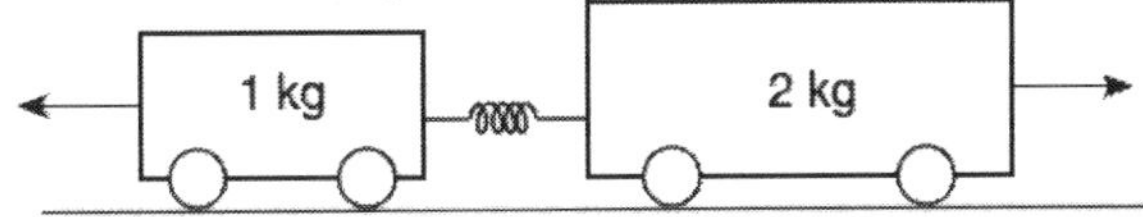

   If the average force on the 1-kilogram cart is 1 newton, what is the average force on the 2-kilogram cart?
   1. 1 N
   2. 0.0 N
   3. 0.5 N
   4. 4 N

5. What is the speed of a 1.0×10^3-kilogram car that has a momentum of 2.0×10^4 kilogram·meters per second east?
   1. 5.0×10^{-2} m/s
   2. 2.0×10^1 m/s
   3. 1.0×10^4 m/s
   4. 2.0×10^7 m/s

6. A motorcycle being driven on a dirt path hits a rock. It's 60-kilogram cyclist is projected over the handlebars at 20 meters per second into a haystack. if the cyclist is brought to rest in 0.50 second, the magnitude of the average force exerted on the cyclist by the haystack is
   1. 6.0×10^1 N
   2. 5.9×10^2 N
   3. 1.2×10^3 N
   4. 2.4×10^3 N

7. A 70-kilogram hockey player skating east on an ice rink is hit by a 0.1-kilogram hockey puck moving toward the west. The puck exerts a 50-newton force toward the west on the player. Determine the magnitude of the force that the player exerts on the puck during this collision.

8. Which situation will produce the greatest change of momentum for a 1.0-kilogram cart?
   1. accelerating it from rest to 3.0 m/s
   2. accelerating it from 2.0 m/s to 4.0 m/s
   3. applying a net force of 5.0 N for 2.0 s
   4. applying a net force of 10.0 N for 0.5 s

9. A 0.149-kilogram baseball, initially moving at 15 meters per second, is brought to rest in 0.040 second by a baseball glove on a catcher's hand. The magnitude of the average force exerted on the ball by the glove is
   1. 2.2 N
   2. 2.9 N
   3. 17 N
   4. 56 N

Name:____________________ Period: __________

Momentum-Impulse

10. Calculate the magnitude of the impulse applied to a 0.75-kilogram cart to change its velocity from 0.50 meter per second east to 2.00 meters per second east. [Show all work, including the equation and substitution with units.]

11. Which is a scalar quantity?
    1. acceleration
    2. momentum
    3. speed
    4. displacement

12. A 0.45-kilogram football traveling at a speed of 22 meters per second is caught by an 84-kilogram stationary receiver. If the football comes to rest in the receiver's arms, the magnitude of the impulse imparted to the receiver by the ball is
    1. 1800 N·s
    2. 9.9 N·s
    3. 4.4 N·s
    4. 3.8 N·s

13. A force of 6.0 newtons changes the momentum of a moving object by 3.0 kilogram·meters per second. How long did the force act on the mass?
    1. 1.0 s
    2. 2.0 s
    3. 0.25 s
    4. 0.50 s

14. A 1000-kilogram car traveling due east at 15 meters per second is hit from behind and receives a forward impulse of 6000 newton-seconds. Determine the magnitude of the car's change in momentum due to this impulse.

15. Cart A has a mass of 2 kilograms and a speed of 3 meters per second. Cart B has a mass of 3 kilograms and a speed of 2 meters per second. Compared to the inertia and magnitude of momentum of cart A, cart B has
    1. the same inertia and a smaller magnitude of momentum
    2. the same inertia and the same magnitude of momentum
    3. greater inertia and a smaller magnitude of momentum
    4. greater inertia and the same magnitude of momentum

16. A 6.0-kilogram block, sliding to the east across a horizontal, frictionless surface with a momentum of 30 kilogram·meters per second, strikes an obstacle. The obstacle exerts an impulse of 10 newton·seconds to the west on the block. The speed of the block after the collision is
    1. 1.7 m/s
    2. 3.3 m/s
    3. 5.0 m/s
    4. 20 m/s

17. A 60-kilogram student jumps down from a laboratory counter. At the instant he lands on the floor his speed is 3 meters per second. If the student stops in 0.2 second, what is the average force of the floor on the student?
    1. 1×10^{-2} N
    2. 1×10^{2} N
    3. 9×10^{2} N
    4. 4 N

18. A 2.0-kilogram laboratory cart is sliding across a horizontal frictionless surface at a constant velocity of 4.0 meters per second east. What will be the cart's velocity after a 6.0-newton westward force acts on it for 2.0 seconds?
    1. 2.0 m/s east
    2. 2.0 m/s west
    3. 10 m/s east
    4. 10 m/s west

Name: ______________________________ Period: ____________

Momentum-Impulse

19. A 40-kilogram mass is moving across a horizontal surface at 5.0 meters per second. What is the magnitude of the net force required to bring the mass to a stop in 8.0 seconds?
    1. 1.0 N
    2. 5.0 N
    3. 25 N
    4. 40 N

20. A 0.15-kilogram baseball moving at 20 meters per second is stopped by a catcher in 0.010 second. The average force stopping the ball is
    1. 3.0×10^{-2} N
    2. 3.0×10^{0} N
    3. 3.0×10^{1} N
    4. 3.0×10^{2} N

21. A 2.0-kilogram body is initially traveling at a velocity of 40 meters per second east. If a constant force of 10 newtons due east is applied to the body for 5.0 seconds, the final speed of the body is
    1. 15 m/s
    2. 25 m/s
    3. 65 m/s
    4. 130 m/s

22. A 75-kilogram hockey player is skating across the ice at a speed of 6.0 meters per second. What is the magnitude of the average force required to stop the player in 0.65 second?
    1. 120 N
    2. 290 N
    3. 690 N
    4. 920 N

23. A bicycle and its rider have a combined mass of 80 kg and a speed of 6 m/s. What is the magnitude of the average force needed to bring the bicycle and its rider to a stop in 4.0 seconds?
    1. 1.2×10^{2} N
    2. 3.2×10^{2} N
    3. 4.8×10^{2} N
    4. 1.9×10^{3} N

24. A 5-kilogram block slides along a horizontal, frictionless surface at 10 meters per second for 4 seconds. The magnitude of the block's momentum is
    1. 200 kg·m/s
    2. 50 kg·m/s
    3. 20 kg·m/s
    4. 12.5 kg·m/s

25. Calculate the time required for a 6000-newton net force to stop a 1200-kilogram car initially traveling at 10 meters per second. [Show all work, including the equation and substitution with units.]

26. Which term identifies a scalar quantity?
    1. displacement
    2. momentum
    3. velocity
    4. time

27. A baseball bat exerts an average force of 600 newtons east on a ball, imparting an impulse of 3.6 newton•seconds east to the ball. Calculate the amount of time the baseball bat is in contact with the ball. [Show all work, including the equation and substitution with units.]

28. An air bag is used to safely decrease the momentum of a driver in a car accident. The air bag reduces the magnitude of the force acting on the driver by
    1. increasing the length of time the force acts on the driver
    2. decreasing the distance over which the force acts on the driver
    3. increasing the rate of acceleration of the driver
    4. decreasing the mass of the driver

29. A 3.0-kilogram object is acted upon by an impulse having a magnitude of 15 newton•seconds. What is the magnitude of the object's change in momentum due to this impulse?
    1. 5.0 kg•m/s
    2. 15 kg•m/s
    3. 3.0 kg•m/s
    4. 45 kg•m/s

Name:__________________________ Period: __________

Momentum-Impulse

30. A 1.5-kilogram cart initially moves at 2.0 meters per second. It is brought to rest by a constant net force in 0.30 second. What is the magnitude of the net force?
    1. 0.40 N
    2. 0.90 N
    3. 10 N
    4. 15 N

31. A 0.0600-kilogram ball traveling at 60.0 meters per second hits a concrete wall. What speed must a 0.0100-kilogram bullet have in order to hit the wall with the same magnitude of momentum as the ball?
    1. 3.60 m/s
    2. 6.00 m/s
    3. 360 m/s
    4. 600 m/s

Name: ______________________ Period: __________

Momentum-Conservation

1. A 1.2-kilogram block and a 1.8-kilogram block are initially at rest on a frictionless, horizontal surface. When a compressed spring between the blocks is released, the 1.8-kilogram block moves to the right at 2.0 meters per second, as shown.

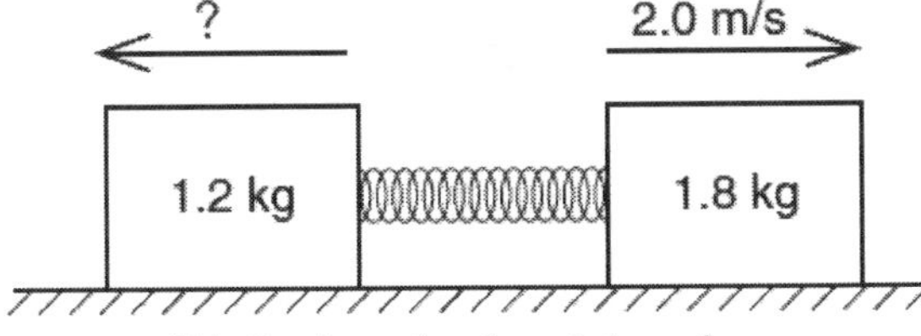

What is the speed of the 1.2-kilogram block after the spring is released?
1. 1.4 m/s
2. 2.0 m/s
3. 3.0 m/s
4. 3.6 m/s

Base your answers to questions 2 and 3 on the information below.

An 8.00-kilogram ball is fired horizontally from a 1.00×10^3-kilogram cannon initially at rest. After having been fired, the momentum of the ball is 2.40×10^3 kilogram·meters per second east. [Neglect friction.]

2. Calculate the magnitude of the cannon's velocity after the ball is fired. [Show all work, including the equation and substitution with units.]

3. Identify the direction of the cannon's velocity after the ball is fired.

4. Ball A of mass 5.0 kilograms moving at 20 meters per second collides with ball B of unknown mass moving at 10 meters per second in the same direction. After the collision, ball A moves at 10 meters per second and ball B at 15 meters per second, both still in the same direction. What is the mass of ball B?
1. 6.0 kg
2. 2.0 kg
3. 10 kg
4. 12 kg

5. In the diagram below, scaled vectors represent the momentum of each of two masses, A and B, sliding toward each other on a frictionless, horizontal surface.

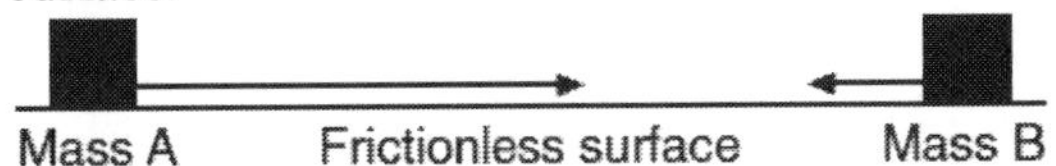

Which scaled vector best represents the momentum of the system after the masses collide?

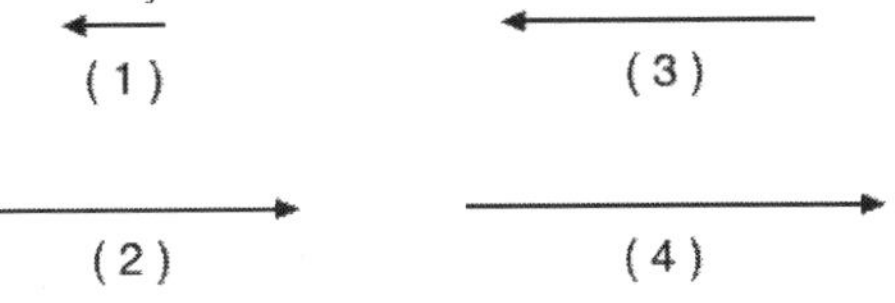

6. At the circus, a 100-kilogram clown is fired 15 meters per second from a 500-kilogram cannon. What is the recoil speed of the cannon?
1. 75 m/s
2. 15 m/s
3. 3.0 m/s
4. 5.0 m/s

7. A woman with horizontal velocity v_1 jumps off a dock into a stationary boat. After landing in the boat, the woman and the boat move with velocity v_2. Compared to velocity v_1, velocity v_2 has
1. the same magnitude and the same direction
2. the same magnitude and the opposite direction
3. smaller magnitude and the same direction
4. larger magnitude and the same direction

Name:____________________ Period: ________

Momentum-Conservation

8. On a snow-covered road, a car with a mass of 1.1×10^3 kilograms collides head-on with a van having a mass of 2.5×10^3 kilograms traveling at 8.0 meters per second. As a result of the collision, the vehicles lock together and immediately come to rest. Calculate the speed of the car immediately before the collision. [Neglect friction.] [Show all work, including the equation and substitution with units.]

9. A 3.0-kilogram steel block is at rest on a frictionless horizontal surface. A 1.0-kilogram lump of clay is propelled horizontally at 6.0 meters per second toward the block as shown in the diagram below.

Steel block

Clay

6.0 m/s

1.0 kg

3.0 kg

Frictionless surface

Upon collision, the clay and steel block stick together and move to the right with a speed of
1. 1.5 m/s
2. 2.0 m/s
3. 3.0 m/s
4. 6.0 m/s

10. A 1.0-kilogram laboratory cart moving with a velocity of 0.50 meter per second due east collides with and sticks to a similar cart initially at rest. After the collision, the two carts move off together with a velocity of 0.25 meter per second due east. The total momentum of this frictionless system is
1. zero before the collision
2. zero after the collision
3. the same before and after the collision
4. greater before the collision than after the collision.

11. Which two quantities can be expressed using the same units?
1. energy and force
2. impulse and force
3. momentum and energy
4. impulse and momentum

12. A 3.1-kilogram gun initially at rest is free to move. When a 0.015-kilogram bullet leaves the gun with a speed of 500 meters per second, what is the speed of the gun?
1. 0.0 m/s
2. 2.4 m/s
3. 7.5 m/s
4. 500 m/s

Base your answers to questions 13 and 14 on the information below. Show all work, including the equation and substitution with units.

A 1200-kilogram car moving at 12 meters per second collides with a 2300-kilogram car that is waiting at rest at a traffic light. After the collision, the cars lock together and slide. Eventually, the combined cars are brought to rest by a force of kinetic friction as the rubber tires slide across the dry, level asphalt road surface.

13. Calculate the speed of the locked-together cars immediately after the collision.

14. Calculate the magnitude of the frictional force that brings the locked-together cars to rest.

Name: ______________________________ Period: __________

Momentum-Conservation

15. The diagram below represents two masses before and after they collide. Before the collision, mass m_A is moving to the right with speed v, and mass m_B is at rest. Upon collision, the two masses stick together.

Before Collision

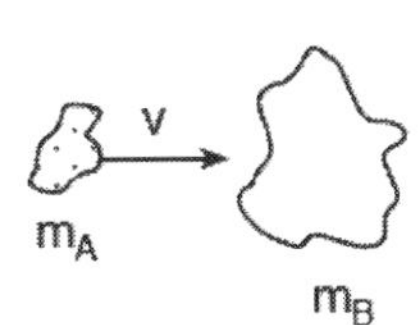

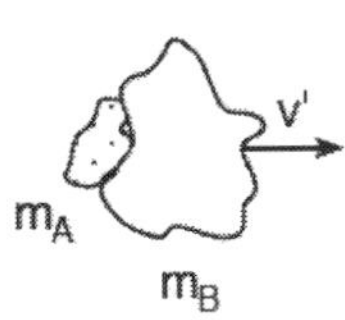

Which expression represents the speed, v', of the masses after the collision? [Assume no outside forces are acting on m_A or m_B.]

(1) $\dfrac{m_A + m_B v}{m_A}$

(2) $\dfrac{m_A + m_B}{m_A v}$

(3) $\dfrac{m_B v}{m_A + m_B}$

(4) $\dfrac{m_A v}{m_A + m_B}$

---

16. In the diagram below, a block of mass M initially at rest on a frictionless horizontal surface is struck by a bullet of mass m moving with a horizontal velocity v.

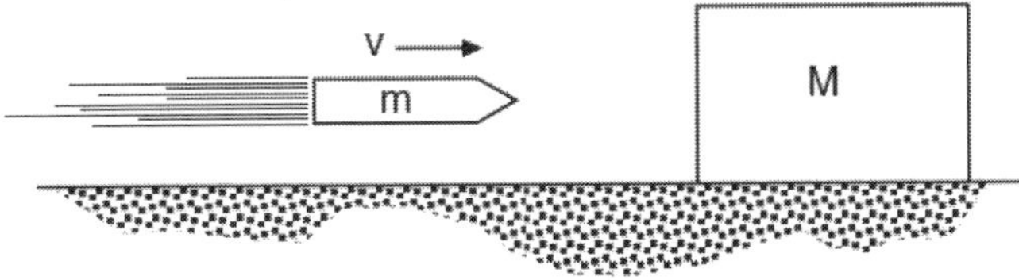

What is the velocity of the bullet-block system after the bullet embeds itself in the block?

(1) $\left(\dfrac{M+v}{M}\right)m$

(2) $\left(\dfrac{m+M}{m}\right)v$

(3) $\left(\dfrac{m+v}{M}\right)m$

(4) $\left(\dfrac{m}{m+M}\right)v$

---

17. When a 1.0-kilogram cart moving with a speed of 0.50 meter per second on a horizontal surface collides with a second 1.0-kilogram cart initially at rest, the carts lock together. What is the speed of the combined carts after the collision? [Neglect friction.]
    1. 1.0 m/s
    2. 0.50 m/s
    3. 0.25 m/s
    4. 0 m/s

18. The diagram below shows an 8.0-kilogram cart moving to the right at 4.0 meters per second about to make a head-on collision with a 4.0-kilogram cart moving to the left at 6.0 meters per second.

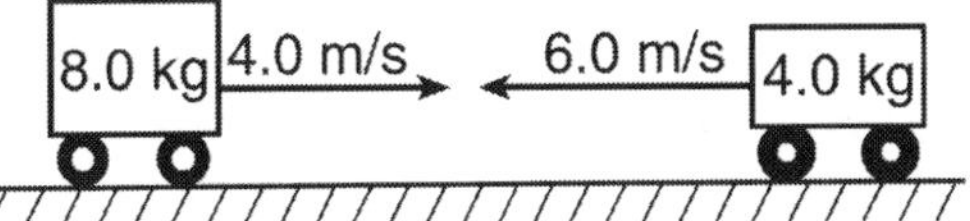

Frictionless, horizontal surface

After the collision, the 4.0-kilogram cart moves to the right at 3.0 meters per second. What is the velocity of the 8.0-kilogram cart after the collision?
    1. 0.50 m/s left
    2. 0.50 m/s right
    3. 5.5 m/s left
    4. 5.5 m/s right

Name: ______________________________ Period: __________

Momentum-Conservation

19. A 7.28-kilogram bowling ball traveling 8.50 meters per second east collides head-on with a 5.45-kilogram bowling ball traveling 10.0 meters per second west. Determine the magnitude of the total momentum of the two-ball system after the collision.

Name: ______________________________ Period: ____________

WEP-Work and Power

1. The work done in accelerating an object along a frictionless horizontal surface is equal to the change in the object's
   1. momentum
   2. velocity
   3. potential energy
   4. kinetic energy

2. The graph below represents the relationship between the work done by a student running up a flight of stairs and the time of ascent.

   Work vs. Time

   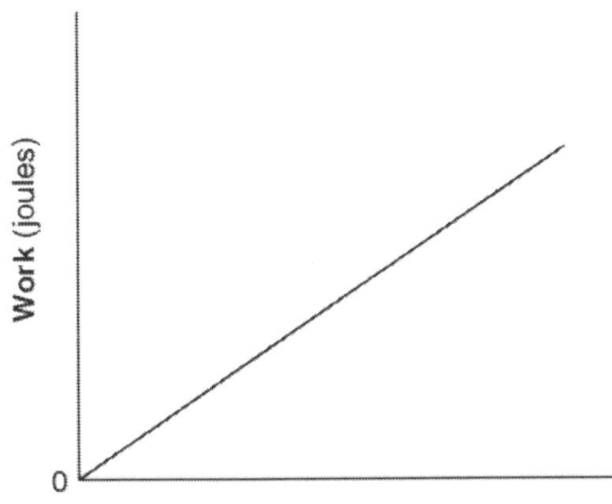

   What does the slope of this graph represent?
   1. impulse
   2. momentum
   3. speed
   4. power

3. A student does 60 joules of work pushing a 3.0-kilogram box up the full length of a ramp that is 5.0 meters long. What is the magnitude of the force applied to the box to do this work?
   1. 20 N
   2. 15 N
   3. 12 N
   4. 4.0 N

4. A boat weighing 900 newtons requires a horizontal force of 600 newtons to move it across the water at 15 meters per second. The boat's engine must provide energy at the rate of
   1. 2.5×10^{-2} J
   2. 4.0×10^{1} W
   3. 7.5×10^{3} J
   4. 9.0×10^{3} W

5. A motor used 120 watts of power to raise a 15-newton object in 5.0 seconds. Through what vertical distance was the object raised?
   1. 1.6 m
   2. 8.0 m
   3. 40 m
   4. 360 m

6. The diagram below shows points A, B, and C at or near Earth's surface. As a mass is moved from A to B, 100 joules of work are done against gravity.

   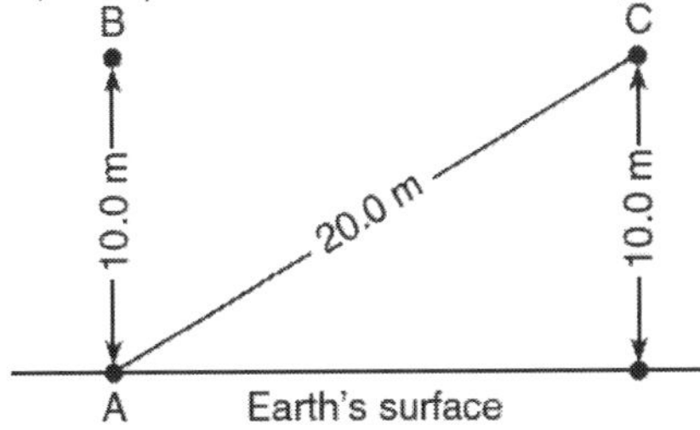

   What is the amount of work done against gravity as an identical mass is moved from A to C?
   1. 100 J
   2. 173 J
   3. 200 J
   4. 273 J

7. One watt is equivalent to one
   1. N·m
   2. N/m
   3. J·s
   4. J/s

8. Two weightlifters, one 1.5 meters tall and one 2.0 meters tall, raise identical 50-kilogram masses above their heads. Compared to the work done by the weightlifter who is 1.5 meters tall, the work done by the weightlifter who is 2.0 meters tall is
   1. less
   2. greater
   3. the same

9. A 40-kilogram student runs up a staircase to a floor that is 5.0 meters higher than her starting point in 7.0 seconds. The student's power output is
   1. 29 W
   2. 280 W
   3. 1.4×10^{3} W
   4. 1.4×10^{4} W

Name: ______________________________ Period: ____________

WEP-Work and Power

10. A 3.0-kilogram block is initially at rest on a frictionless, horizontal surface. The block is moved 8.0 meters in 2.0 seconds by the application of a 12-newton horizontal force, as shown in the diagram below.

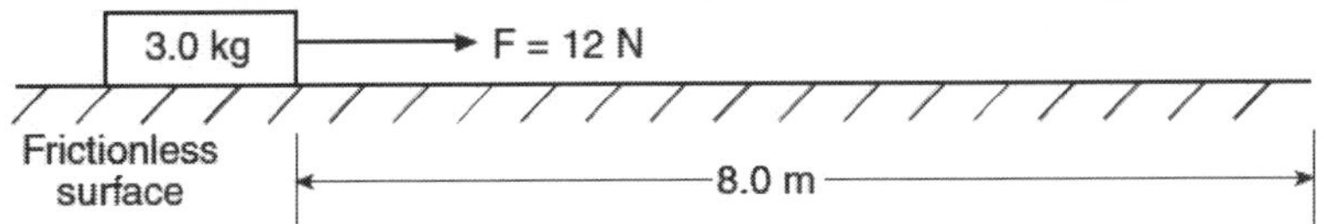

What is the average power developed while moving the block?
1. 24 W
2. 32 W
3. 48 W
4. 96 W

11. The graph below shows the relationship between the work done by a student and the time of ascent as the student runs up a flight of stairs.

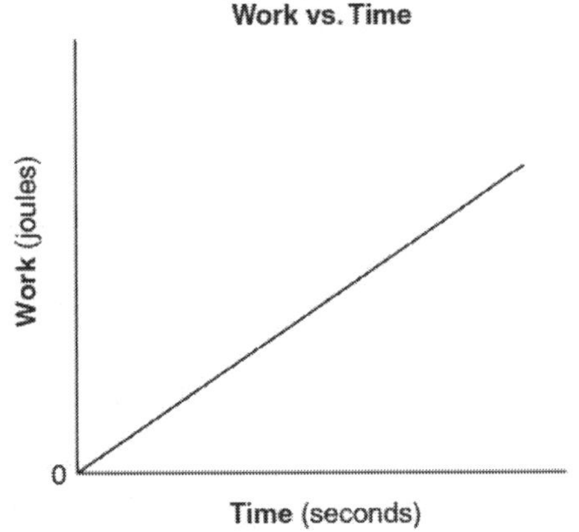

The slope of the graph would have units of
1. joules
2. seconds
3. watts
4. newtons

12. The amount of work done against friction to slide a box in a straight line across a uniform, horizontal floor depends most on the
1. time taken to move the box
2. distance the box is moved
3. speed of the box
4. direction of the box's motion

13. What is the average power developed by a motor as it lifts a 400-kilogram mass at a constant speed through a vertical distance of 10 meters in 8 seconds?
1. 320 W
2. 500 W
3. 4,900 W
4. 32,000 W

14. The work done in lifting an apple one meter near Earth's surface is approximately
1. 1 J
2. 0.01 J
3. 100 J
4. 1000 J

15. A 70-kilogram cyclist develops 210 watts of power while pedaling at a constant velocity of 7.0 meters per second east. What average force is exerted eastward on the bicycle to maintain this constant speed?
1. 490 N
2. 30 N
3. 3.0 N
4. 0 N

16. A 95-kilogram student climbs 4.0 meters up a rope in 3.0 seconds. What is the power output of the student?
1. 130 W
2. 380 W
3. 1200 W
4. 3700 W

17. Which is an SI unit for work done on an object?
1. $\frac{kg \bullet m^2}{s^2}$
2. $\frac{kg \bullet m^2}{s}$
3. $\frac{kg \bullet m}{s}$
4. $\frac{kg \bullet m}{s^2}$

Name:______________________ Period: __________

WEP-Work and Power

18. As shown in the diagram below, a child applies a constant 20-newton force along the handle of a wagon which makes a 25° angle with the horizontal.

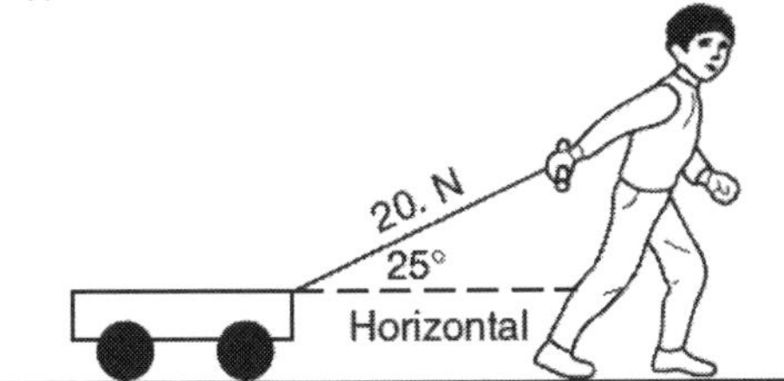

How much work does the child do in moving the wagon a horizontal distance of 4.0 meters?
1. 5.0 J
2. 34 J
3. 73 J
4. 80 J

19. Which graph best represents the relationship between the power required to raise an elevator and the speed at which the elevator rises?

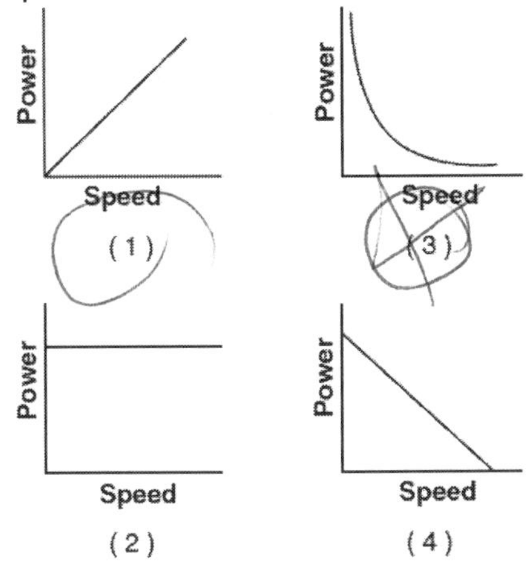

20. A 110-kilogram bodybuilder and his 55-kilogram friend run up identical flights of stairs. The bodybuilder reaches the top in 4.0 seconds while his friend takes 2.0 seconds. Compared to the power developed by the bodybuilder while running up the stairs, the power developed by his friend is
1. the same
2. twice as much
3. half as much
4. four times as much

21. Which quantity is a vector?
1. impulse
2. power
3. speed
4. time

22. Which quantity is a measure of the rate at which work is done?
1. energy
2. power
3. momentum
4. velocity

23. What is the average power required to raise a 1.81×10^4-newton elevator 12.0 meters in 22.5 seconds?
1. 8.04×10^2 W
2. 9.65×10^3 W
3. 2.17×10^5 W
4. 4.89×10^6 W

24. A 15.0-kilogram mass is moving at 7.50 meters per second on a horizontal, frictionless surface. What is the total work that must be done on the mass to increase its speed to 11.5 meters per second?
1. 120 J
2. 422 J
3. 570 J
4. 992 J

25. A truck weighing 3.0×10^4 newtons was driven up a hill that is 1.6×10^3 meters long to a level area that is 8.0×10^2 meters above the starting point. If the trip took 480 seconds, what was the *minimum* power required?
1. 5.0×10^4 W
2. 1.0×10^5 W
3. 1.2×10^{10} W
4. 2.3×10^{10} W

26. A joule is equivalent to a
1. N·m
2. N·s
3. N/m
4. N/s

27. What is the power output of an electric motor that lifts a 2.0-kilogram block 15 meters vertically in 6.0 seconds?
1. 5.0 J
2. 5.0 W
3. 49 J
4. 49 W

Name:__________________________________ Period: ____________

WEP-Work and Power

Base your answers to questions 28 and 29 on the information and diagram below.

A 10-kilogram block is pushed across a floor by a horizontal force of 50 newtons. The block moves from point A to point B in 3.0 seconds.

28. Using a scale of 1.0 centimeter = 1.0 meter, determine the magnitude of the displacement of the block as it moves from point A to point B.

29. Calculate the power required to move the block from point A to point B in 3.0 seconds. [Show all work, including the equation and substitution with units.]

30. How much work is required to lift a 10-newton weight from 4.0 meters to 40 meters above the surface of Earth?
    1. 2.5 J
    2. 3.6 J
    3. 3.6×10^2 J
    4. 4.0×10^2 J

31. Student A lifts a 50-newton box from the floor to a height of 0.40 meter in 2.0 seconds. Student B lifts a 40-newton box from the floor to a height of 0.50 meter in 1.0 second. Compared to student A, student B does
    1. the same work but develops more power
    2. the same work but develops less power
    3. more work but develops less power
    4. less work but develops more power

32. The graph below represents the relationship between the work done by a person and time.

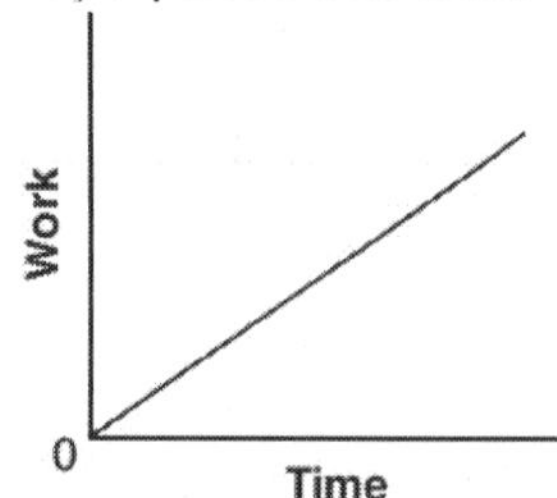

Identify the physical quantity represented by the slope of the graph.

33. A 60-kilogram student climbs a ladder a vertical distance of 4.0 meters in 8.0 seconds. Approximately how much total work is done against gravity by the student during the climb?
    1. 2.4×10^3 J
    2. 2.9×10^2 J
    3. 2.4×10^2 J
    4. 3.0×10^1 J

34. A box is pushed to the right with a varying horizontal force. The graph below represents the relationship between the applied force and the distance the box moves.

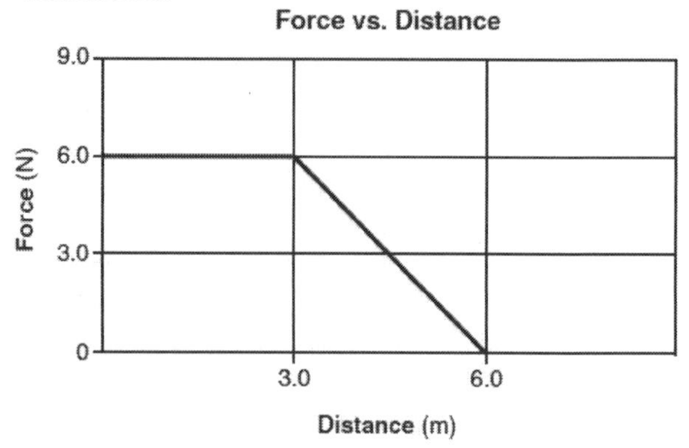

What is the total work done in moving the box 6.0 meters?
    1. 9.0 J
    2. 18 J
    3. 27 J
    4. 36 J

Name: ______________________________ Period: __________

WEP-Work and Power

35. The total work done in lifting a typical high school physics textbook a vertical distance of 0.10 meter is approximately
    1. 0.15 J
    2. 1.5 J
    3. 15 J
    4. 150 J

36. Through what vertical distance is a 50-newton object moved if 250 joules of work is done against the gravitational field of Earth?
    1. 2.5 m
    2. 5.0 m
    3. 9.8 m
    4. 25 m

37. A small electric motor is used to lift a 0.50-kilogram mass at constant speed. If the mass is lifted a vertical distance of 1.5 meters in 5.0 seconds, the average power developed by the motor is
    1. 0.15 W
    2. 1.5 W
    3. 3.8 W
    4. 7.5 W

38. What is the maximum amount of work that a 6000-watt motor can do in 10 seconds?
    1. 6.0×10^1 J
    2. 6.0×10^2 J
    3. 6.0×10^3 J
    4. 6.0×10^4 J

39. How much work is done by the force lifting a 0.1-kilogram hamburger vertically upward at constant velocity 0.3 meter from a table?
    1. 0.03 J
    2. 0.1 J
    3. 0.3 J
    4. 0.4 J

40. The watt·second is a unit of
    1. power
    2. energy
    3. potential difference
    4. electric field strength

41. Which quantity has both a magnitude and direction?
    1. energy
    2. impulse
    3. power
    4. work

42. Two elevators, A and B, move at constant speed. Elevator B moves with twice the speed of elevator A. Elevator B weighs twice as much as elevator A. Compared to the power needed to lift elevator A, the power needed to lift elevator B is
    1. the same
    2. twice as great
    3. half as great
    4. four times as great

43. What is the maximum height to which a motor having a power rating of 20.4 watts can lift a 5.00-kilogram stone vertically in 10 seconds?
    1. 0.0416 m
    2. 0.408 m
    3. 4.16 m
    4. 40.8 m

44. If a motor lifts a 400-kilogram mass a vertical distance of 10 meters in 8.0 seconds, the minimum power generated by the motor is
    1. 3.2×10^2 W
    2. 5.0×10^2 W
    3. 4.9×10^3 W
    4. 3.2×10^4 W

45. The graph below represents the work done against gravity by a student as she walks up a flight of stairs at constant speed.

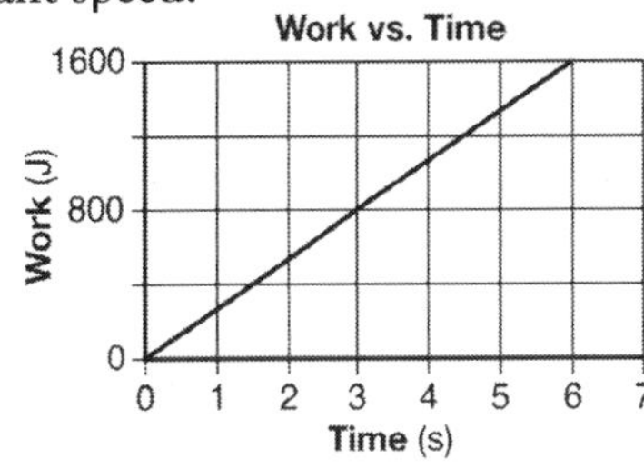

Compared to the power generated by the student after 2.0 seconds, the power generated by the student after 4.0 seconds is
    1. the same
    2. twice as great
    3. half as great
    4. four times as great

Name: ______________________________ Period: __________

WEP-Work and Power

46. Which graph best represents the greatest amount of work?

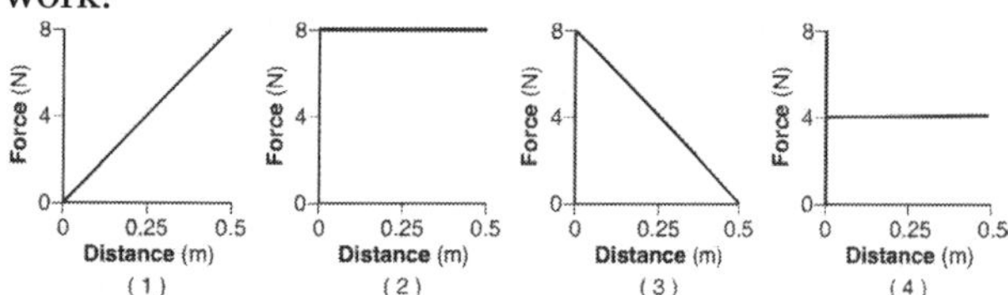

47. Calculate the average power required to lift a 490-newton object a vertical distance of 2.0 meters in 10 seconds. [Show all work, including the equation and substitution with units.]

48. Which combination of fundamental units can be used to express the amount of work done on an object?
   1. kg·m/s
   2. kg·m/s^2
   3. kg·m^2/s^2
   4. kg·m^2/s^3

49. The graph below represents the relationship between the force exerted on an elevator and the distance the elevator is lifted.

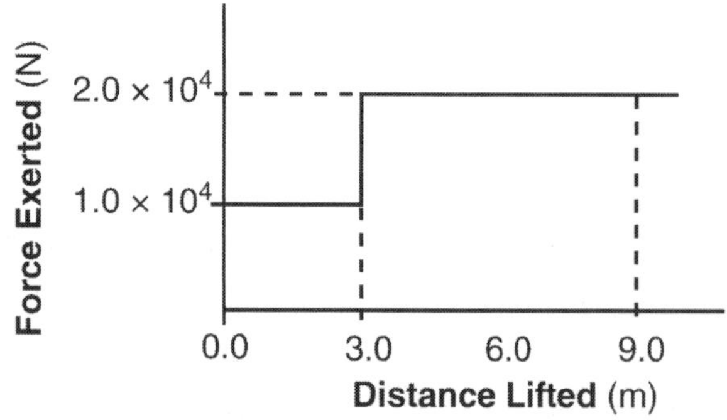

How much total work is done by the force in lifting the elvator from 0.0 m to 9.0 m?
   1. 9.0×10^4 J
   2. 1.2×10^5 J
   3. 1.5×10^5 J
   4. 1.8×10^5 J

50. An electric motor has a rating of 4.0×10^2 watts. How much time will it take for this motor to lift a 50-kilogram mass a vertical distance of 8.0 meters? [Assume 100% efficiency.]
   1. 0.98 s
   2. 9.8 s
   3. 98 s
   4. 980 s

51. Calculate the minimum power output of an electric motor that lifts a 1.30×10^4-newton elevator car vertically upward at a constant speed of 1.50 meters per second. [Show all work, including the equation and substitution with units.]

Name: ____________________ Period: __________

WEP-Springs

Base your answers to questions 1 through 3 on the information and data table below.

In an experiment, a student applied various forces to a spring and measured the spring's corresponding elongation. The table below shows his data.

Force (newtons)	Elongation (meters)
0	0
1.0	0.30
3.0	0.67
4.0	1.00
5.0	1.30
6.0	1.50

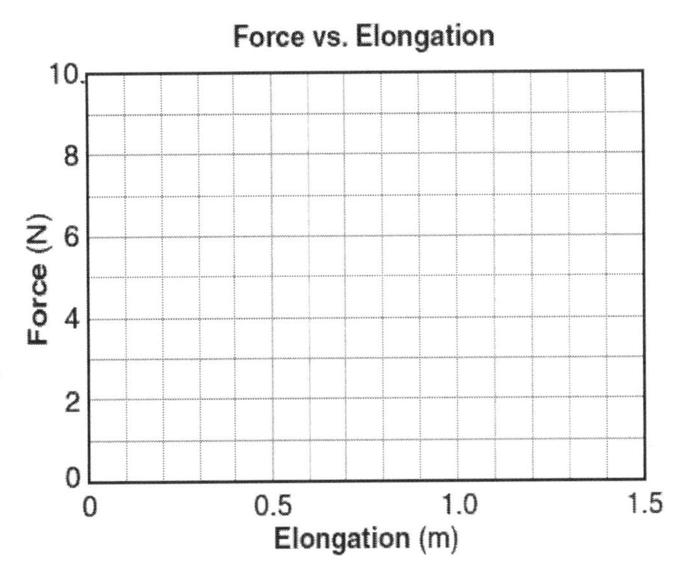

1. On the grid at right, plot the data points for force versus elongation.
2. Draw the best-fit line
3. Using your graph, calculate the spring constant of the spring. [Show all work, including the equation and substitution with units.]

4. A 10-newton force is required to hold a stretched spring 0.20 meter from its rest position. What is the potential energy stored in the stretched spring?
   1. 1.0 J
   2. 2.0 J
   3. 5.0 J
   4. 50 J

5. A 5-newton force causes a spring to stretch 0.2 meter. What is the potential energy stored in the stretched spring?
   1. 1 J
   2. 0.5 J
   3. 0.2 J
   4. 0.1 J

6. The spring of a toy car is wound by pushing the car backward with an average force of 15 newtons through a distance of 0.50 meter. How much elastic potential energy is stored in the car's spring during this process?
   1. 1.9 J
   2. 7.5 J
   3. 30 J
   4. 56 J

Base your answers to questions 7 and 8 on the information and graph below.

The graph represents the relationship between the force applied to each of two springs, A and B, and their elongations.

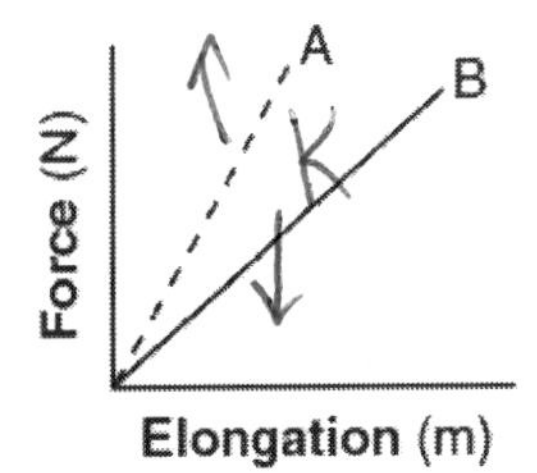

7. What physical quantity is represented by the slope of each line?

8. A 1.0-kilogram mass is suspended from each spring. If each mass is at rest, how does the potential energy stored in spring A compare to the potential energy stored in spring B?

Name: ________________________ Period: ____________

WEP-Springs

9. A spring scale reads 20 newtons as it pulls a 5.0-kilogram mass across a table. What is the magnitude of the force exerted by the mass on the spring scale?
   1. 49 N
   2. 20 N
   3. 5.0 N
   4. 4.0 N

Base your answers to questions 10 through 12 on the information and diagram below.

A mass, M, is hung from a spring and reaches equilibrium at position B. The mass is then raised to position A and released. The mass oscillates between positions A and C. [Neglect friction.]

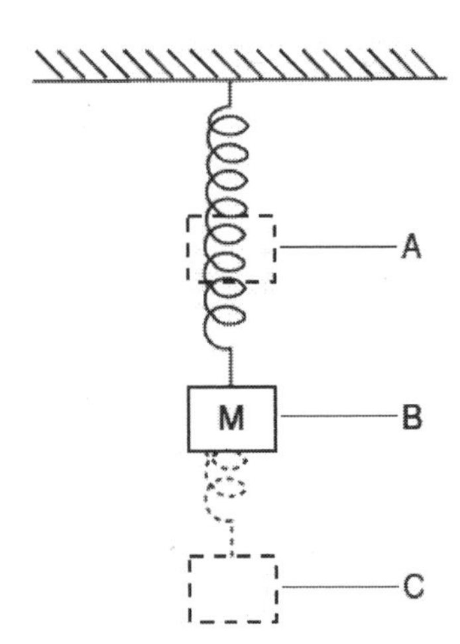

10. At which position, A, B, or C, is mass M located when the kinetic energy of the system is at a maximum? Explain your choice.

11. At which position, A, B, or C, is mass M located when the gravitational potential energy of the system is at a maximum? Explain your choice.

12. At which position, A, B, or C, is mass M located when the elastic potential energy of the system is at a maximum? Explain your choice.

13. A spring with a spring constant of 80 newtons per meter is displaced 0.30 meter from its equilibrium position. The potential energy stored in the spring is
   1. 3.6 J
   2. 7.2 J
   3. 12 J
   4. 24 J

14. The diagram below shows a 0.1-kilogram apple attached to a branch of a tree 2 meters above a spring on the ground below.

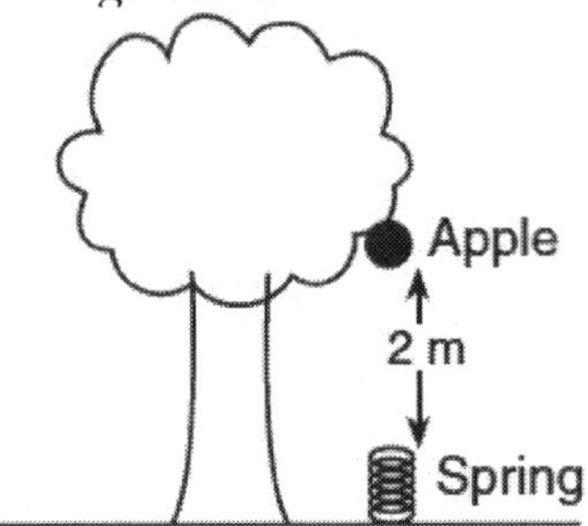

The apple falls and hits the spring, compressing it 0.1 meter from its rest position. If all of the gravitational potential energy of the apple on the tree is transferred to the spring when it is compressed, what is the spring constant of this spring?
   1. 10 N/m
   2. 40 N/m
   3. 100 N/m
   4. 400 N/m

15. The graph below shows elongation as a function of the applied force for two springs, A and B.

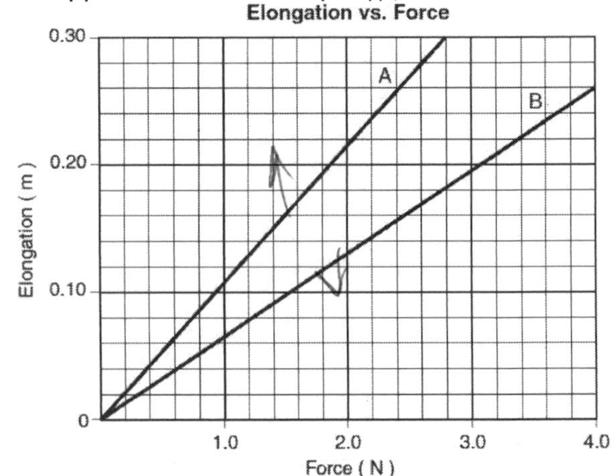

Compared to the spring constant for spring A, the spring constant for spring B is
   1. smaller
   2. larger
   3. the same

Name: ______________________________ Period: ____________

WEP-Springs

16. As shown in the diagram below, a 0.50-meter-long spring is stretched from its equilibrium position to a length of 1.00 meter by a weight.

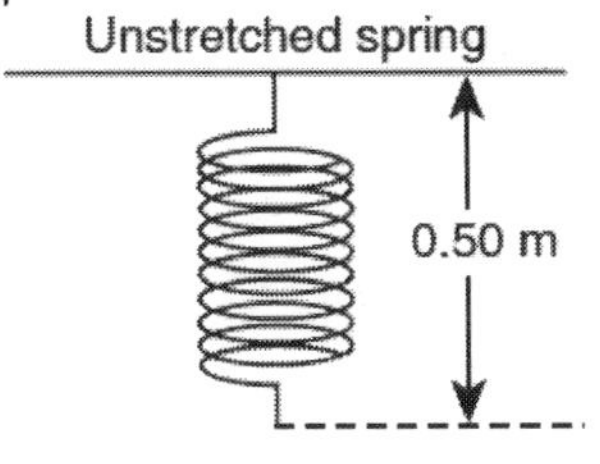

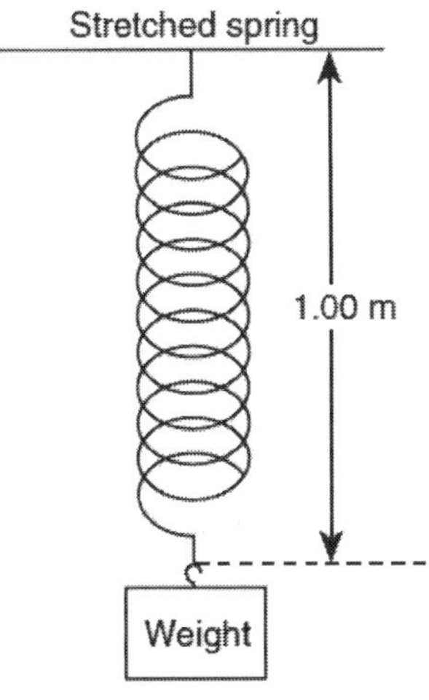

If 15 joules of energy are stored in the stretched spring, what is the value of the spring constant?

1. 30 N/m
2. 60 N/m
3. 120 N/m
4. 240 N/m

Base your answers to questions 17 through 19 on the information and data table below.

A student performed an experiment in which the weight attached to a suspended spring was varied and the resulting total length of the spring measured. The data for the experiment are in the table below.

Attached Weight (N)	Total Spring Length (m)
0.98	0.37
1.96	0.42
2.94	0.51
3.92	0.59
4.91	0.64

Using the information in the data table, construct a graph on the grid at right by following the directions below.

17. Plot the data points for the attached weight versus total spring length.

18. Draw the line or curve of best fit.

19. Using your graph, determine the length of the spring before any weight was attached.

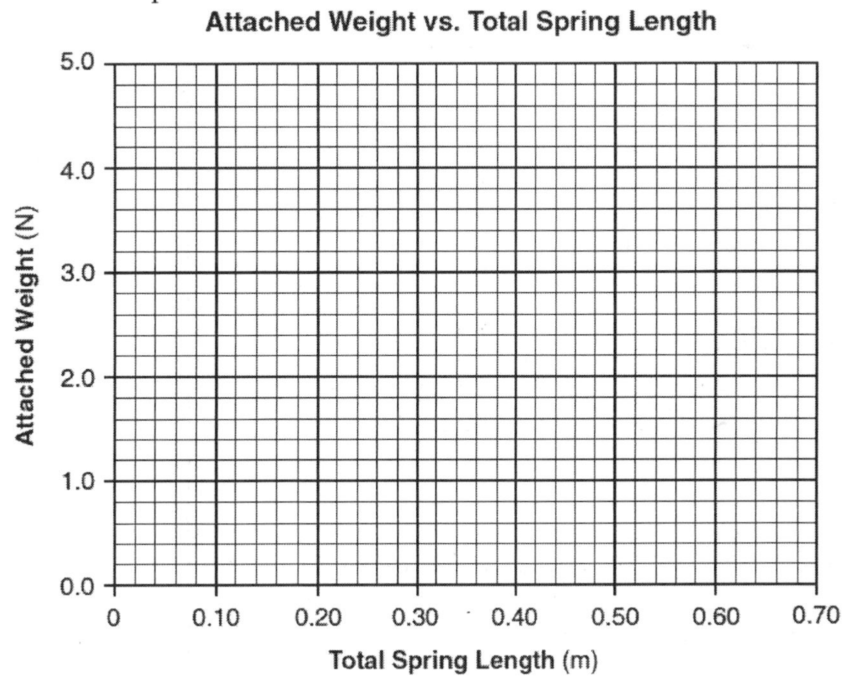

Name:____________________ Period: __________

WEP-Springs

Base your answers to questions 20 and 21 on the information and diagram below.

A pop-up toy has a mass of 0.020 kilogram and a spring constant of 150 newtons per meter. A force is applied to the toy to compress the spring 0.050 meter.

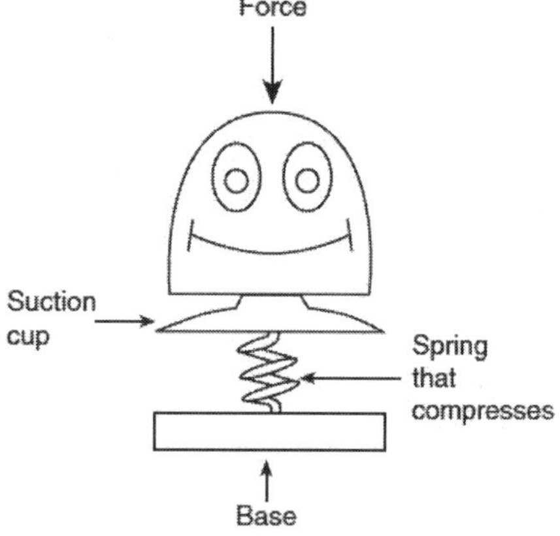

20. Calculate the potential energy stored in the compressed spring. [Show all work, including the equation and substitution with units.]

21. The toy is activated and all the compressed spring's potential energy is converted to gravitational potential energy. Calculate the maximum vertical height to which the toy is propelled. [Show all work, including the equation and substitution with units.]

22. A 10-newton force compresses a spring 0.25 meter from its equilibrium position. Calculate the spring constant of this spring. [Show all work, including the equation and substitution with units.]

23. The diagram below represents a spring hanging vertically that stretches 0.075 meter when a 5.0-newton block is attached. The spring-block system is at rest in the position shown.

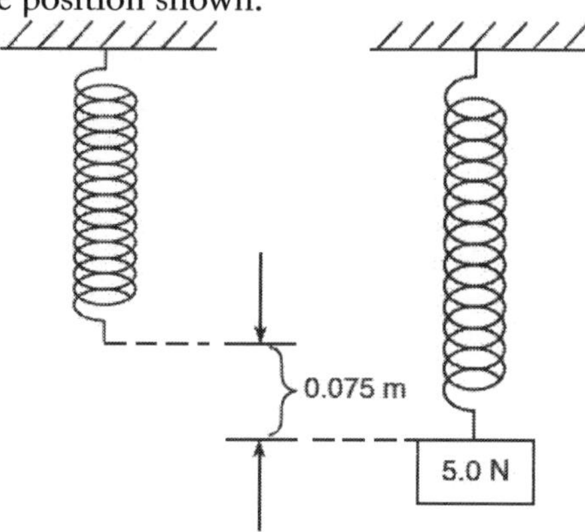

The value of the spring constant is
1. 38 N/m
2. 67 N/m
3. 130 N/m
4. 650 N/m

24. A spring with a spring constant of 4.0 newtons per meter is compressed by a force of 1.2 newtons. What is the total elastic potential energy stored in this compressed spring?
1. 0.18 J
2. 0.36 J
3. 0.60 J
4. 4.8 J

25. The spring in a scale in the produce department of a supermarket stretches 0.025 meter when a watermelon weighing 1.0×10^2 newtons is placed on the scale. The spring constant for this spring is
1. 3.2×10^5 N/m
2. 4.0×10^3 N/m
3. 2.5 N/m
4. 3.1×10^{-2} N/m

26. When a 1.53-kilogram mass is placed on a spring with a spring constant of 30.0 newtons per meter, the spring is compressed 0.500 meter. How much energy is stored in the spring?
1. 3.75 J
2. 7.50 J
3. 15.0 J
4. 30.0 J

Name: ____________________ Period: __________

WEP-Springs

Base your answers to questions 27 through 30 on the information and data table below.

The spring in a dart launcher has a spring constant of 140 newtons per meter. The launcher has six power settings, 0 through 5, with each successive setting having a spring compression 0.020 meter beyond the previous setting. During testing, the launcher is aligned to the vertical, the spring is compressed, and a dart is fired upward. The maximum vertical displacement of the dart in each test trial is measured. The results of the testing are shown in the table below.

Data Table

Power Setting	Spring Compression (m)	Dart's Maximum Vertical Displacement (m)
0	0.000	0.00
1	0.020	0.29
2	0.040	1.14
3	0.060	2.57
4	0.080	4.57
5	0.100	7.10

Directions (27-28): Using the information in the data table, construct a graph on the grid below.

27. Plot the data points for the dart's maximum vertical displacement versus spring compression.

28. Draw the line or curve of best fit.

29. Using information from your graph, calculate the energy provided by the compressed spring that causes the dart to achieve a maximum vertical displacement of 3.50 meters. [Show all work, including the equation and substitution with units.]

30. Determine the magnitude of the force, in newtons, needed to compress the spring 0.040 meter.

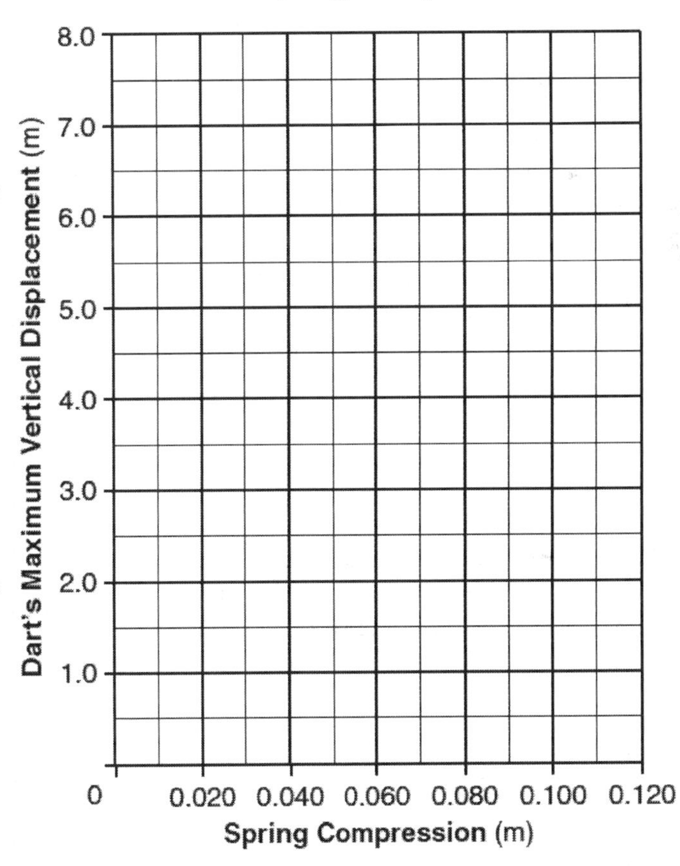

Name:______________________________ Period: __________

WEP-Springs

Base your answers to questions 31 and 32 on the information below.

In a laboratory investigation, a student applied various downward forces to a verticle spring. The applied forces and the corresponding elongations of the spring from its equilibrium position are recorded in the data table below.

Data Table

Force (N)	Elongation (m)
0	0
0.5	0.010
1.0	0.018
1.5	0.027
2.0	0.035
2.5	0.046

31. Construct a graph on the grid below. Mark an appropriate scale on the axis labeled "Force (N)," plot the data points for force versus elongation, and draw the best-fit line or curve.

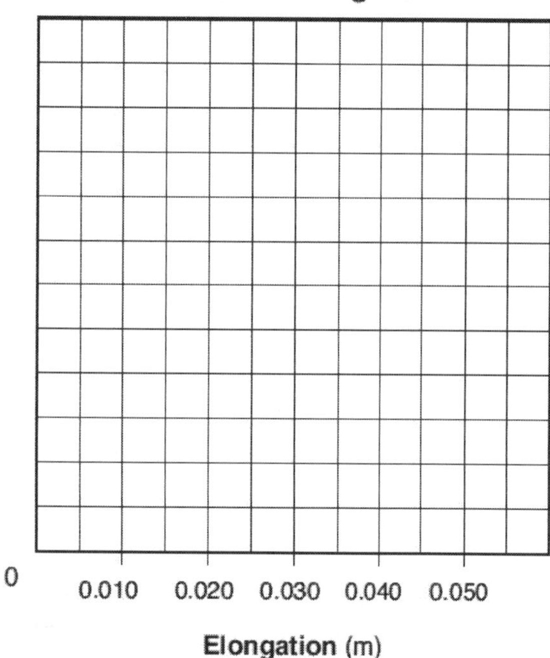

32. Using your graph, calculate the spring constant of this spring. [Show all work, including the equation and substitution with units.]

33. When a mass is placed on a spring with a spring constant of 15 newtons per meter, the spring is compressed 0.25 meter. How much elastic potential energy is stored in the spring?
    1. 0.47 J
    2. 0.94 J
    3. 1.9 J
    4. 3.8 J

34. The potential energy stored in a compressed spring is to the change in the spring's length as the kinetic energy of a moving body is to the body's
    1. speed
    2. mass
    3. radius
    4. acceleration

35. The diagram below shows a toy cart possessing 16 joules of kinetic energy traveling on a frictionless, horizontal surface toward a horizontal spring.

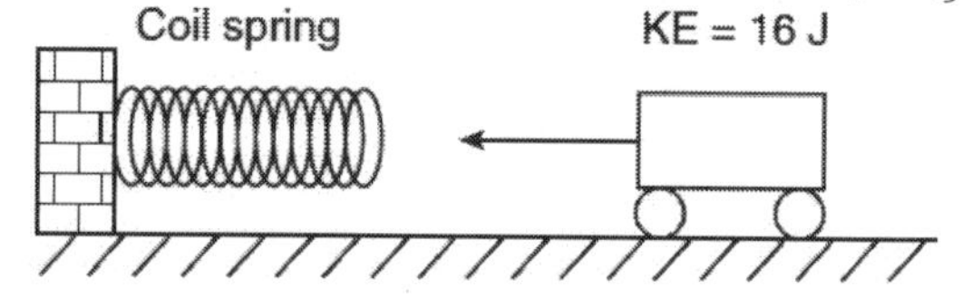

If the cart comes to rest after compressing the spring a distance of 1.0 meter, what is the spring constant of the spring?
    1. 32 N/m
    2. 16 N/m
    3. 8.0 N/m
    4. 4.0 N/m

36. A spring in a toy car is compressed a distance, x. When released, the spring returns to its original length, transferring its energy to the car. Consequently, the car having mass m moves with speed v.

Derive the spring constant, k, of the car's spring in terms of m, x, and v. [Assume an ideal mechanical system with no loss of energy.]

Name:________________________ Period: ____________

WEP-Springs

37. The graph below represents the relationship between the force applied to a spring and spring elongation for four different springs.

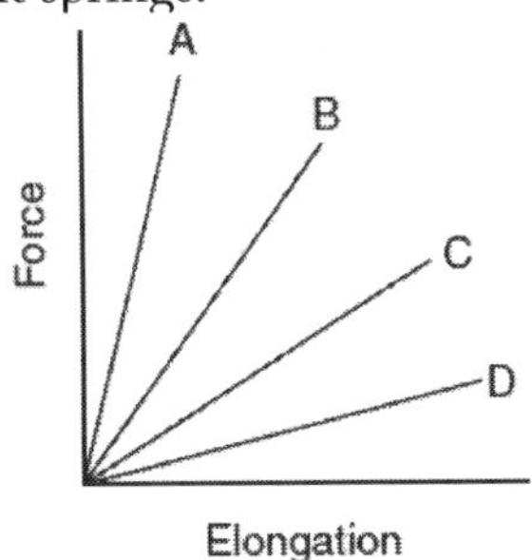

Which spring has the greatest spring constant?

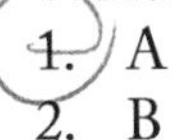

1. A
2. B
3. C
4. D

Base your answers to questions 38 and 39 on the information below.

A vertically hung spring has a spring constant of 150 newtons per meter. A 2.00-kilogram mass is suspended from the spring and allowed to come to rest.

38. Calculate the elongation of the spring produced by the suspended 2.00-kilogram mass. [Show all work, including the equation and substitution with units.]

39. Calculate the total elastic potential energy stored in the spring due to the suspended 2.00-kilogram mass. [Show all work, including the equation and substitution with units.]

40. Which graph best represents the relationship between the elastic potential energy stored in a spring and its elongation from equilibrium?

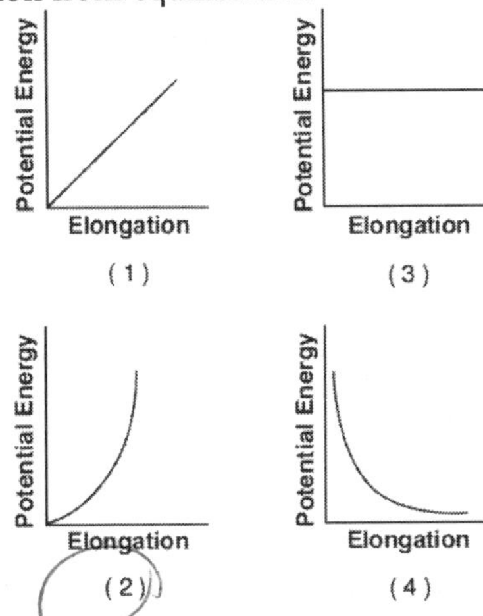

41. A child does 0.20 joule of work to compress the spring in a pop-up toy. If the mass of the toy is 0.010 kilogram, what is the maximum vertical height that the toy can reach after the spring is released?
   1. 20 m
   2. 2.0 m
   3. 0.20 m
   4. 0.020 m

42. A vertical spring 0.100 meter long is elongated to a length of 0.119 meter when a 1.00-kilogram mass is attached to the bottom of the spring. The spring constant of this spring is
   1. 9.8 N/m
   2. 82 N/m
   3. 98 N/m
   4. 520 N/m

43. An unstretched spring has a length of 10 centimeters. When the spring is stretched by a force of 16 newtons, its length is increased to 18 centimeters. What is the spring constant of this spring?
   1. 0.89 N/cm
   2. 2.0 N/cm
   3. 1.6 N/cm
   4. 1.8 N/cm

44. When a spring is compressed 2.50×10^{-2} meter from its equilibrium position, the total potential energy stored in the spring is 1.25×10^{-2} joule. Calculate the spring constant of the spring.

Name: ______________________________ Period: __________

WEP-Springs

Base your answers to questions 45 and 46 on the information below.

A student produced various elongations of a spring by applying a series of forces to the spring. The graph at right represents the relationship between the applied force and the elongation of the spring.

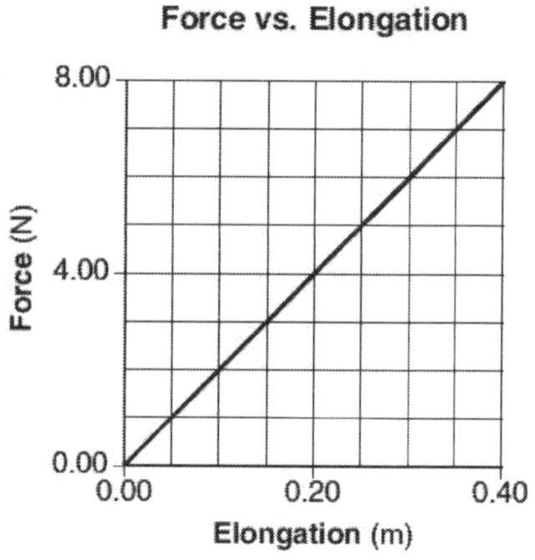

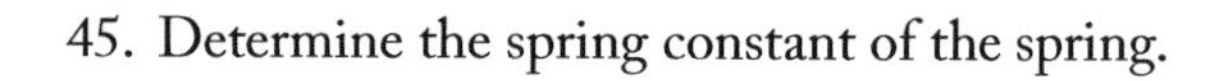

45. Determine the spring constant of the spring.

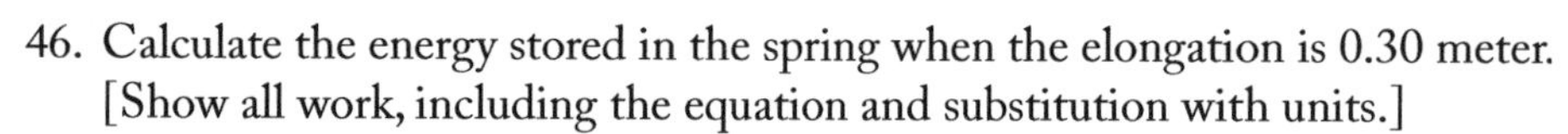

46. Calculate the energy stored in the spring when the elongation is 0.30 meter. [Show all work, including the equation and substitution with units.]

47. A spring gains 2.34 joules of elastic potential energy as it is compressed 0.250 meter from its equilibrium position. What is the spring constant of this spring?
    1. 9.36 N/m
    2. 18.7 N/m
    3. 37.4 N/m
    4. 74.9 N/m

48. A vertical spring has a spring constant of 100 newtons per meter. When an object is attached to the bottom of the spring, the spring changes from its unstretched length of 0.50 meter to a length of 0.65 meter. The magnitude of the weight of the attached object is
    1. 1.1 N
    2. 15 N
    3. 50 N
    4. 65 N

Name: ______________________________ Period: ____________

WEP-Energy

1. A 1-kilogram rock is dropped from a cliff 90 meters high. After falling 20 meters, the kinetic energy of the rock is approximately
   1. 20 J
   2. 200 J
   3. 700 J
   4. 900 J

2. If the speed of a car is doubled, the kinetic energy of the car is
   1. quadrupled
   2. quartered
   3. doubled
   4. halved

3. A constant force is used to keep a block sliding at constant velocity along a rough horizontal track. As the block slides, there could be an increase in its
   1. gravitational potential energy, only
   2. internal energy, only
   3. gravitational potential energy and kinetic energy
   4. internal energy and kinetic energy

4. As an object falls freely, the kinetic energy of the object
   1. decreases
   2. increases
   3. remains the same

5. An object weighing 15 newtons is lifted from the ground to a height of 0.22 meter. The increase in the object's gravitational potential energy is approximately
   1. 310 J
   2. 32 J
   3. 3.3 J
   4. 0.34 J

6. A 0.50-kilogram ball is thrown vertically upward with an initial kinetic energy of 25 joules. Approximately how high will the ball rise? [Neglect air resistance.]
   1. 2.6 m
   2. 5.1 m
   3. 13 m
   4. 25 m

7. A 45-kilogram boy is riding a 15-kilogram bicycle with a speed of 8 meters per second. What is the combined kinetic energy of the boy and the bicycle?
   1. 240 J
   2. 480 J
   3. 1440 J
   4. 1920 J

8. The work done in moving a block across a rough surface and the heat energy gained by the block can both be measured in
   1. watts
   2. degrees
   3. newtons
   4. joules

Base your answers to questions 9 through 11 on the information below.

A 50-kilogram child running at 6 meters per second jumps onto a stationary 10-kilogram sled. The sled is on a level frictionless surface.

9. Calculate the speed of the sled with the child after she jumps onto the sled. [Show all work, including the equation and substitution with units.]

10. Calculate the kinetic energy of the sled with the child after she jumps onto the sled. [Show all work, including the equation and substitution with units.]

11. After a short time, the moving sled with the child aboard reaches a rough level surface that exerts a constant frictional force of 54 newtons on the sled. How much work must be done by friction to bring the sled with the child to a stop?

Name:______________________________ Period: __________

WEP-Energy

12. Which graph best represents the relationship between the kinetic energy, KE, and the velocity of an objct accelerating in a straight line?

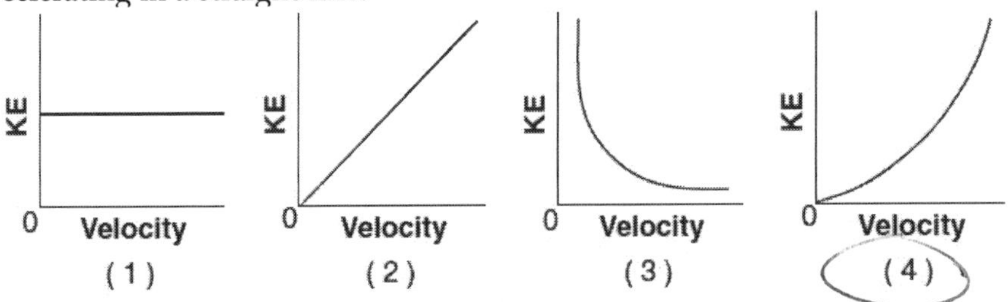

13. The graph below represents the kinetic energy, gravitational potential energy, and total mechanical energy of a moving block.

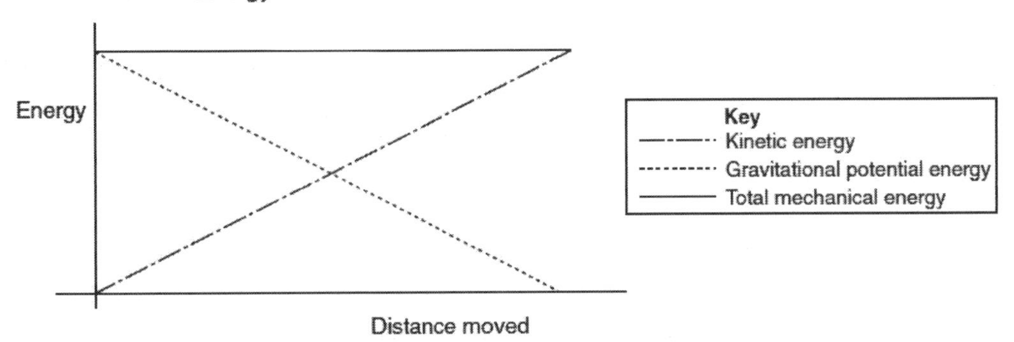

Which best describes the motion of the block?
1. accelerating on a flat horizontal surface
2. sliding up a frictionless incline
3. falling freely
4. being lifted at constant velocity

Base your answers to questions 14 through 16 on the information and diagram below.

A 1000-kilogram empty cart moving with a speed of 6 meters per second is about to collide with a stationary loaded cart having a total mass of 5000 kilograms, as shown. After the collision, the carts lock and move together. [Assume friction is negligible.]

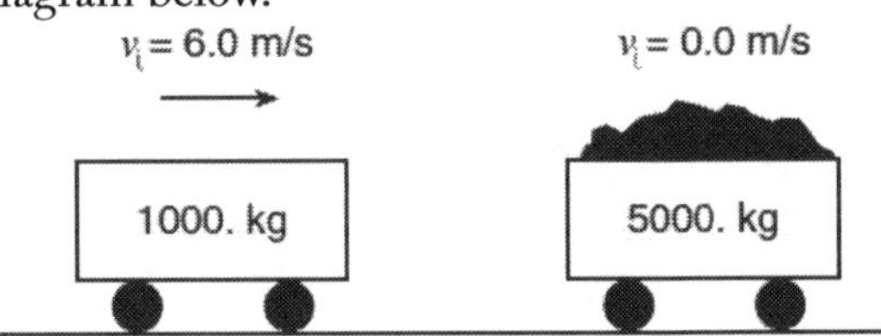

14. Calculate the speed of the combined carts after the collision. [Show all work, including the equation and substitution with units.]

15. Calculate the kinetic energy of the combined carts after the collision. [Show all work, including the equation and substitution with units.]

16. How does the kinetic energy of the combined carts after the collision compare to the kinetic energy of the carts before the collision?

Name: ______________________________ Period: ____________

WEP-Energy

17. When a force moves an object over a rough, horizontal surface at a constant velocity, the work done against friction produces an increase in the object's
    1. weight
    2. momentum
    3. potential energy
    4. internal energy

Base your answers to questions 18 through 21 on the information below.

The driver of a car made an emergency stop on a straight horizontal road. The wheels locked and the car skidded to a stop. The marks made by the rubber tires on the dry asphalt are 16 meters long, and the car's mass is 1200 kilograms.

18. Determine the weight of the car

19. Calculate the magnitude of the frictional force the road applied to the car in stopping it. [Show all work, including the equation and substitution with units.]

20. Calculate the work done by the frictional force in stopping the car. [Show all work, including the equation and substitution with units.]

21. Assuming that energy is conserved, calculate the speed of the car before the brakes were applied. [Show all work, including the equation and substitution with units.]

22. A 60-kilogram runner has 1920 joules of kinetic energy. At what speed is she running?
    1. 5.66 m/s
    2. 8.00 m/s
    3. 32.0 m/s
    4. 64.0 m/s

23. As a block slides across a table, its speed decreases while its temperature increases. Which two changes occur in the block's energy as it slides?
    1. a decrease in kinetic energy and an increase in internal energy
    2. an increase in kinetic energy and a decrease in internal energy
    3. a decrease in both kinetic energy and internal energy
    4. an increase in both kinetic energy and internal energy

24. If the direction of a moving car changes and its speed remains constant, which quantity must remain the same?
    1. velocity
    2. momentum
    3. displacement
    4. kinetic energy

25. What is the gravitational potential energy with respect to the surface of the water of a 75.0-kilogram diver located 3.00 meters above the water?
    1. 2.17×10^4 J
    2. 2.21×10^3 J
    3. 2.25×10^2 J
    4. 2.29×10^1 J

26. As a ball falls freely (without friction) toward the ground, its total mechanical energy
    1. decreases
    2. increases
    3. remains the same

27. The gravitational potential energy, with respect to Earth, this is possessed by an object is dependent on the object's
    1. acceleration
    2. momentum
    3. position
    4. speed

Name:________________________ Period: ____________

WEP-Energy

Base your answers to questions 28 and 29 on the information below.

A boy pushes his wagon at constant speed along a level sidewalk. The graph below represents the relationship between the horizontal force exerted by the boy and the distance the wagon moves.

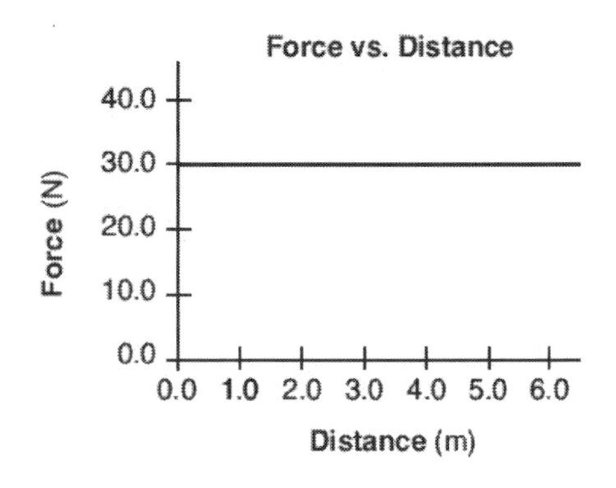

28. What is the total work done by the boy in pushing the wagon 4.0 meters?
    1. 5.0 J
    2. 7.5 J
    3. 120 J
    4. 180 J

29. As the boy pushes the wagon, what happens to the wagon's energy?
    1. Gravitational potential energy increases.
    2. Gravitational potential energy decreases.
    3. Internal energy increases.
    4. Internal energy decreases.

30. A 1.0-kilogram book resting on the ground is moved 1.0 meter at various angles relative to the horizontal. In which direction does the 1.0-meter displacement produce the greatest increase in the book's gravitational potential energy?

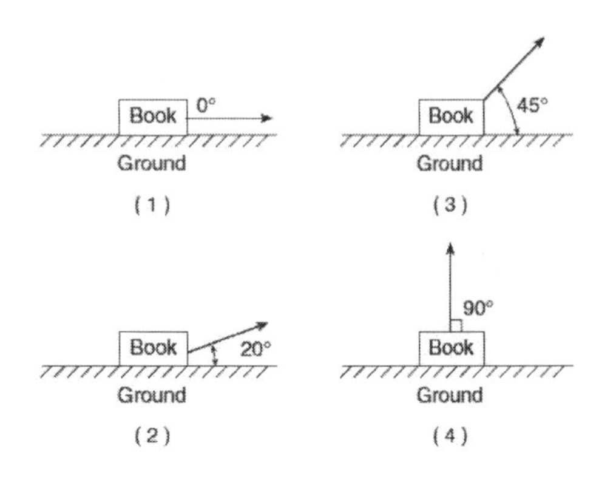

31. As shown in the diagram below, a student exerts an average force of 600 newtons on a rope to lift a 50-kilogram crate a vertical distance of 3 meters.

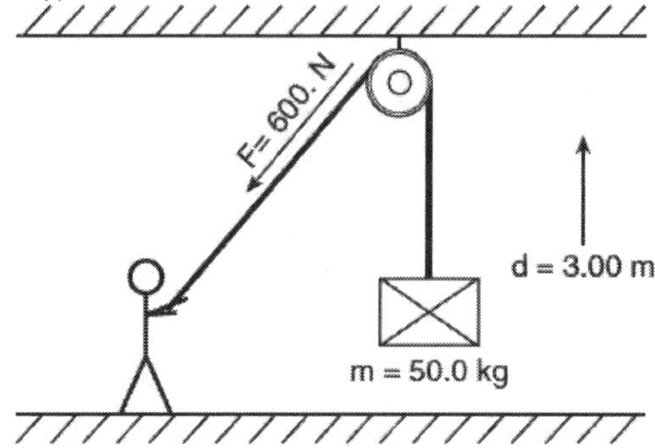

Compared to the work done by the student, the gravitational potential energy gained by the crate is
1. exactly the same
2. 330 J less
3. 330 J more
4. 150 J more

32. A book sliding across a horizontal tabletop slows until it comes to rest. Describe what change, if any, occurs in the book's kinetic energy and internal energy as it slows.

33. A pendulum is pulled to the side and released from rest. Which graph best represents the relationship between the gravitational potential energy of the pendulum and its displacement from its point of release?

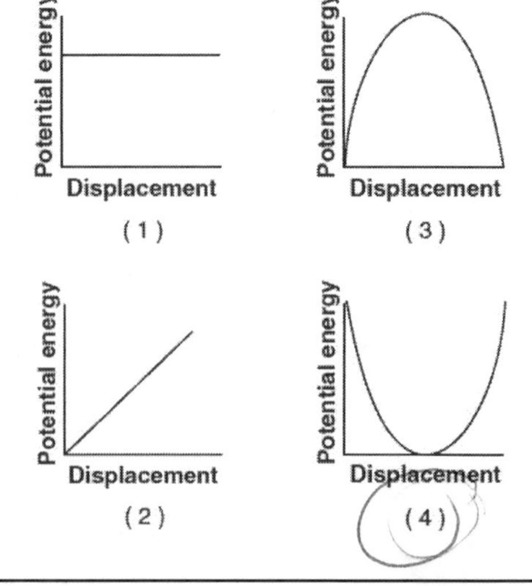

Name: ______________________________ Period: __________

WEP-Energy

Base your answers to questions 34 through 36 on the information and diagram below.

A 250-kilogram car is initially at rest at point A on a roller coaster track. The car carries a 75-kilogram passenger and is 20 meters above the ground at point A. [Neglect friction.]

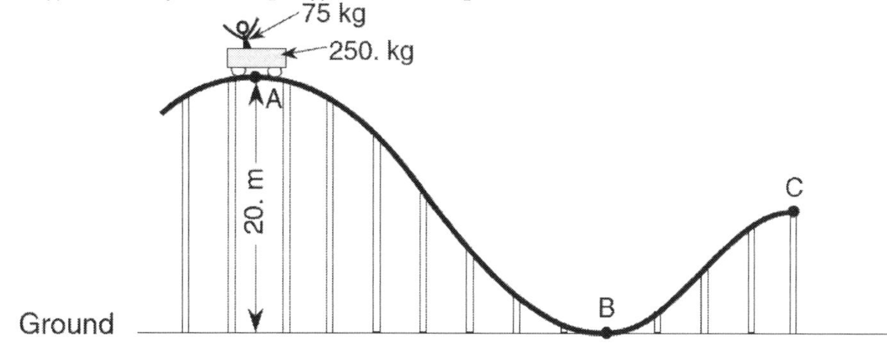

34. Calculate the total gravitational potential energy, relative to the ground, of the car and the passenger at point A. [Show all work, including the equation and substitution with units.]

35. Calculate the speed of the car and passenger at point B. [Show all work, including the equation and substitution with units.]

36. Compare the total mechanical energy of the car and passenger at points A, B, and C.

37. Which graph best represents the relationship between the gravitational potential energy of an object near the surface of Earth and its height above Earth's surface?

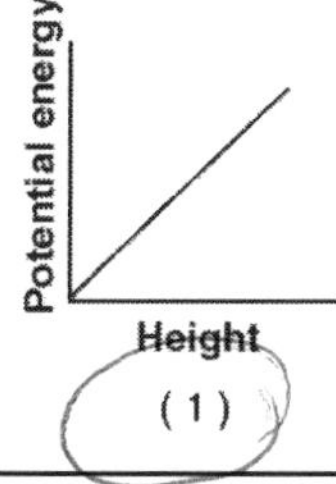

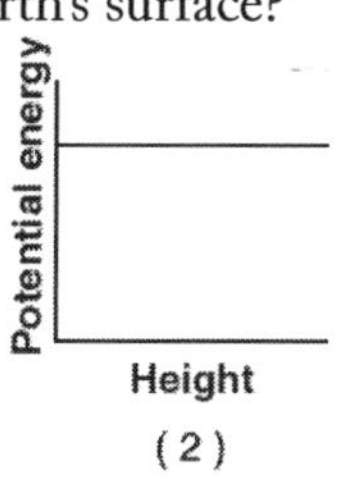

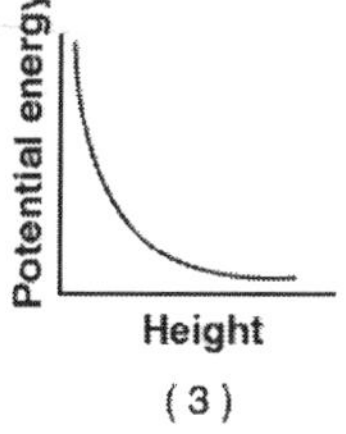

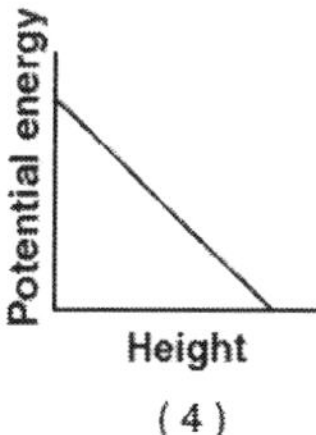

38. A horizontal force of 5.0 newtons acts on a 3.0-kilogram mass over a distance of 6.0 meters along a horizontal, frictionless surface. What is the change in kinetic energy of the mass during its movement over the 6.0-meter distance?
    1. 6.0 J
    2. 15 J
    3. 30 J
    4. 90 J

39. As a ball falls freely toward the ground, its total mechanical energy
    1. decreases
    2. increases
    3. remains the same

Name: ______________________ Period: __________

WEP-Energy

Base your answers to questions 40 through 42 on the information below.

A roller coaster car has a mass of 290 kilograms. Starting from rest, the car acquires 3.13×10^5 joules of kinetic energy as it descends to the bottom of a hill in 5.3 seconds.

40. Calculate the height of the hill. [Neglect friction. Show all work, including the equation and substitution with units.]

41. Calculate the speed of the roller coaster car at the bottom of the hill. [Show all work, including the equation and substitution with units.]

42. Calculate the magnitude of the average acceleration of the roller coaster car as it descends to the bottom of the hill. [Show all work, including the equation and substitution with units.]

43. The table below lists the mass and speed of each of four objects.

Data Table

Objects	Mass (kg)	Speed (m/s)
A	1.0	4.0
B	2.0	2.0
C	0.5	4.0
D	4.0	1.0

Which two objects have the same kinetic energy?
1. A and D
2. B and D
3. A and C
4. B and C

44. A 1.00-kilogram ball is dropped from the top of a building. Just before striking the ground, the ball's speed is 12.0 meters per second. What was the ball's gravitational potential energy, relative to the ground, at the instant it was dropped? [Neglect friction.]
1. 6.00 J
2. 24.0 J
3. 72.0 J
4. 144 J

45. A person weighing 6.0×10^2 newtons rides an elevator upward at an average speed of 3.0 meters per second for 5.0 seconds. How much does this person's gravitational potential energy increase as a result of this ride?
1. 3.6×10^2 J
2. 1.8×10^3 J
3. 3.0×10^3 J
4. 9.0×10^3 J

46. Which combination of fundamental units can be used to express energy?
1. kg·m/s
2. kg·m^2/s
3. kg·m/s^2
4. kg·m^2/s^2

47. A ball is dropped from the top of a cliff. Which graph best represents the relationship between the ball's total energy and elapsed time as the ball falls to the ground? [Neglect friction.]

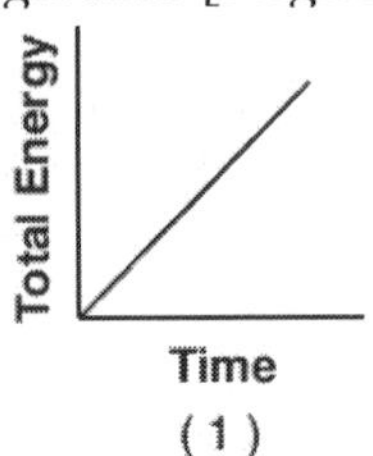

(1)

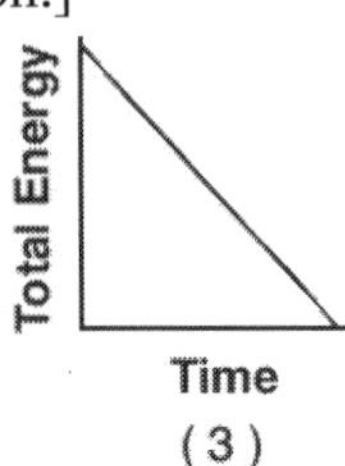

(3)

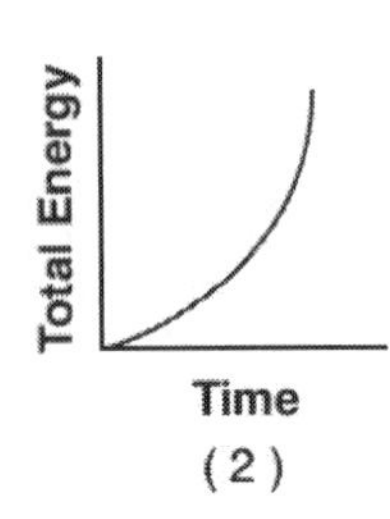

(2)

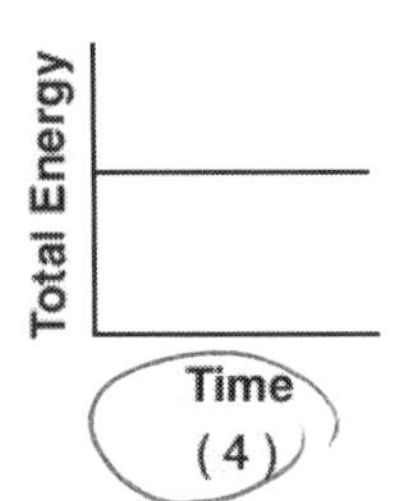

(4)

Name:________________________________ Period: __________

WEP-Energy

Base your answers to questions 48 through 50 on the information and diagram below.

A 3.0-kilogram object is placed on a frictionless track at point A and released from rest. (Assume the gravitational potential energy of the system to be zero at point C.)

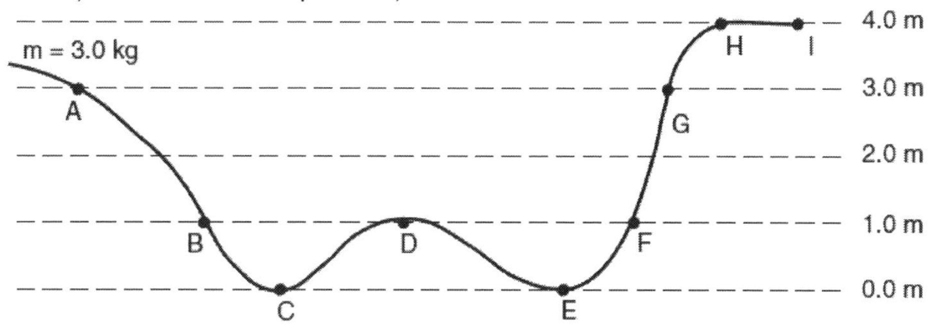

48. Calculate the gravitational potential energy of the object at point A. [Show all work, including the equation and substitution with units.]

49. Calculate the kinetic energy of the object at point B. [Show all work, including the equation and substitution with units.]

50. Which letter represents the farthest point on the track that the object will reach?

---

51. An object is thrown vertically upward. Which pair of graphs best represents the object's kinetic energy and gravitational potential energy as functions of displacement while it rises? PE↑ KE↓

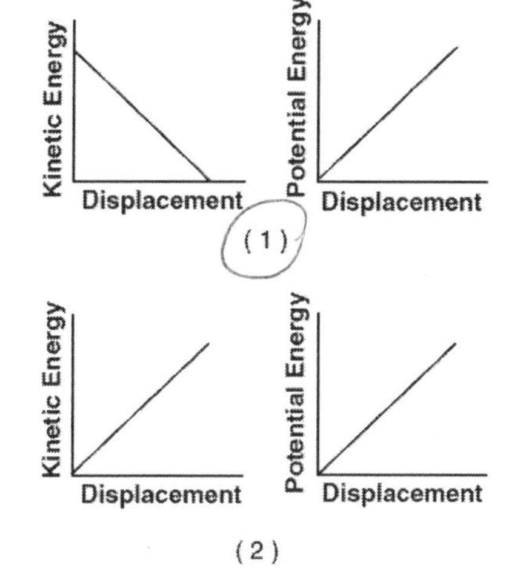

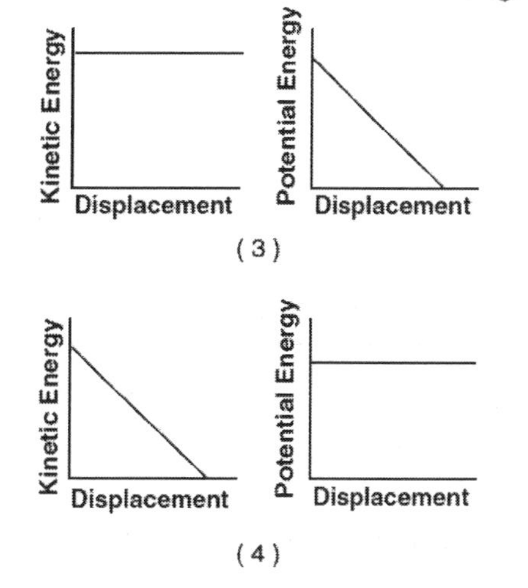

Name:______________________________ Period: ____________

WEP-Energy

Base your answers to questions 52 through 55 on the graph below, which represents the relationship between vertical height and gravitational potential energy for an object near Earth's surface.

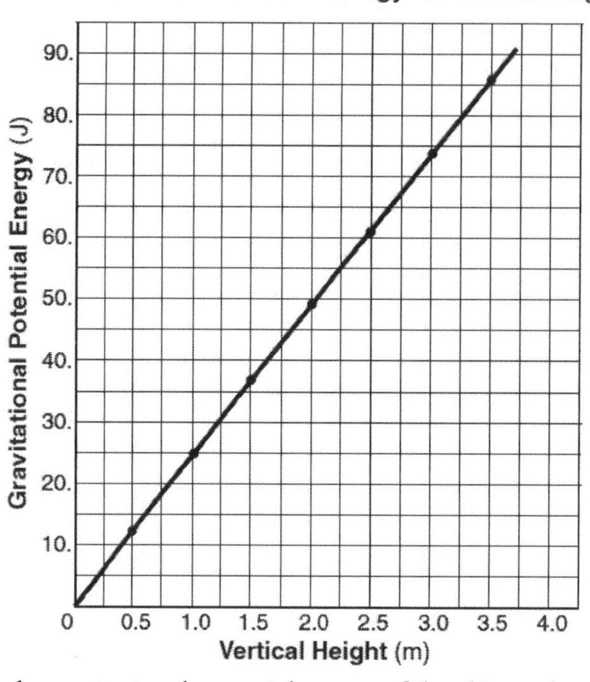

52. Based on the graph, what is the gravitational potential energy of the object when it is 2.25 meters above the surface of Earth?

53. Using the graph, calculate the mass of the object. [Show all work, including the equation and substitution with units.]

54. What physical quantity does the slope of the graph represent?

55. Using a straightedge, draw a line on the graph to represent the relationship between gravitational potential energy and vertical height for an object having a greater mass.

---

56. An object falls freely near Earth's surface. Which graph best represents the relationship between the object's kinetic energy and its time of fall?

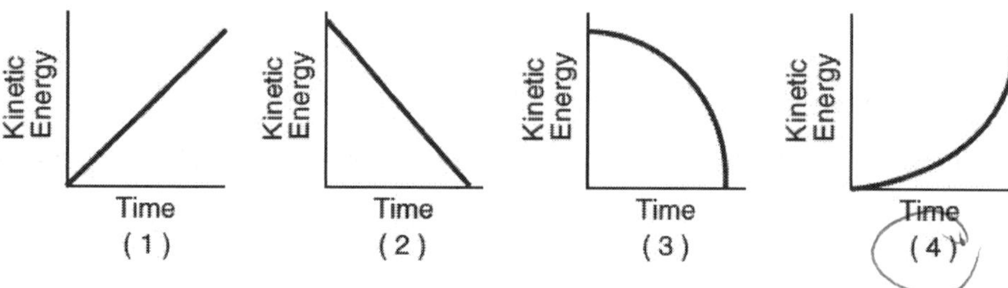

Name: ______________________________ Period: ____________

WEP-Energy

57. A wound spring provides the energy to propel a toy car across a level floor. At time t_i, the car is moving at speed v_i across the floor and the spring is unwinding, as shown below. At time t_f, the spring has fully unwound and the car has coasted to a stop.

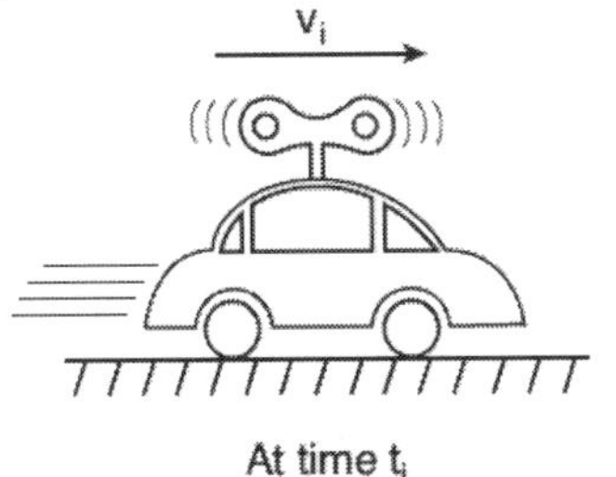

$v_f = 0$

At time t_f

Which statement best describes the transformation of energy that occurs between times t_i and t_f?

1. Gravitational potential energy at t_i is converted to internal energy at t_f.
2. Elastic potential energy at t_i is converted to kinetic energy at t_f.
3. Both elastic potential energy and kinetic energy at t_i are converted to internal energy at t_f.
4. Both kinetic energy and internal energy at t_i are converted to elastic potential energy at t_f.

58. A wooden crate is pushed at constant speed across a level wooden floor. Which graph best represents the relationship between the total mechanical energy of the crate and the duration of time the crate is pushed?

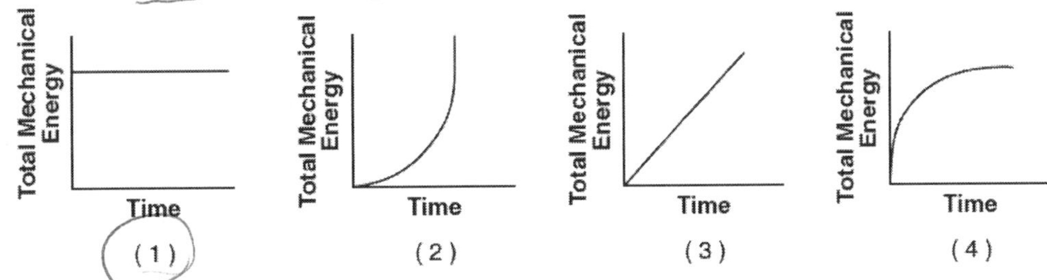

59. Which graph represents the relationship between the gravitational potential energy (GPE) of an object near the surface of Earth and its height above the surface of Earth?

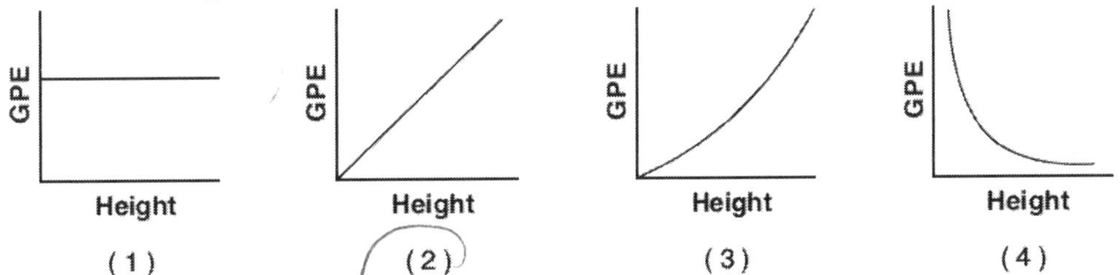

Base your answers to questions 60 and 61 on the information below.

A 75-kilogram athlete jogs 1.8 kilometers along a straight road in 1.2×10^3 seconds.

60. Determine the average speed of the athlete in meters per second.

61. Calculate the average kinetic energy of the athlete. [Show all work, including the equation and substitution with units.]

Name: ____________________ Period: __________

WEP-Energy

62. A 6.8-kilogram block is sliding down a horizontal, frictionless surface at a constant speed of 6.0 meters per second. The kinetic energy of the block is approximately
    1. 20 J
    2. 41 J
    3. 120 J
    4. 240 J

63. If the speed of a moving object is doubled, the kinetic energy of the object is
    1. halved
    2. doubled
    3. unchanged
    4. quadrupled

64. The diagram below represents a 155-newton box on a ramp. Applied force F causes the box to slide from point A to point B.

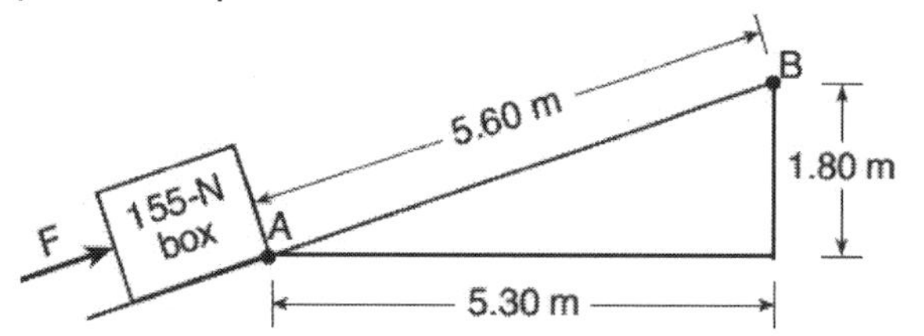

What is the total amount of gravitational potential energy gained by the box?
    1. 28.4 J
    2. 279 J
    3. 868 J
    4. 2740 J

65. A car travels at constant speed v up a hill from point A to point B, as shown in the diagram below.

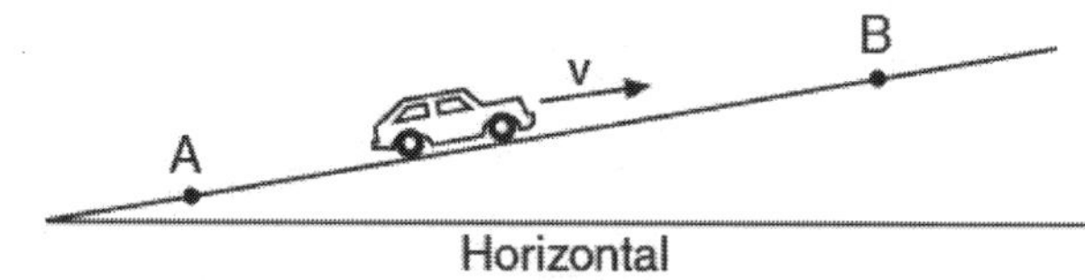

As the car travels from A to B, its gravitational potential energy
    1. increases and its kinetic energy decreases
    2. increases and its kinetic energy remains the same
    3. remains the same and its kinetic energy decreases
    4. remains the same and its kinetic energy remains the same

66. A student makes a simple pendulum by attaching a mass to the free end of a 1.50-meter length of string suspended from the ceiling of her physics classroom. She pulls the mass up to her chin and releases it from rest, allowing the pendulum to swing in its curved path. Her classmates are surprised that the mass doesn't reach her chin on the return swing, even though she does not move. Explain why the mass does *not* have enough energy to return to its starting position and hit the girl on the chin.

Base your answers to questions 67 and 68 on the information below. [Show all work, including the equation and substitution with units.]

A 65-kilogram pole vaulter wishes to vault to a height of 5.5 meters.

67. Calculate the minimum amount of kinetic energy the vaulter needs to reach this height if air friction is neglected and all the vaulting energy is derived from kinetic energy.

68. Calculate the speed the vaulter must attain to have the necessary kinetic energy.

69. Which pair of quantities can be expressed using the same units?
    1. work and kinetic energy
    2. power and momentum
    3. impulse and potential energy
    4. acceleration and weight

Name: ______________________________ Period: ____________

WEP-Energy

70. A car with mass *m* possesses a momentum of magnitude *p*. Which expression correctly represents the kinetic energy, KE, of the car in terms of *m* and *p*?
    1. KE=p/2m
    2. $KE=mp^2/2$
    3. KE=mp/2
    4. $KE=p^2/2m$

71. Which statement best explains why a "wet saw" used to cut through fine optical crystals is constantly lubricated with oil?
    1. Lubrication decreases friction and minimizes the increase of internal energy.
    2. Lubrication decreases friction and maximizes the increase of internal energy.
    3. Lubrication increases friction and minimizes the increase of internal friction.
    4. Lubrication increases friction and maximizes the increase of internal energy.

72. A car, initially traveling at 30 meters per second, slows uniformly as it skids to a stop after the brakes are applied. On the axes below, sketch a graph showing the relationship between the kinetic energy of the car as it is being brought to a stop and the work done by friction in stopping the car.

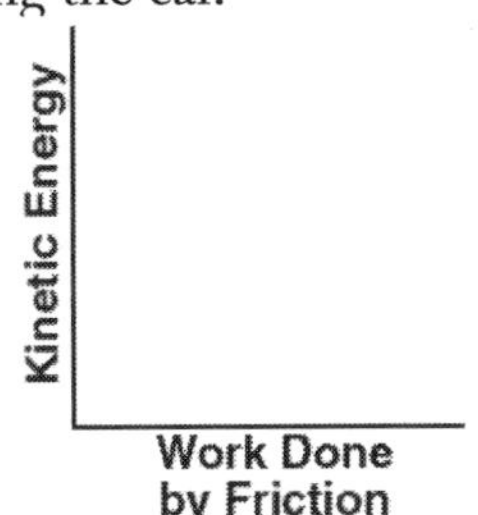

73. The work done on a slingshot is 40 joules to pull back a 0.10-kilogram stone. If the slingshot projects the stone straight up in the air, what is the maximum height to which the stone will rise? [Neglect friction.]
    1. 0.41 m
    2. 41 m
    3. 410 m
    4. 4.1 m

74. Calculate the kinetic energy of a particle with a mass of 3.34×10^{-27} kg and a speed of 2.89×10^{5} meters per second.

75. A 55-kilogram diver falls freely from a diving platform that is 3.00 meters above the surface of the water in a pool. When she is 1.00 meter above the water, what are her gravitational potential energy and kinetic energy with respect to the water's surface.
    1. PE=1620 J and KE=0 J
    2. PE=1080 J and KE=540 J
    3. PE=810 J and KE=810 J
    4. PE=540 J and KE=1080 J

76. As a box is pushed 30 meters across a horizontal floor by a constant horizontal force of 25 newtons, the kinetic energy of the box increases by 300 joules. How much total internal energy is produced during this process?
    1. 150 J
    2. 250 J
    3. 450 J
    4. 750 J

77. The diagram below shows an ideal simple pendulum.

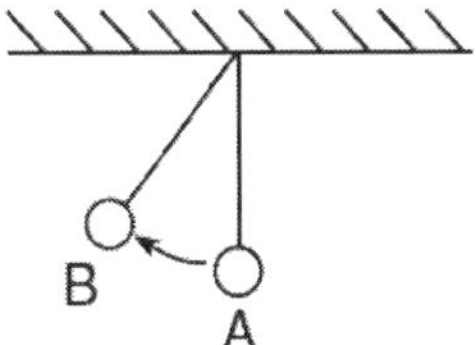

As the pendulum swings from position A to position B, what happens to its total mechanical energy? [Neglect friction.]
    1. It decreases.
    2. It increases.
    3. It remains the same.

78. Which situation describes a system with *decreasing* gravitational potential energy.
    1. a girl stretching a horizontal spring
    2. a bicyclist riding up a steep hill
    3. a rocket rising vertically from Earth
    4. a boy jumping down from a tree limb

79. A 2.0-kilogram block sliding down a ramp from a height of 3.0 meters above the ground reaches the ground with a kinetic energy of 50 joules. The total work done by friction on the block as it slides down the ramp is approximately
    1. 6 J
    2. 9 J
    3. 18 J
    4. 44 J

Name: ______________________________ Period: ____________

WEP-Energy

80. Which statement describes the kinetic energy and total mechanical energy of a block as it is pulled at constant speed up an incline?
    1. Kinetic energy decreases and total mechanical energy increases.
    2. Kinetic energy decreases and total mechanical energy remains the same.
    3. Kinetic energy remains the same and total mechanical energy increases.
    4. Kinetic energy remains the same and total mechanical energy remains the same.

81. A 75-kilogram bicyclist coasts down a hill at a constant speed of 12 meters per second. What is the kinetic energy of the bicyclist?
    1. 4.5×10^2 J
    2. 9.0×10^2 J
    3. 5.4×10^3 J
    4. 1.1×10^4 J

82. An electrical generator in a science classroom makes a lightbulb glow when a student turns a hand crank on the generator. During its operation, this generator converts
    1. chemical energy to electrical energy
    2. mechanical energy to electrical energy
    3. electrical energy to mechanical energy
    4. electrical energy to chemical energy

83. A block weighing 40 newtons is released from rest on an incline 8 meters above the horizontal, as shown in the diagram below.

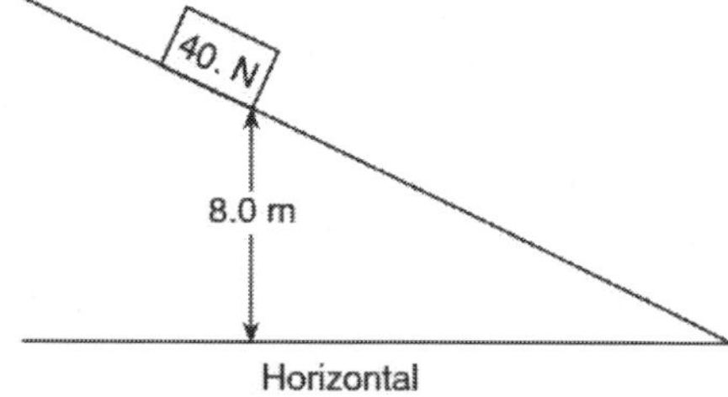

If 50 joules of heat is generated as the block slides down the incline, the maximum kinetic energy of the block at the bottom of the incline is
    1. 50 J
    2. 270 J
    3. 320 J
    4. 3100 J

84. A child, starting from rest at the top of a playground slide, reaches a speed of 7.0 meters per second at the bottom of the slide. What is the vertical height of the slide? [Neglect friction.]
    1. 0.71 m
    2. 1.4 m
    3. 2.5 m
    4. 3.5 m

85. Two students of equal weight go from the first floor to the second floor. The first student uses an elevator and the second student walks up a flight of stairs. Compared to the gravitational potential energy gained by the first student, the gravitational potential energy gained by the second student is
    1. less
    2. greater
    3. the same

86. During an emergency stop, a 1.5×10^3-kilogram car lost a total of 3.0×10^5 joules of kinetic energy. What was the speed of the car at the moment the brakes were applied?
    1. 10 m/s
    2. 14 m/s
    3. 20 m/s
    4. 25 m/s

87. While riding a chairlift, a 55-kilogram skier is raised a vertical distance of 370 meters. What is the total change in the skier's gravitational potential energy?
    1. 5.4×10^1 J
    2. 5.4×10^2 J
    3. 2.0×10^4 J
    4. 2.0×10^5 J

88. A book of mass m falls freely from rest to the floor from the top of a desk of height h. What is the speed of the book upon striking the floor?
    1. $\sqrt{2gh}$
    2. 2gh
    3. mgh
    4. mh

89. A box at the top of a rough incline possesses 981 joules more gravitational potential energy than it does at the bottom. As the box slides to the bottom of the incline, 245 joules of heat is produced. Determine the kinetic energy of the box at the bottom of the incline.

Name:______________________________ Period: __________

WEP-Energy

90. A car uses its brakes to stop on a level road. During this process, there must be a conversion of kinetic energy into
    1. light energy
    2. nuclear energy
    3. gravitational potential energy
    4. internal energy

91. A pendulum is made from a 7.50-kilogram mass attached to a rope connected to the ceiling of a gymnasium. The mass is pushed to the side until it is at position A, 1.5 meters higher than its equilibrium position. After it is released from rest at position A, the pendulum moves freely back and forth between positions A and B, as shown in the diagram below.

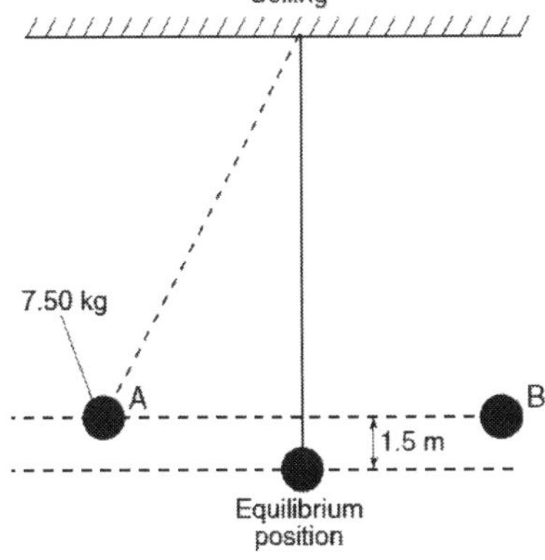

What is the total amount of kinetic energy that the mass has as it swings freely through its equilibrium position? [Neglect friction.]
    1. 11 J
    2. 94 J
    3. 110 J
    4. 920 J

Base your answers to questions 92 through 95 on the information below.

A runner accelerates uniformly from rest to a speed of 8 meters per second. The kinetic energy of the runner was determined at 2-meter-per-second intervals and recorded in the data table.

Speed (m/s)	Kinetic Energy (J)
0.00	0.00
2.00	140
4.00	560
6.00	1260
8.00	2240

Using the information in the data table, construct a graph on the grid following the directions below.

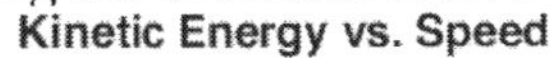

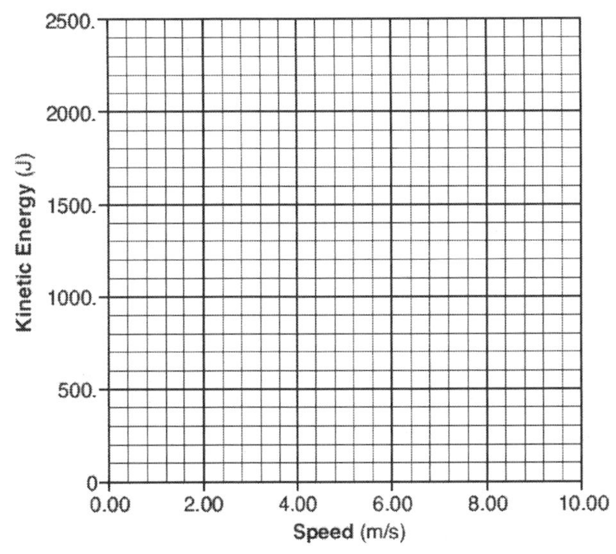

92. Plot the data points for kinetic energy of the runner versus his speed.

93. Draw the line or curve of best fit.

94. Calculate the mass of the runner [Show all work, including the equation and substitution with units.]

95. A soccer player having less mass than the runner also accelerates uniformly from rest to a speed of 8 meters per second. Compare the kinetic energy of the less massive soccer player to the kinetic energy of the more massive runner when both are traveling at the same speed.

Name:______________________ Period: __________

WEP-Energy

96. Which graph represents the relationship between the kinetic energy and the speed of a freely falling object?

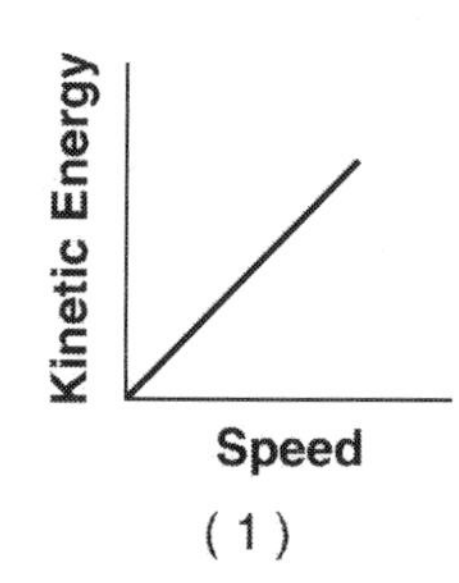

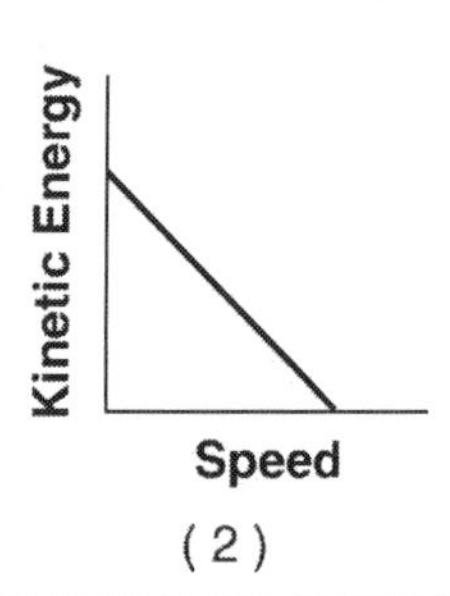

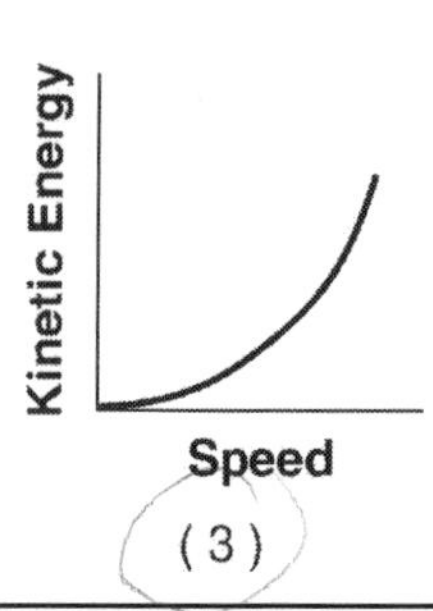

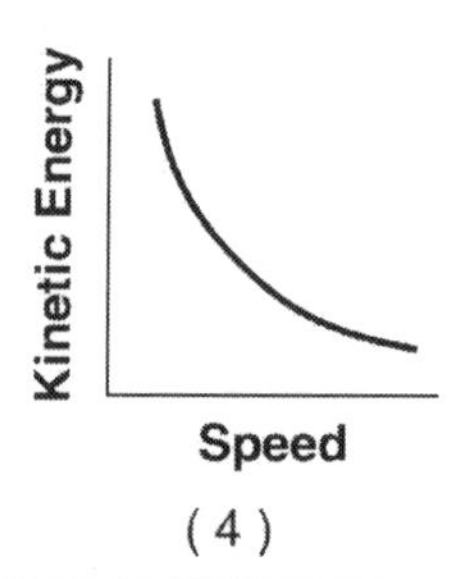

Base your answers to questions 97 through 100 on the information below and diagram at right.

A 30.4-newton force is used to slide a 40.0-newton crate a distance of 6.00 meters at constant speed along an incline to a vertical height of 3.00 meters.

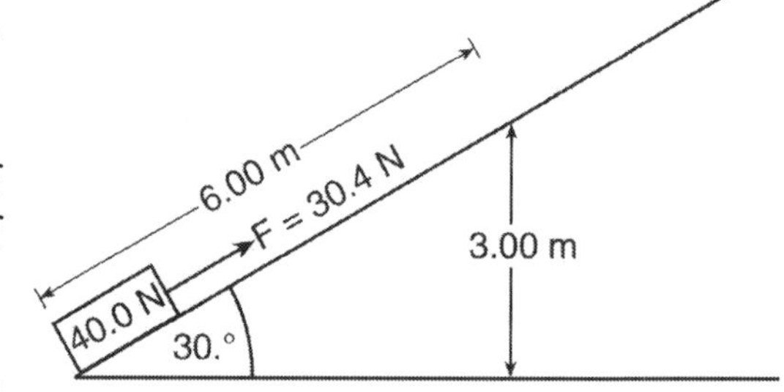

97. Determine the total work done by the 30.4-newton force in sliding the crate along the incline.

98. Calculate the total increase in the gravitational potential energy of the crate after it has slid 6.00 meters along the incline. [Show all work, including the equation and substitution with units.]

99. State what happens to the kinetic energy of the crate as it slides along the incline.

100. State what happens to the internal energy of the crate as it slides along the incline.

101. When a teacher shines a light on a photocell attached to a fan, the blades of the fan turn. The brighter the light shone on the photocell, the faster the blades turn. Which energy conversion is illustrated by this demonstration?
1. light -> thermal -> mechanical
2. light -> nuclear -> thermal
3. light -> electrical -> mechanical
4. light -> mechanical -> chemical

102. In the diagram below, an ideal pendulum released from position A swings freely to position B.

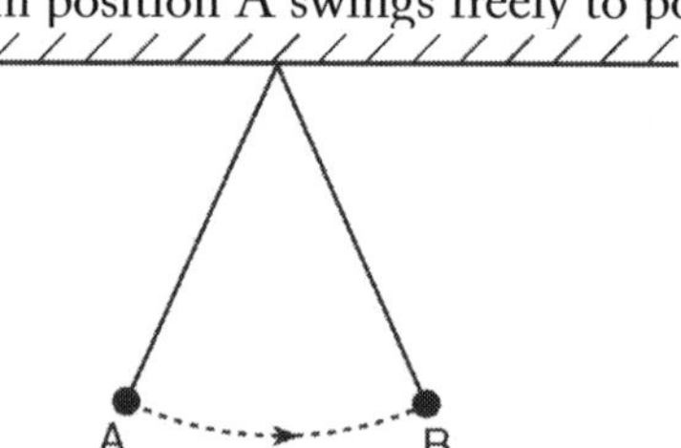

As the pendulum swings from A to B, its total mechanical energy
1. decreases, then increases
2. increases, only
3. increases, then decreases
4. remains the same

Name: ______________________________ Period: ____________

WEP-Energy

103. Two pieces of flint rock produce a visible spark when they are struck together. During this process, mechanical energy is converted into
   1. nuclear energy and electromagnetic energy
   2. internal energy and nuclear energy
   3. electromagnetic energy and internal energy
   4. elastic potential energy and nuclear energy

104. Which graph best represents an object in equilibrium moving in a straight line?

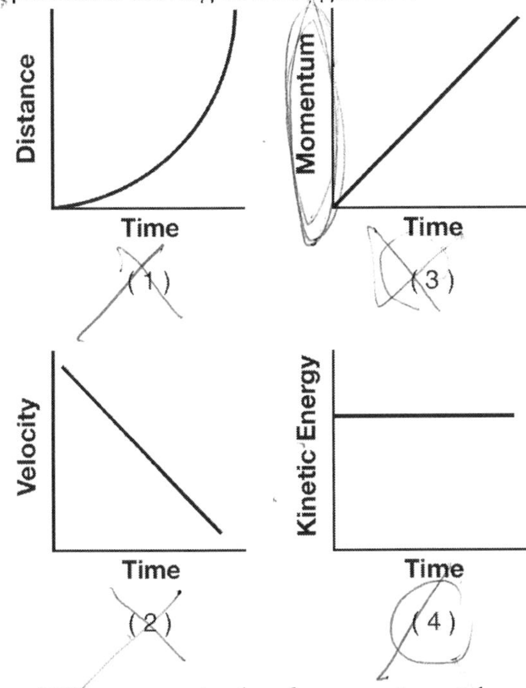

105. When a mass is placed on a spring with a spring constant of 60.0 newtons per meter, the spring is compressed 0.500 meter. How much energy is stored in the spring?
   1. 60.0 J
   2. 30.0 J
   3. 15.0 J
   4. 7.50 J

106. A shopping cart slows as it moves along a level floor. Which statement describes the energies of the cart?
   1. The kinetic energy increases and the gravitational potential energy remains the same.
   2. The kinetic energy increases and the gravitational potential energy decreases.
   3. The kinetic energy decreases and the gravitational potential energy remains the same.
   4. The kinetic energy decreases and the gravitational potential energy increases.

107. A 25-gram paper cup falls from rest off the edge of a tabletop 0.90 meter above the floor. If the cup has 0.20 joule of kinetic energy when it hits the floor, what is the total amount of energy converted into internal (thermal) energy during the cup's fall?
   1. 0.02 J
   2. 0.22 J
   3. 2.2 J
   4. 220 J

108. A 3.00-newton force causes a spring to stretch 60 centimeters. Calculate the spring constant of this spring. [Show all work, including the equation and substitution with units.]

109. Regardless of the method used to generate electrical energy, the amount of energy provided by the source is always greater than the amount of electrical energy produced. Explain why there is a difference between the amount of energy provided by the source and the amount of electrical energy produced.

110. Which quantities are scalar?
   1. speed and work
   2. velocity and force
   3. distance and acceleration
   4. momentum and power

111. Which energy transformation occurs in an operating electric motor?
   1. electrical --> mechanical
   2. mechanical --> electrical
   3. chemical --> electrical
   4. electrical --> chemical

Name: ________________________________ Period: ____________

WEP-Energy

112. A block slides across a rough, horizontal table-top. As the block comes to rest, there is an increase in the block-tabletop system's
   1. gravitational potential energy
   2. elastic potential energy
   3. kinetic energy
   4. internal (thermal) energy

113. A compressed spring in a toy is used to launch a 5.00-gram ball. If the ball leaves the toy with an initial horizontal speed of 5.00 meters per second, the minimum amount of potential energy stored in the compressed spring was
   1. 0.0125 J
   2. 0.0250 J
   3. 0.0625 J
   4. 0.125 J

Name: ______________________________ Period: ____________

Electrostatics-Charge

1. The diagram below represents two electrically charged identical-sized metal spheres, A and B.

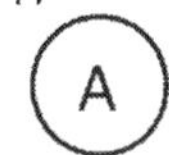

$+2.0 \times 10^{-7}$ C $+1.0 \times 10^{-7}$ C

If the spheres are brought into contact, which sphere will have a net gain of electrons?
  1. A, only
  2. B, only
  3. both A and B
  4. neither A nor B

2. Metal sphere A has a charge of -2 units and an identical metal sphere, B, has a charge of -4 units. If the spheres are brought into contact with each other and then separated, the charge on sphere B will be
  1. 0 units
  2. -2 units
  3. -3 units
  4. +4 units

3. If an object has a net negative charge of 4.0 coulombs, the object possesses
  1. 6.3×10^{18} more electrons than protons
  2. 2.5×10^{19} more electrons than protons
  3. 6.3×10^{18} more protons than electrons
  4. 2.5×10^{19} more protons than electrons

Base your answers to questions 4 and 5 on the information below

A lightweight sphere hangs by an insulating thread. A student wishes to determine if the sphere is neutral or electrostatically charged. She has a negatively charged hard rubber rod and a positively charged glass rod. She does not touch the sphere with the rods, but runs tests by bringing them near the sphere one at a time.

4. Describe the test result that would prove that the sphere is neutral

5. Describe the test result that would prove that the sphere is positively charged.

6. Oil droplets may gain electrical charges as they are projected through a nozzle. Which quantity of charge is *not* possible on an oil droplet?
  1. 8.0×10^{-19} C
  2. 4.8×10^{-19} C
  3. 3.2×10^{-19} C
  4. 2.6×10^{-19} C

7. A positive test charge is placed between an electron, *e*, and a proton, *p*, as shown in the diagram below.

A

Test charge

(e)D  B(p)

C

When the test charge is released, it will move toward
  1. A
  2. B
  3. C
  4. D

8. A metal sphere has a net negative charge of 1.1×10^{-6} coulomb. Approximately how many more electrons than protons are on the sphere?
  1. 1.8×10^{12}
  2. 5.7×10^{12}
  3. 6.9×10^{12}
  4. 9.9×10^{12}

9. A positively charged glass rod attracts object X. The net charge of object X
  1. may be zero or negative
  2. may be zero or positive
  3. must be negative
  4. must be positive

10. The charge-to-mass ratio of an electron is
  1. 5.69×10^{-12} C/kg
  2. 1.76×10^{-11} C/kg
  3. 1.76×10^{11} C/kg
  4. 5.69×10^{12} C/kg

11. What is the magnitude of the charge, in coulombs, of a lithium nucleus containing three protons and four neutrons?

Name: ______________________ Period: __________

Electrostatics-Charge

12. The diagram below shows three neutral metal spheres, x, y, and z, in contact and on insulating stands.

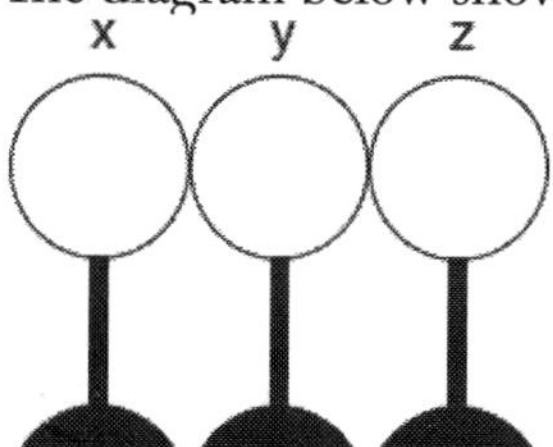

Which diagram best represents the charge distribution on the spheres when a positively charged rod is brought near sphere x, but does not touch it.

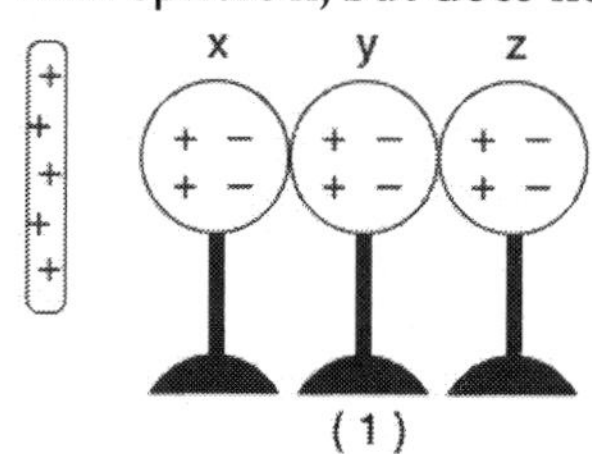

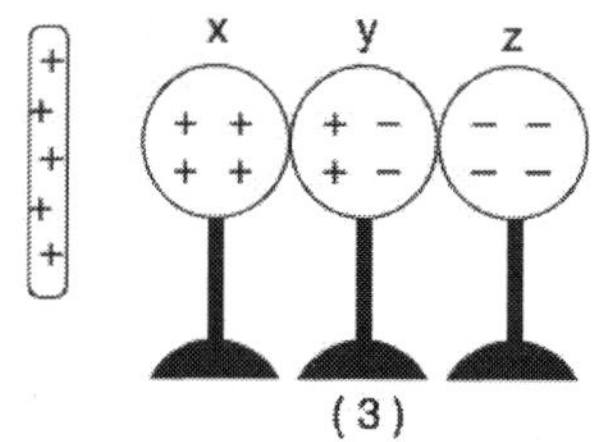

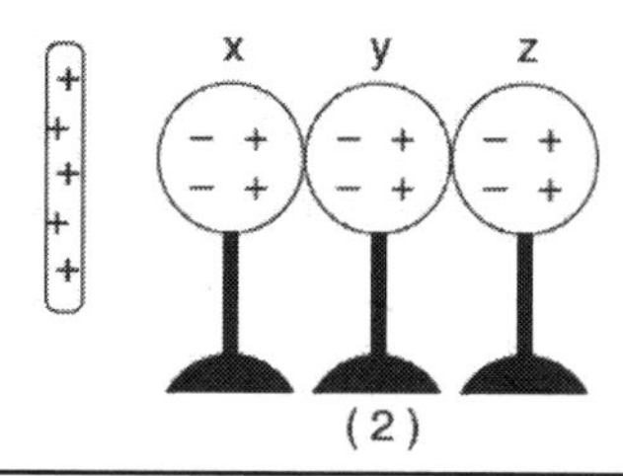

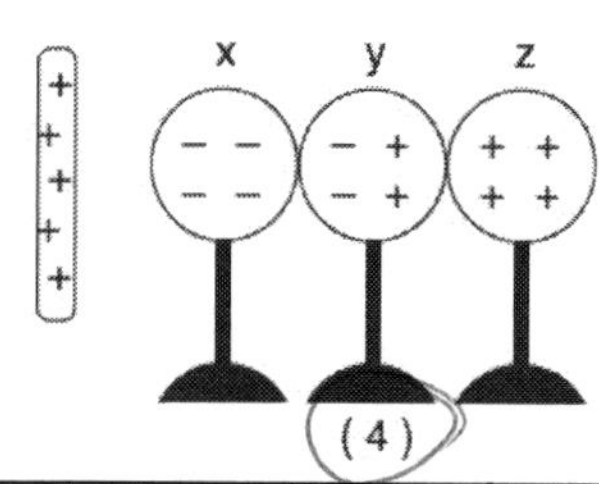

13. What is the net electrical charge on a magnesium ion that is formed when a neutral magnesium atom loses two electrons?
    1. -3.2×10^{-19} C
    2. -1.6×10^{-19} C
    3. $+1.6 \times 10^{-19}$ C
    4. $+3.2 \times 10^{-19}$ C

14. A negatively charged plastic comb is brought close to, but does not touch, a small piece of paper. If the comb and the paper are attracted to each other, the charge on the paper
    1. may be negative or neutral
    2. may be positive or neutral
    3. must be negative
    4. must be positive

15. An object possessing an excess of 6.0×10^{6} electrons has a net charge of
    1. 2.7×10^{-26} C
    2. 5.5×10^{-24} C
    3. 3.8×10^{-13} C
    4. 9.6×10^{-13} C

16. When a neutral metal sphere is charged by contact with a positively charged glass rod, the sphere
    1. loses electrons
    2. gains electrons
    3. loses protons
    4. gains protons

17. Which quantity of excess electric charge could be found on an object?
    1. 6.25×10^{-19} C
    2. 4.8×10^{-19} C
    3. 6.25 elementary charges
    4. 1.60 elementary charges

18. A particle could have a charge of
    1. 0.8×10^{-19} C
    2. 1.2×10^{-19} C
    3. 3.2×10^{-19} C
    4. 4.1×10^{-19} C

Name: ______________________________ Period: ____________

Electrostatics-Charge

19. A dry plastic rod is rubbed with wool cloth and then held near a thin stream of water from a faucet. The path of the stream of water is changed, as represented in the diagram below.

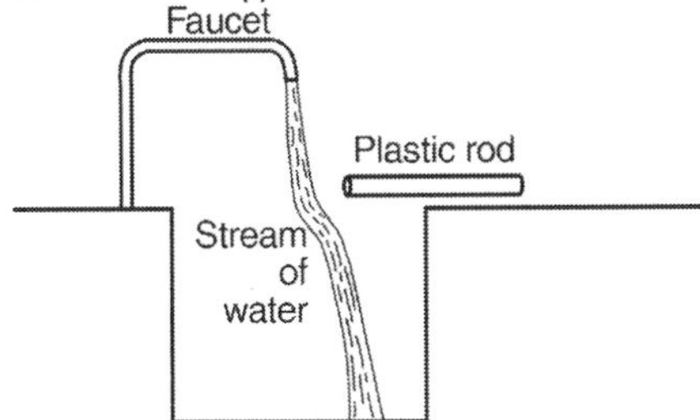

Which force causes the path of the stream of water to change due to the plastic rod?

1. nuclear
2. magnetic
3. electrostatic
4. gravitational

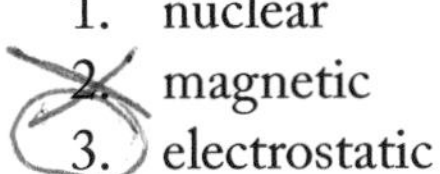

20. Which net charge could be found on an object?

1. $+4.80 \times 10^{-19}$ C
2. $+2.40 \times 10^{-19}$ C
3. -2.40×10^{-19} C
4. -5.60×10^{-19} C

21. Two identically-sized metal spheres, A and B, are on insulating stands, as shown in the diagram below. Sphere A possesses an excess of 6.3×10^{10} electrons and sphere B is neutral.

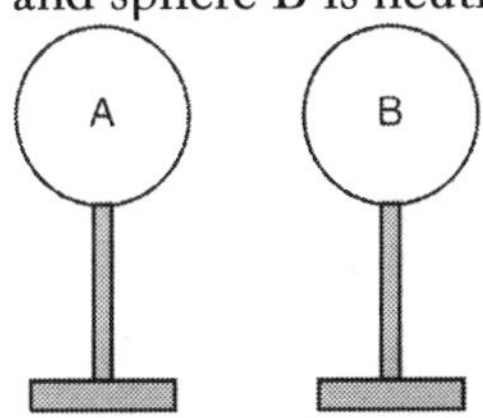

Insulating stands

Which diagram best represents the charge distribution on sphere B?

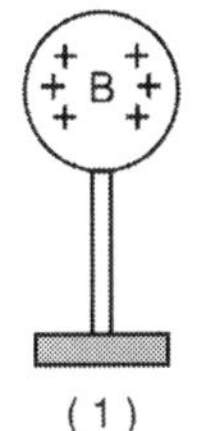

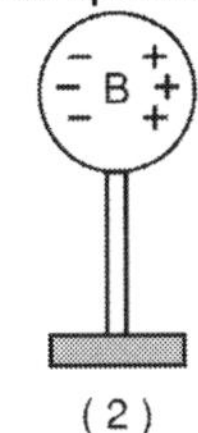

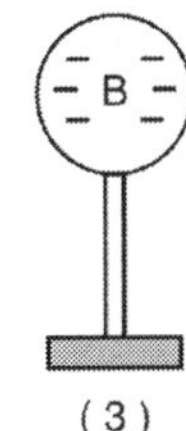

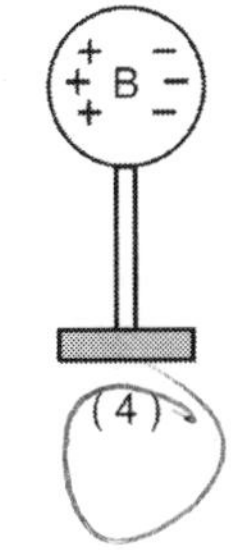

22. Two identically-sized metal spheres on insulating stands are positioned as shown below. The charge on sphere A is -4.0×10^{-6} coulomb and the charge on sphere B is -8.0×10^{-6} coulomb.

-4.0×10^{-6} C -8.0×10^{-6} C

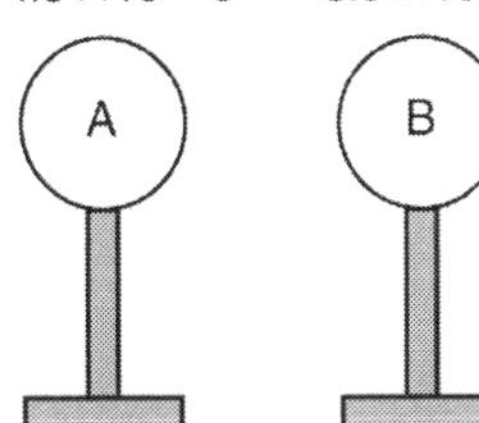

The two spheres are touched together and then separated. The total number of excess electrons on sphere A after the separation is

1. 2.5×10^{13}
2. 3.8×10^{13}
3. 5.0×10^{13}
4. 7.5×10^{13}

Name: ______________________________ Period: ____________

Electrostatics-Coulomb's Law

1. Which graph best represents the electrostatic force between an alpha particle with a charge of +2 elementary charges and a positively charged nucleus as a function of their distance of separation?

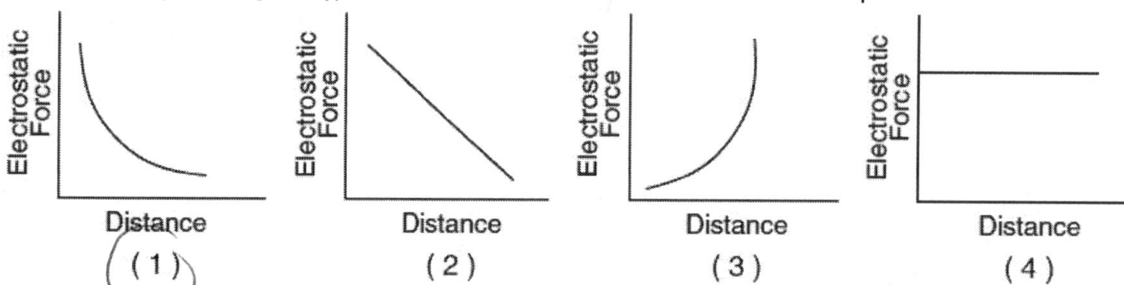

2. In the diagram below, two positively charged spheres, A and B, of masses m_A and m_B are located a distance d apart.

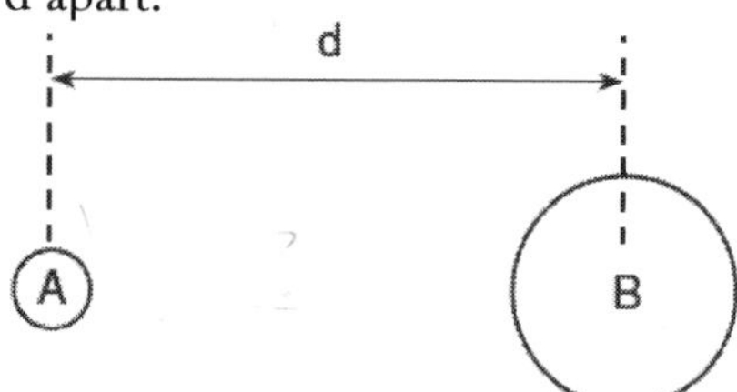

Which diagram best represents the directions of the gravitational force, F_g, and the electrostatic force, F_e, acting on sphere A due to the mass and charge of sphere B? [Vectors are not drawn to scale.]

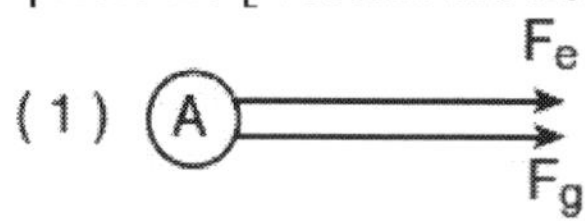

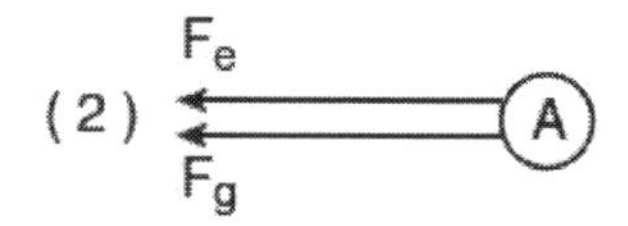

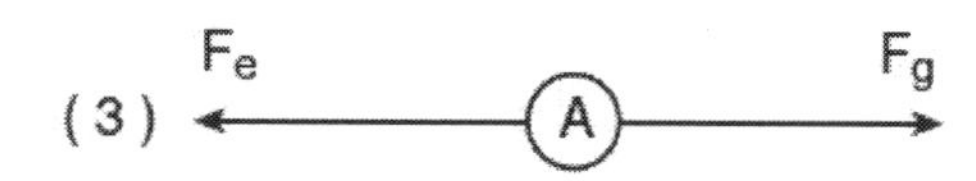

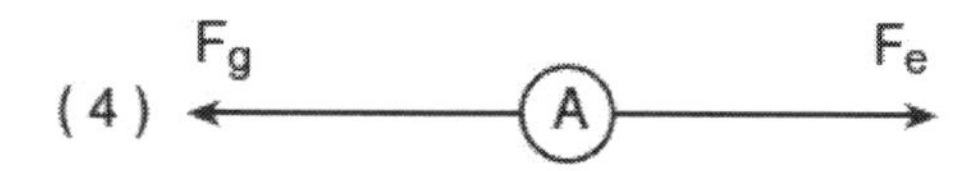

3. What is the magnitude of the electrostatic force between two electrons separated by a distance of 1.00×10^{-8} meter?
   1. 2.56×10^{-22} N
   2. 2.30×10^{-20} N
   3. 2.30×10^{-12} N
   4. 1.44×10^{-1} N

4. Two metal spheres, A and B, possess charges of 1.0 microcoulomb and 2.0 microcoulombs, respectively. In the diagram below, arrow F represents the electrostatic force exerted on sphere B by sphere A.

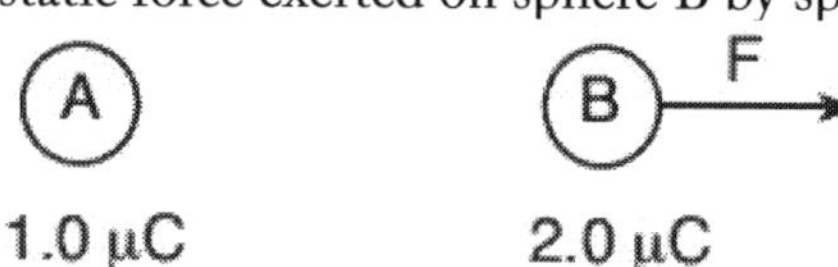

Which arrow represents the magnitude and direction of the electrostatic force exerted on sphere A by sphere B?

F ← (1)

F → (3)

2F ← (2)

2F → (4)

5. Two small identical metal spheres, A and B, on insulated stands, are each given a charge of $+2.0 \times 10^{-6}$ coulomb. The distance between the spheres is 2.0×10^{-1} meter. Calculate the magnitude of the electrostatic force that the charge on sphere A exerts on the charge on the sphere B. [Show all work, including the equation and substitution with units.]

Name:______________________________ Period: ____________

Electrostatics-Coulomb's Law

Base your answers to questions 6 through 9 on the information below.

A force of 6.0×10^{-15} newton due south and a force of 8.0×10^{-15} newton due east act concurrently on an electron, e^-.

6. On the diagram, draw a force diagram to represent the two forces acting on the electron. (The electron is represented by a dot.) Use a metric ruler and the scale of 1.0 centimeter = 1.0×10^{-15} newton. Begin each vector at the dot representing the electron and label its magnitude in newtons.

e⁻

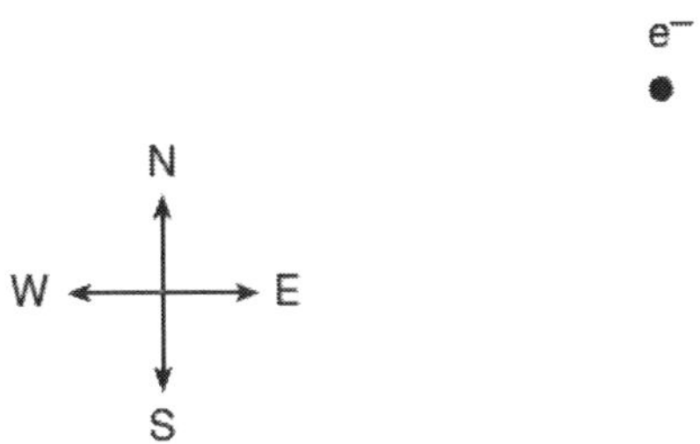

7. Determine the resultant force on the electron graphically. Label the resultant vector R.

8. Determine the magnitude of the resultant vector R.

9. Determine the angle between the resultant and the 6.0×10^{-15}-newton vector.

---

10. The distance between an electron and a proton is varied. Which pair of graphs best represents the relationship between gravitational force, F_g, and distance, r, and the relationship between electrostatic force, F_e, and distance, r, for these particles?

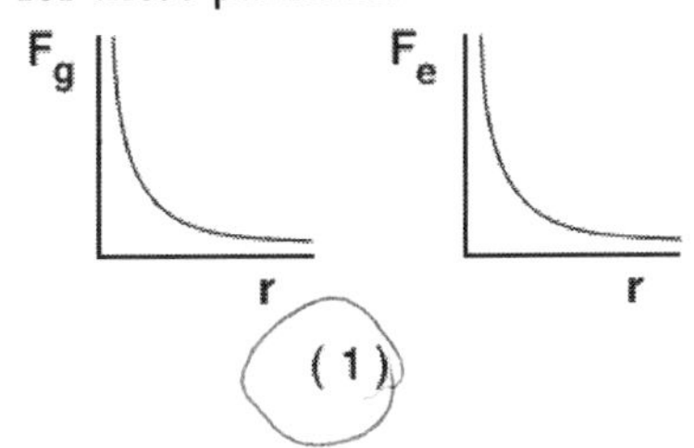

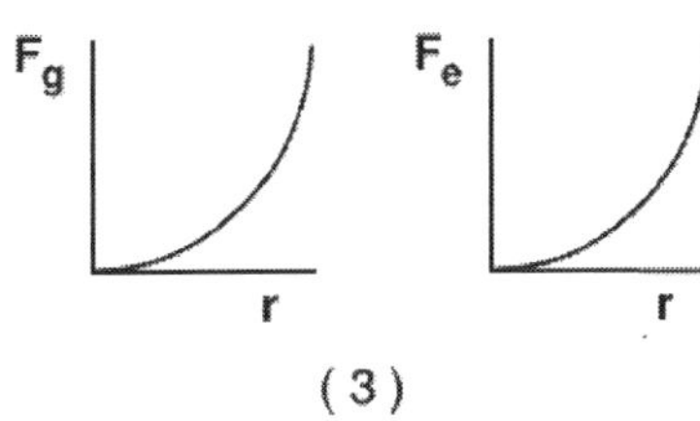

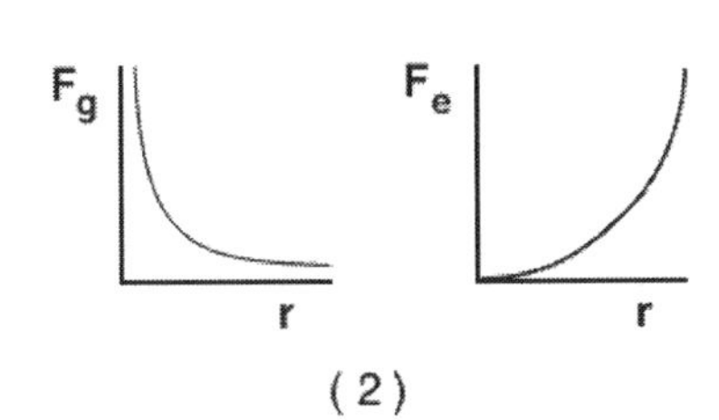

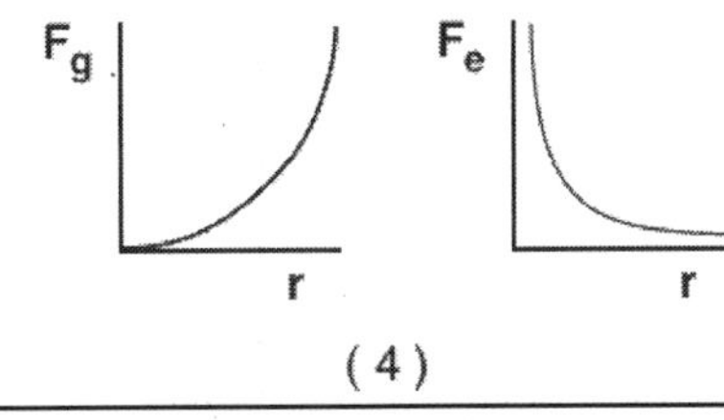

Name: ______________________________ Period: __________

Electrostatics-Coulomb's Law

11. Which graph best represents the relationship between the magnitude of the electrostatic force and the distance between two oppositely charged particles?

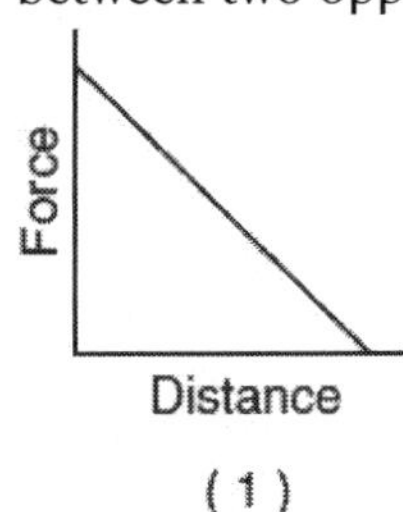

(1)

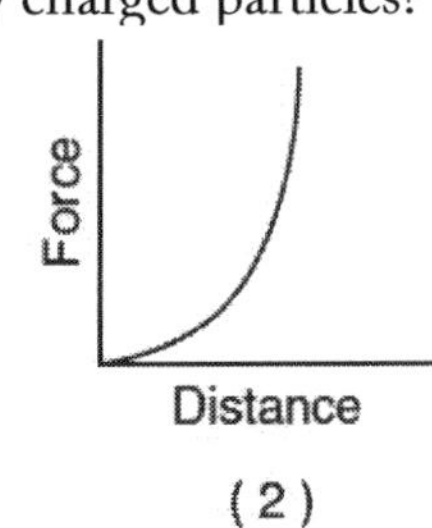

(2)

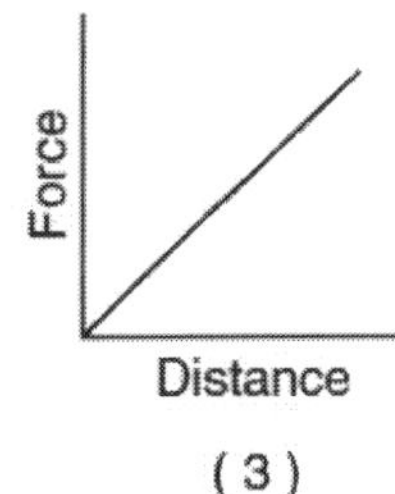

(3)

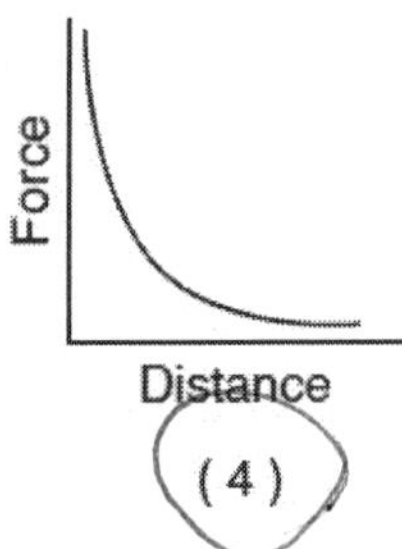

(4)

12. The diagram below shows two identical metal spheres, A and B, separated by a distance d. Each sphere has mass m and possesses charge q.

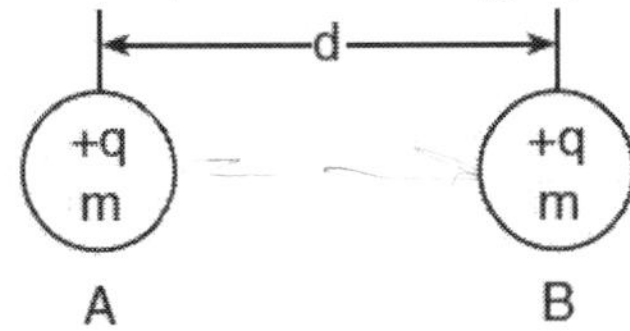

Which diagram best represents the electrostatic force F_e and the gravitational force F_g acting on sphere B due to sphere A?

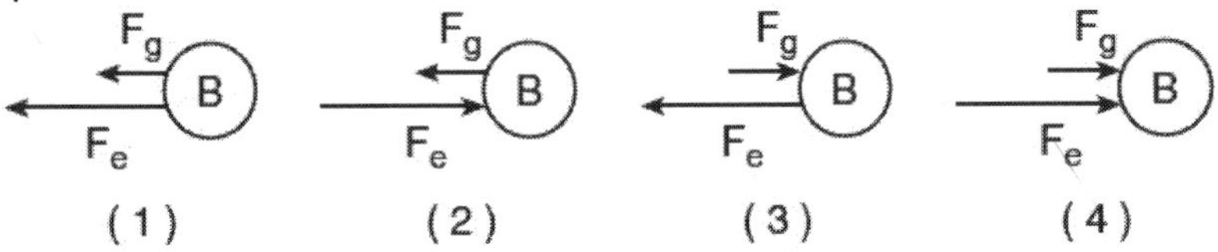

13. The diagram below shows the arrangement of three small spheres, A, B, and C, having charges of 3q, q, and q, respectively. Spheres A and C are located distance r from sphere B.

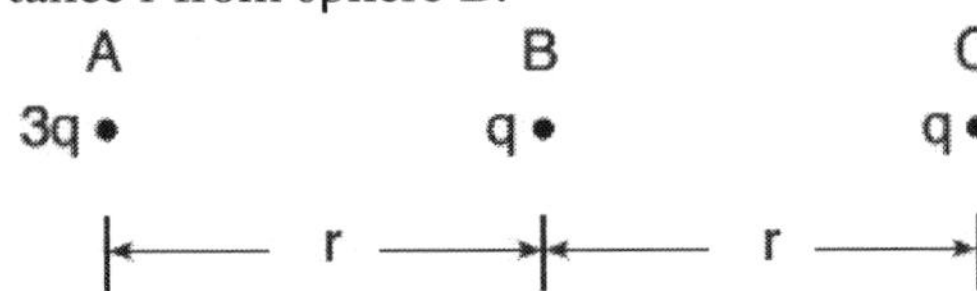

Compared to the magnitude of the electrostatic force exerted by sphere B on sphere C, the magnitude of the electrostatic force exerted by sphere A on sphere C is

1. the same
2. twice as great
3. 3/4 as great
4. 3/2 as great

14. If the distance separating an electron and a proton is halved, the magnitude of the electrostatic force between these charges particles will be

1. unchanged
2. doubled
3. quartered
4. quadrupled

15. Two similar metal spheres, A and B, have charged of $+2.0 \times 10^{-6}$ coulomb and $+1.0 \times 10^{-6}$ coulomb, respectively, as shown in the diagram below.

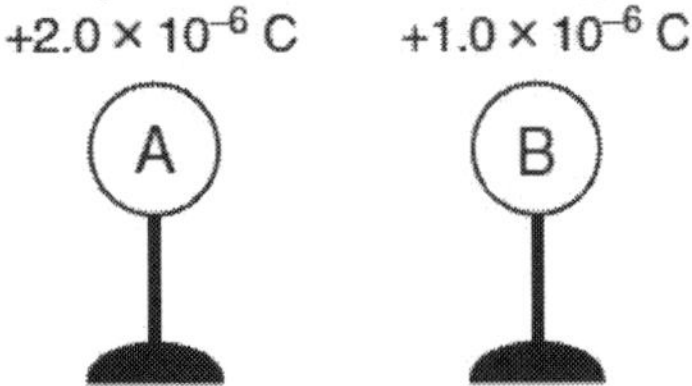

The magnitude of the electrostatic force on A due to B is 2.4 newtons. What is the magnitude of the electrostatic force on B due to A?

1. 1.2 N
2. 2.4 N
3. 4.8 N
4. 9.6 N

16. Two protons are located one meter apart. Compared to the gravitational force of attraction between the two protons, the electrostatic force between the protons is

1. stronger and repulsive
2. weaker and repulsive
3. stronger and attractive
4. weaker and attractive

Name:______________________________ Period: ____________

Electrostatics-Coulomb's Law

17. The diagram below shows two small metal spheres, A and B. Each sphere possesses a net charge of 4.0×10^{-6} coulomb. The spheres are separated by a distance of 1.0 meter.

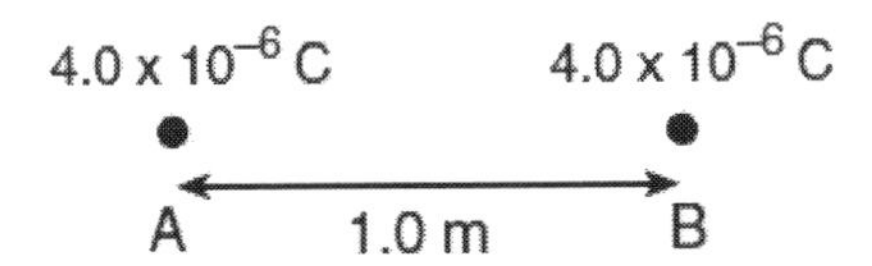

Which combination of charged spheres and separation distance produces an electrostatic force of the same magnitude as the electrostatic force between spheres A and B?

(1)
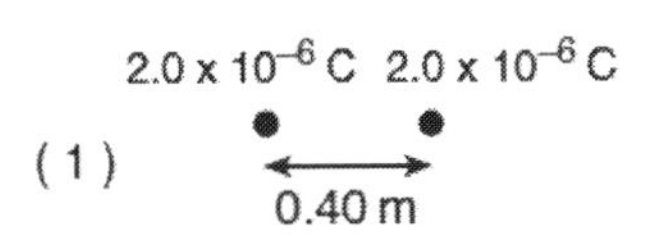

(2)
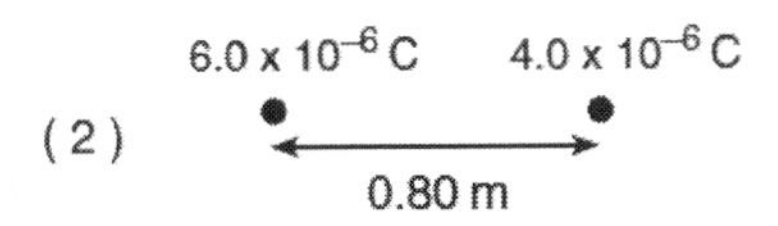

(3)
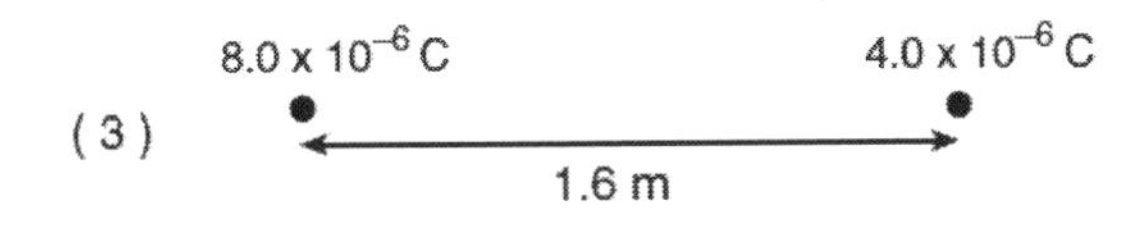

(4)
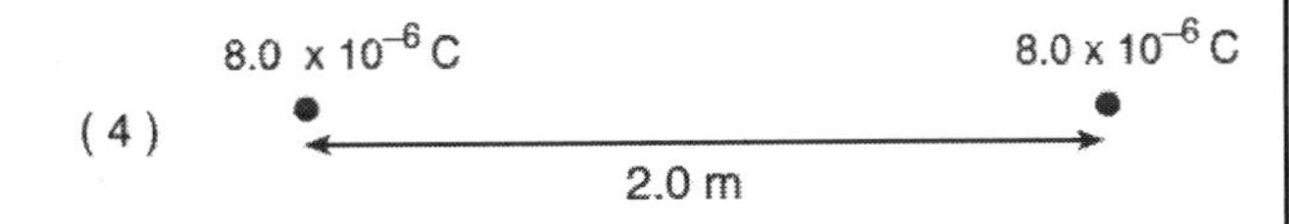

18. A balloon is rubbed against a student's hair and then touched to a wall. The balloon "sticks" to the wall due to
    1. electrostatic forces between the particles of the balloon
    2. magnetic forces between the particles of the wall
    3. electrostatic forces between the particles of the balloon and the particles of the wall
    4. magnetic forces between the particles of the balloon and the particles of the wall

19. The magnitude of the electrostatic force between two point charges is F. If the distance between the charges is doubled, the electrostatic force between the charges will become
    1. F/4
    2. 2F
    3. F/2
    4. 4F

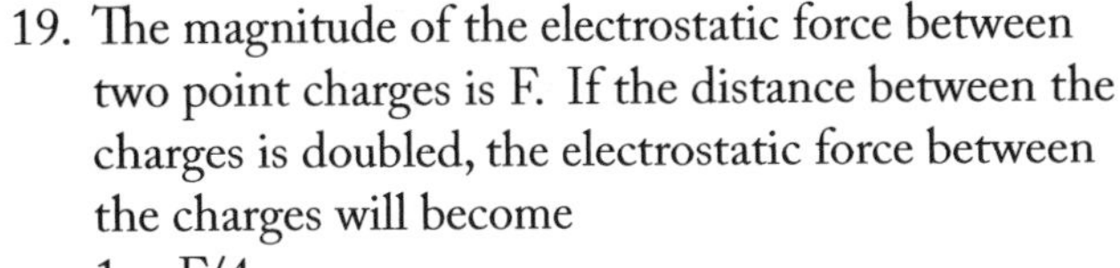

20. In the diagram below, a positive test charge is located between two charged spheres, A and B. Sphere A has a charge of +2q and is located 0.2 meter from the test charge. Sphere B has a charge of -2q and is located 0.1 meter from the test charge.

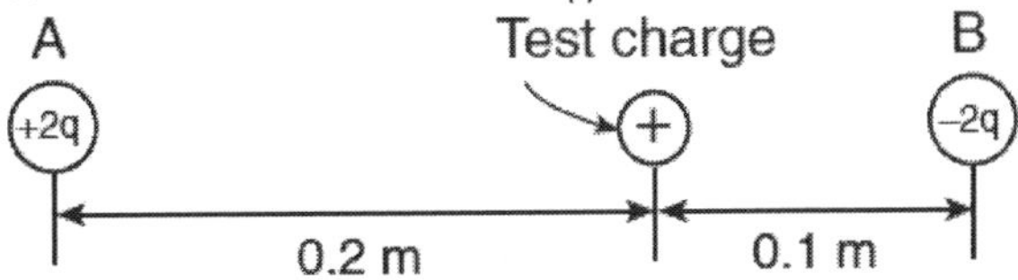

If the magnitude of the force on the test charge due to sphere A is F, what is the magnitude of the force on the test charge due to sphere B?
    1. F/4
    2. 2F
    3. F/2
    4. 4F

21. Two positively charged masses are separated by a distance, r. Which statement best describes the gravitational and electrostatic forces between the two masses?
    1. Both forces are attractive.
    2. Both forces are repulsive.
    3. The gravitational force is repulsive and the electrostatic force is attractive.
    4. The gravitational force is attractive and the electrostatic force is repulsive.

22. A distance of 1.0 meter separates the centers of two small charged spheres. The spheres exert gravitational force F_g and electrostatic force F_e on each other. If the distance between the spheres' centers is increased to 3.0 meters, the gravitational force and electrostatic force, respectively, may be represented as
    1. $F_g/9$ and $F_e/9$
    2. $F_g/3$ and $F_e/3$
    3. $3F_g$ and $3F_e$
    4. $9F_g$ and $9F_e$

Name: ______________________ Period: __________

Electrostatics-Coulomb's Law

Base your answers to questions 23 and 24 on the diagram below and on your knowledge of physics. The diagram represents two small, charged, identical metal spheres, A and B, that are separated by a distance of 2.0 meters.

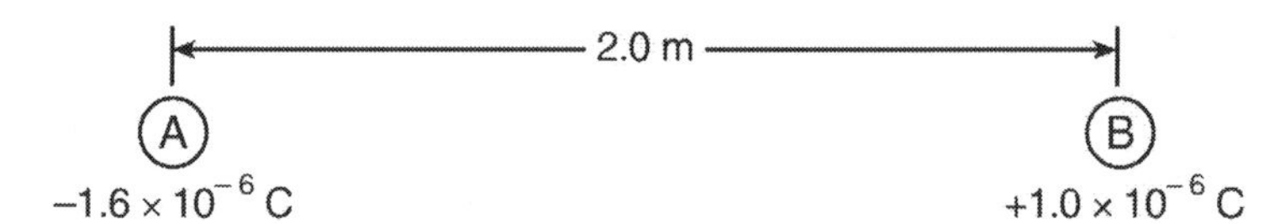

23. What is the magnitude of the electrostatic force exerted by sphere A on sphere B?
    1. 7.2×10^{-3} N
    2. 3.6×10^{-3} N
    3. 8.0×10^{-13} N
    4. 4.0×10^{-13} N

24. If the two spheres were touched together and then separated, the charge on sphere A would be
    1. -3.0×10^{-7} C
    2. -6.0×10^{-7} C
    3. -1.3×10^{-6} C
    4. -2.6×10^{-6} C

25. Two electrons are separated by a distance of 3.00×10^{-6} meter. What are the magnitude and direction of the electrostatic forces each exerts on the other?
    1. 2.56×10^{-17} N away from each other
    2. 2.56×10^{-17} N toward each other
    3. 7.67×10^{-23} N away from each other
    4. 7.67×10^{-23} N toward each other

26. When two point charges of magnitude q_1 and q_2 are separated by a distance r, the magnitude of the electrostatic force between them is F. What would be the magnitude of the electrostatic force between point charges $2q_1$ and $4q_2$ when separated by a distance of 2r?
    1. F
    2. 2F
    3. 16F
    4. 4F

27. When two point charges are a distance d apart, the magnitude of the electrostatic force between them is F. If the distance between the point charges is increased to 3d, the magnitude of the electrostatic force between the two charges will be
    1. F/9
    2. F/3
    3. 2F
    4. 4F

Name: ______________________________ Period: __________

Electrostatics-E Field

1. Which diagram represents the electric field lines between two small electrically charged spheres?

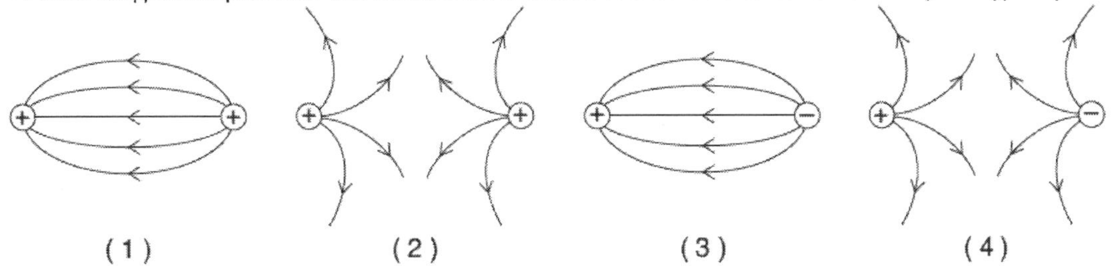

2. Which graph best represents the relationship between the magnitude of the electric field strength, E, around a point charge and the distance, r, from the point charge?

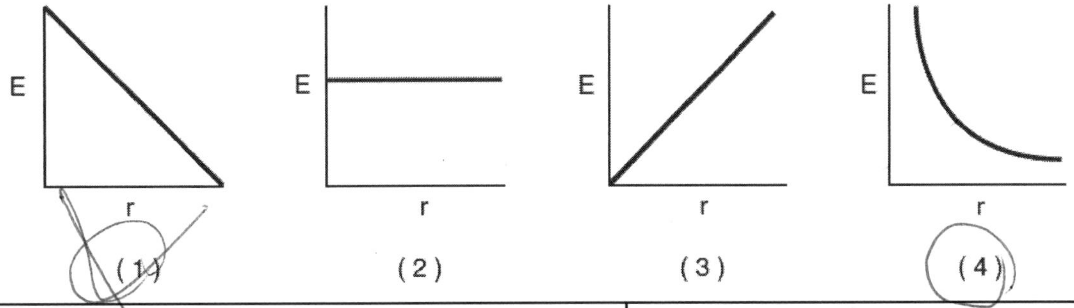

3. The diagram below represents an electron within an electric field between two parallel plates that are charged with a potential difference of 40 volts.

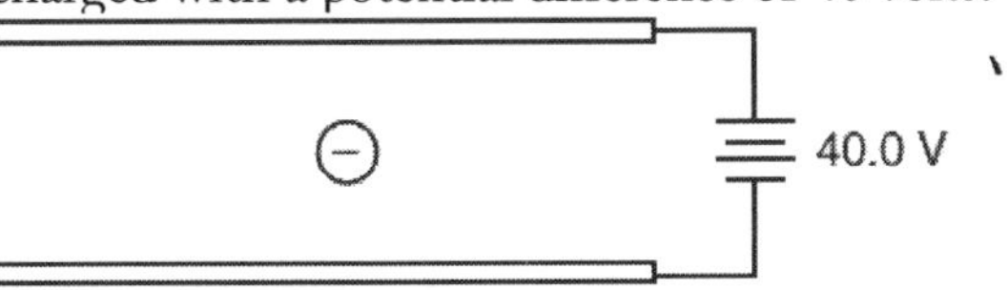

If the magnitude of the electric force on the electron is 2.00×10^{-15} newton, the magnitude of the electric field strength between the charged plates is
1. 3.20×10^{-34} N/C
2. 2.00×10^{-14} N/C
3. 1.25×10^{4} N/C
4. 2.00×10^{16} N/C

4. Two oppositely charged parallel metal plates, 1.00 centimeter apart, exert a force with a magnitude of 3.60×10^{-15} newton on an electron placed between the plates. Calculate the magnitude of the electric field strength between the plates. [Show all work, including the equation and substitution with units.]

5. An electron is located in the electric field between two parallel metal plates as shown in the diagram below.

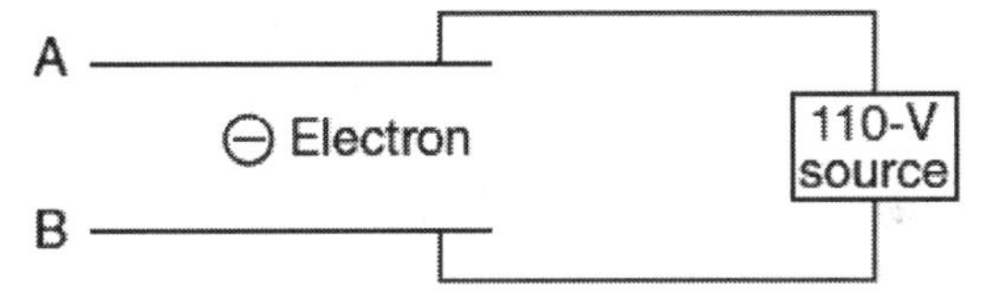

If the electron is attracted to plate A, then plate A is charged
1. positively, and the electric field is directed from plate A toward plate B
2. positively, and the electric field is directed from plate B toward plate A
3. negatively, and the electric field is directed from plate A toward plate B
4. negatively, and the electric field is directed from plate B toward plate A

Name: ____________________ Period: __________

Electrostatics-E Field

6. The diagram below shows a beam of electrons fired through the region between two oppositely charged parallel plates in a cathode ray tube.

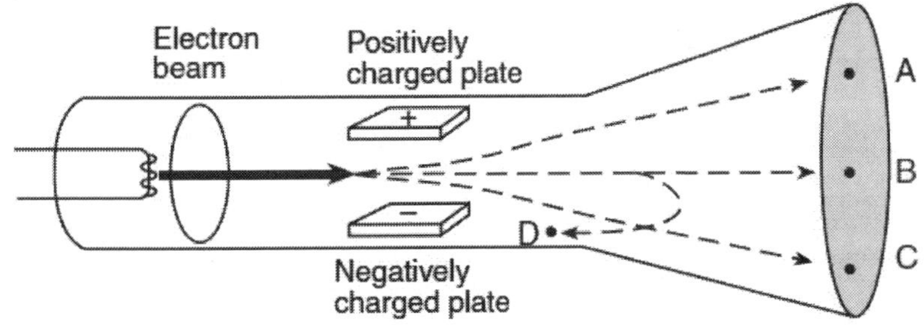

After passing between the charged plates, the electrons will most likely travel path
1. A
2. B
3. C
4. D

7. Two parallel metal plates are connected to a variable source of potential difference. When the potential difference of the source is increased, the magnitude of the electric field strength between the plates increases. The diagram below shows an electron located between the plates.

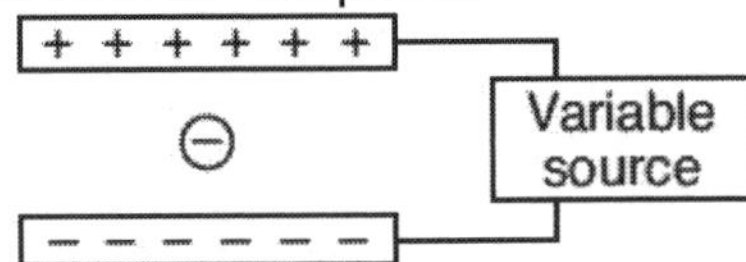

Which graph represents the relationship between the magnitude of the electrostatic force on the electron and the magnitude of the electric field strength between the plates?

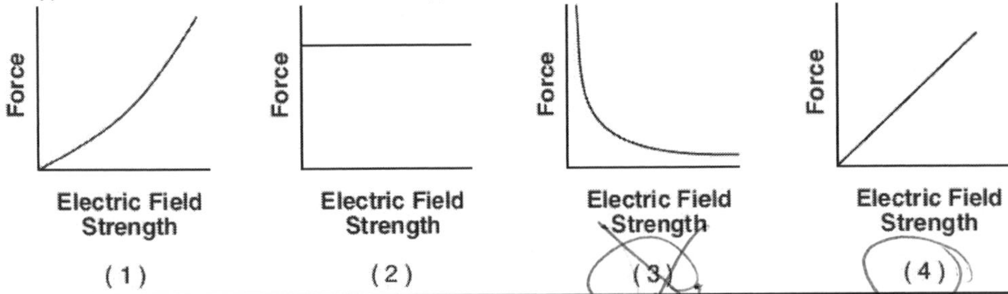

8. On the diagram below, sketch at least four electric field lines with arrowheads that represent the electric field around a negatively charged conducting sphere.

9. Which graph best represents the relationship between the strength of an electric field and distance from a point charge?

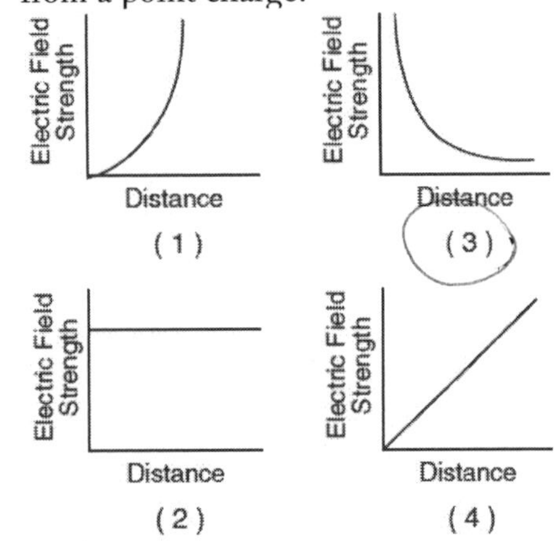

Name:______________________________ Period: ____________

Electrostatics-E Field

Base your answers to questions 11 through 13 on the information below.

The centers of two small charged particles are separated by a distance of 1.2×10^{-4} meter. The charges on the particles are $+8.0 \times 10^{-19}$ coulomb and $+4.8 \times 10^{-19}$ coulomb, respectively.

10. Calculate the magnitude of the electrostatic force between these two particles. [Show all work, including the equation and substitution with units.]

11. On the axes at right, sketch a graph showing the relationship between the magnitude of the electrostatic force between the two charged particles and the distance between the centers of the particles.

12. On the diagram below, draw at least four electric field lines in the region between the two positively charged particles.

8.0×10^{-19} C (+)   (+) 4.8×10^{-19} C

13. A beam of electrons is directed into the electric field between two oppositely charged parallel plates, as shown in the diagram below.

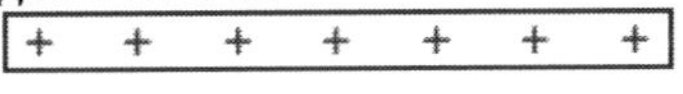

Electron beam

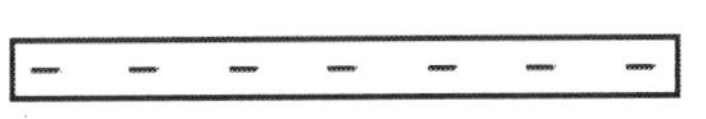

The electrostatic force exerted on the electrons by the electric field is directed

1. into the page
2. out of the page
3. toward the bottom of the page
4. toward the top of the page

14. What is the magnitude of the electric field intensity at a point where a proton experiences an electrostatic force of magnitude 2.30×10^{-25} newton?

1. 3.68×10^{-44} N/C
2. 1.44×10^{-6} N/C
3. 3.68×10^{6} N/C
4. 1.44×10^{44} N/C

15. In the diagram below, P is a point near a negatively charged sphere.

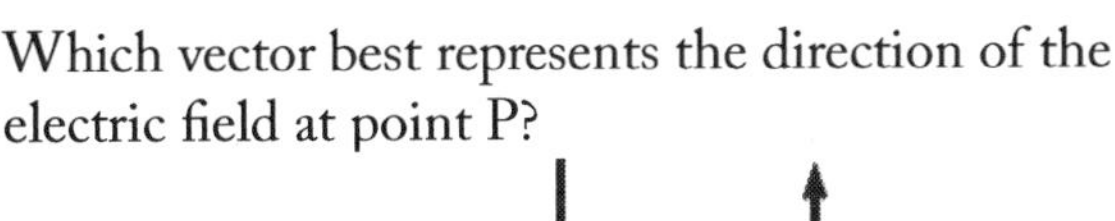

Which vector best represents the direction of the electric field at point P?

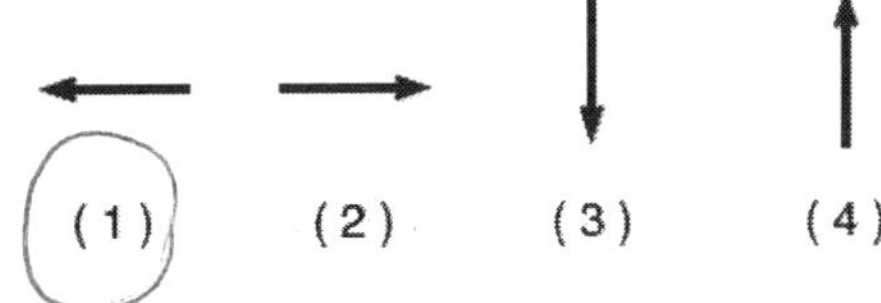

16. An object with a net charge of 4.80×10^{-6} coulomb experiences an electrostatic force having a magnitude of 6.00×10^{-2} newton when placed near a negatively charged metal sphere. What is the electric field strength at this location?

1. 1.25×10^{4} N/C directed away from the sphere
2. 1.25×10^{4} N/C directed toward the sphere
3. 2.88×10^{-8} N/C directed away from the sphere
4. 2.88×10^{-8} N/C directed toward the sphere

Name: ____________________ Period: __________

Electrostatics-E Field

Base your answers to questions 17 and 18 on the information and diagram below.

Two small metallic spheres, A and B, are separated by a distance of 4.0×10^{-1} meter, as shown. The charge on each sphere is $+1.0 \times 10^{-6}$ coulomb. Point P is located near the spheres.

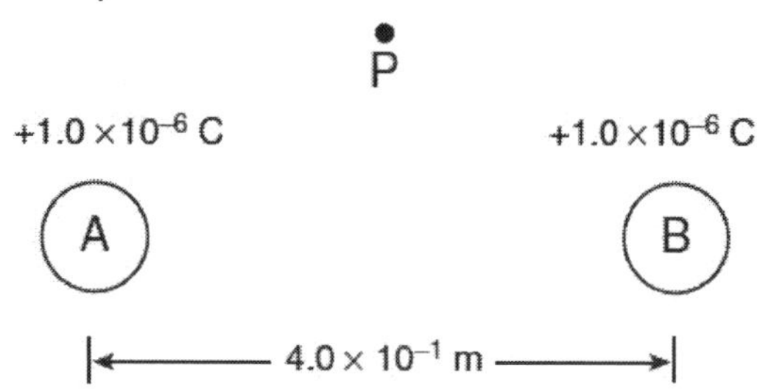

17. What is the magnitude of the electrostatic force between the two charged spheres?
    1. 2.2×10^{-2} N
    2. 5.6×10^{-2} N
    3. 2.2×10^{4} N
    4. 5.6×10^{4} N

18. Which arrow best represents the direction of the resultant electric field at point P due to the charges on spheres A and B?

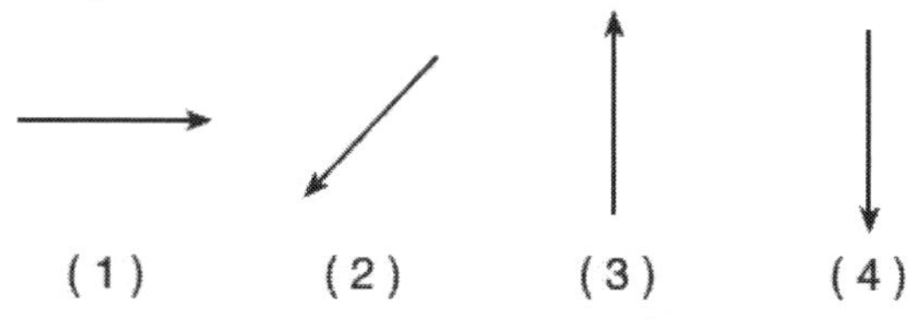

19. The diagram below represents the electric field surrounding two charged spheres, A and B.

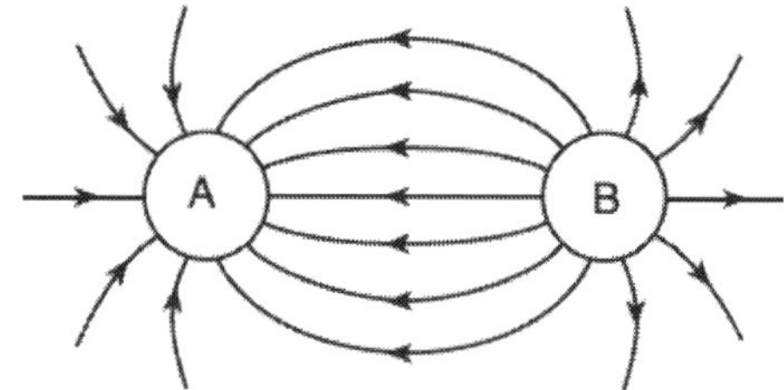

What is the sign of the charge of each sphere?
1. Sphere A is positive and sphere B is negative
2. Sphere A is negative and sphere B is positive
3. Both spheres are positive
4. Both spheres are negative

Base your answers to questions 20 and 21 on the information below.

The magnitude of the electric field strength between two oppositely charged parallel metal places is 2.0×10^{3} newtons per coulomb. Point P is located midway between the plates.

20. On the diagram below, sketch at least five electric field lines to represent the field between the two oppositely charged plates. [Draw an arrowhead on each field line to show the proper direction.]

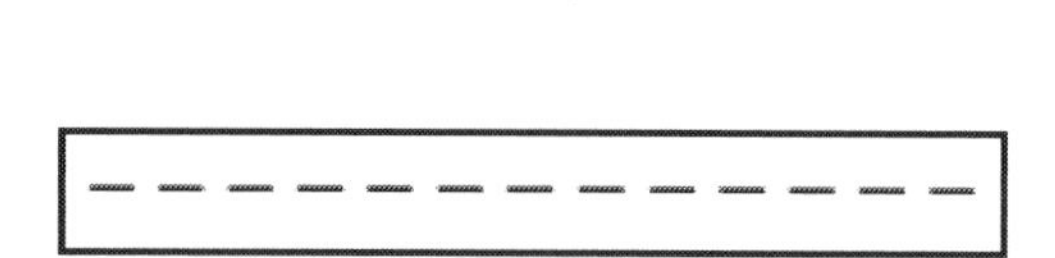

21. An electron is located at point P between the plates. Calculate the magnitude of the force exerted on the electron by the electric field. [Show all work, including the equation and substitution with units.]

22. An electron placed between oppositely charged parallel plates A and B moves toward plate A, as represented in the diagram below.

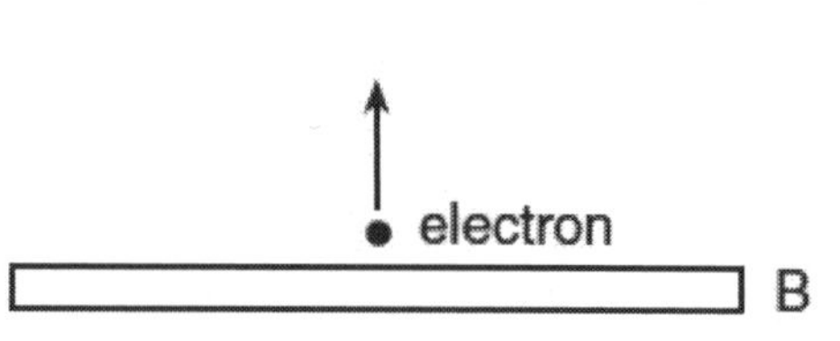

What is the direction of the electric field between the plates?
1. toward plate A
2. toward plate B
3. into the page
4. out of the page

Name: ______________________________ Period: ____________

Electrostatics-E Field

23. A moving electron is deflected by two oppositely charged parallel plates, as shown in the diagram below.

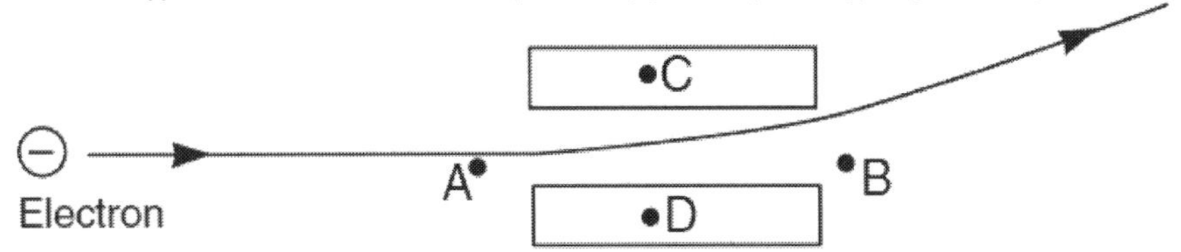

The electric field between the plates is directed from
1. A to B
2. B to A
3. C to D
4. D to C

24. Which diagram represents the electric field between two oppositely charged conducting spheres?

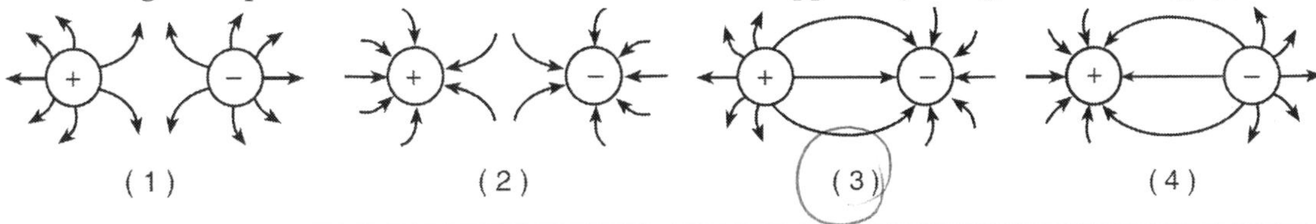

25. A 3.00×10^{-9}-coulomb test charge is placed near a negatively charged metal sphere. The sphere exerts an electrostatic force of magnitude 6.00×10^{-5} newton on the test charge. What is the magnitude and direction of the electric field strength at this location?
1. 2.00×10^{4} N/C directed away from the sphere
2. 2.00×10^{4} N/C directed toward the sphere
3. 5.00×10^{-5} N/C directed away from the sphere
4. 5.00×10^{-5} N/C directed toward the sphere

26. An electron is located in an electric field of magnitude 600 newtons per coulomb. What is the magnitude of the electrostatic force acting on the electron?
1. 3.75×10^{21} N
2. 6.00×10^{2} N
3. 9.60×10^{-17} N
4. 2.67×10^{-22} N

27. A beam of electrons passes through an electric field where the magnitude of the electric field strength is 3.00×10^{3} newtons per coulomb. What is the magnitude of the electrostatic force exerted by the electric field on each electron in the beam?
1. 5.33×10^{-23} N
2. 4.80×10^{-16} N
3. 3.00×10^{3} N
4. 1.88×10^{22} N

28. Two points, A and B, are located within the electric field produced by a -3.0 nanocoulomb charge. Point A is 0.10 meter to the left of the charge and point B is 0.20 meter to the right of the charge, as shown in the diagram below. below.

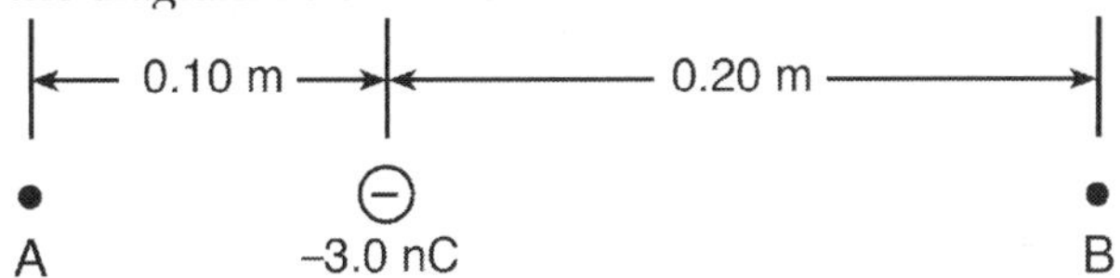

Compared to the magnitude of the electric field strength at point A, the magnitude of the electric field strength at point B is
1. half as great
2. twice as great
3. one-fourth as great
4. four times as great

Name: ______________________ Period: __________

Electrostatics-Potential

1. If 1.0 joule of work is required to move 1.0 coulomb of charge between two points in an electric field, the potential difference between the two points is
   1. 1.0×10^0 V
   2. 9.0×10^9 V
   3. 6.3×10^{18} V
   4. 1.6×10^{-19} V

2. The diagram below represents a positively charged particle about to enter the electric field between two oppositely charged parallel plates.

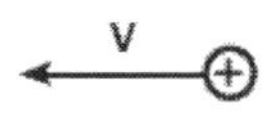

   The electric field will deflect the particle
   1. into the page
   2. out of the page
   3. toward the top of the page
   4. toward the bottom of the page

3. What is the total amount of work required to move a proton through a potential difference of 100 volts?
   1. 1.60×10^{-21} J
   2. 1.60×10^{-17} J
   3. 1.00×10^2 J
   4. 6.25×10^{20} J

4. The diagram below represents two electrons, e_1 and e_2, located between two oppositely charged parallel plates.

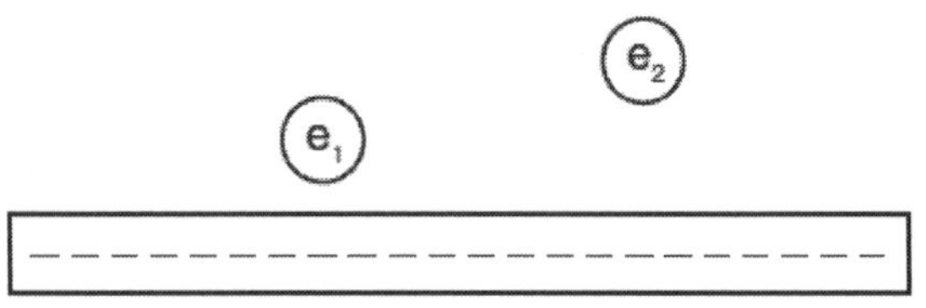

   Compare the magnitude of the force exerted by the electric field on e_1 to the magnitude of the force exerted by the electric field on e_2.

5. Which electrical unit is equivalent to one joule?
   1. volt per meter
   2. ampere·volt
   3. volt per coulomb
   4. coulomb·volt

6. If 60 joules of work is required to move 5.0 coulombs of charge between two points in an electric field, what is the potential difference between these points?
   1. 5.0 V
   2. 12 V
   3. 60 V
   4. 300 V

7. In the diagram below, proton p, neutron n, and electron e are located as shown between two oppositely charged plates.

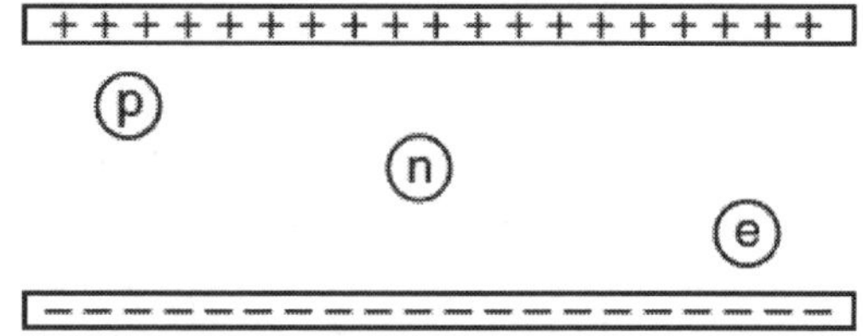

   The magnitude of acceleration will be greatest for the
   1. neutron, because it has the greatest mass
   2. neutron, because it is neutral
   3. electron, because it has the smallest mass
   4. proton, because it is farthest from the negative plate

8. An electron is accelerated through a potential difference of 2.5×10^4 volts in the cathode ray tube of a computer monitor. Calculate the work, in joules, done on the electron. [Show all work, including the equation and substitution with units.]

Name: ______________________________ Period: __________

Electrostatics-Potential

9. Moving 2.5×10^{-6} coulomb of charge from point A to point B in an electric field requires 6.3×10^{-4} joule of work. The potential difference between points A and B is approximately
   1. 1.6×10^{-9} V
   2. 4.0×10^{-3} V
   3. 2.5×10^{2} V
   4. 1.0×10^{14} V

10. The diagram below represents a source of potential difference connected to two large, parallel metal plates separated by a distance of 4.0×10^{-3} meter.

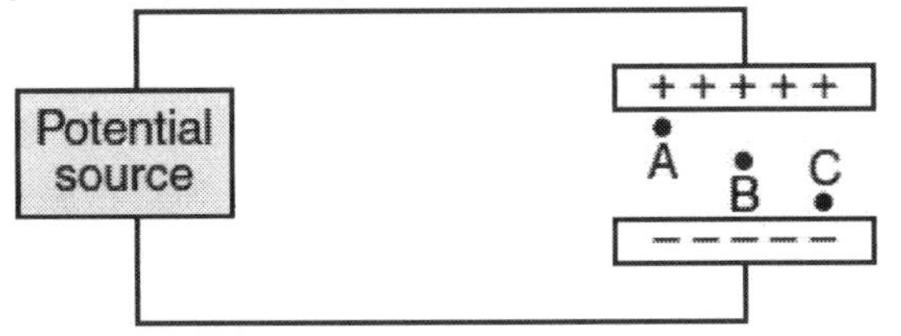

    Which statement best describes the electric field strength between the plates?
   1. It is zero at point B
   2. It is a maximum at point B
   3. It is a maximum at point C
   4. It is the same at points A, B, and C.

11. In an electric field, 0.90 joule of work is required to bring 0.45 coulomb of charge from point A to point B. What is the electric potential difference between points A and B?
   1. 5.0 V
   2. 2.0 V
   3. 0.50 V
   4. 0.41 V

12. A potential difference of 10 volts exists between two points, A and B, within an electric field. What is the magnitude of charge that requires 2.0×10^{-2} joule of work to move it from A to B?
   1. 5.0×10^{2} C
   2. 2.0×10^{-1} C
   3. 5.0×10^{-2} C
   4. 2.0×10^{-3} C

13. If 4.8×10^{-17} joule of work is required to move an electron between two points in an electric field, what is the electric potential difference between these points?
   1. 1.6×10^{-19} V
   2. 4.8×10^{-17} V
   3. 3.0×10^{2} V
   4. 4.8×10^{2} V

Base your answers to questions 14 and 15 on the information below.

A proton starts from rest and gains 8.35×10^{-14} joule of kinetic energy as it accelerates between points A and B in an electric field.

14. What is the final speed of the proton?
   1. 7.07×10^{6} m/s
   2. 1.00×10^{7} m/s
   3. 4.28×10^{8} m/s
   4. 5.00×10^{13} m/s

15. Calculate the potential difference between points A and B in the electric field. [Show all work, including the equation and substitution with units.]

16. Which is a vector quantity?
   1. electric charge
   2. electric field strength
   3. electric potential difference
   4. electric resistance

17. Which object will have the greatest change in electrical energy?
   1. an electron moved through a potential of 2.0 V
   2. a metal sphere with a charge of 1.0×10^{-9} C moved through a potential difference of 2.0 V
   3. an electron moved through a potential of 4.0 V
   4. a metal sphere with a charge of 1.0×10^{-9} C moved through a potential difference of 4.0 V

Name:________________________________ Period: __________

Electrostatics-Potential

Base your answers to questions 18 through 21 on the information and diagram below and on your knowledge of physics.

Two conducting parallel plates 5.0×10^{-3} m apart are charged with a 12-volt potential difference. An electron is located midway between the plates. The magnitude of the electrostatic force on the electron is 3.8×10^{-16} newton.

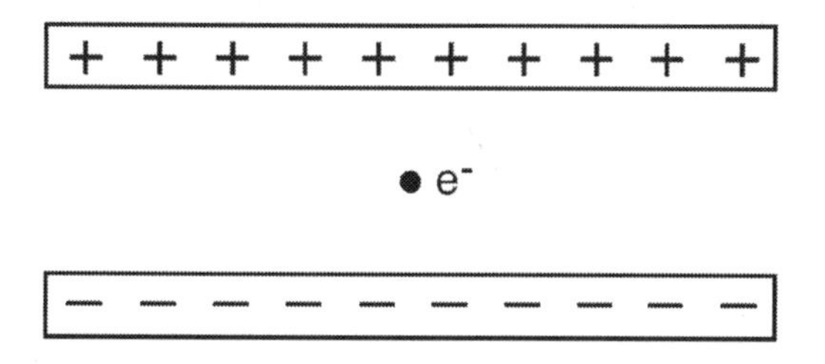

18. On the diagram above, draw at least three field lines to represent the direction of the electric field in the space between the charged plates.

19. Identify the direction of the electrostatic force that the electric field exerts on the electron.

20. Calculate the magnitude of the electric field strength between the plates, in newtons per coulomb. [Show all work including the equation and substitution with units.]

21. Describe what happens to the magnitude of the net electrostatic force on the electron as the electron is moved toward the positive plate.

22. The electronvolt is a unit of
    1. energy
    2. charge
    3. electric field strength
    4. electric potential difference

23. How much work is required to move 3.0 coulombs of electric charge a distance of 0.010 meter through a potential difference of 9.0 volts?
    1. 2.7×10^{3} J
    2. 27 J
    3. 3.0 J
    4. 3.0×10^{-2} J

24. Which combination of units can be used to express electrical energy?
    1. volt/coulomb
    2. coulomb/volt
    3. volt•coulomb
    4. volt•coulomb•second

25. How much work is required to move an electron through a potential difference of 3.00 volts?
    1. 5.33×10^{-20} J
    2. 4.80×10^{-19} J
    3. 3.00 J
    4. 1.88×10^{19} J

Name:______________________________ Period: __________

Circuits-Current

1. What is the current through a wire if 240 coulombs of charge pass through the wire in 2.0 minutes?
   1. 120 A
   2. 2.0 A
   3. 0.50 A
   4. 0.0083 A

2. A 1.5-volt, AAA cell supplies 750 milliamperes of current through a flashlight bulb for 5.0 minutes, while a 1.5-volt, C cell supplies 750 milliamperes of current through the same flashlight bulb for 20 minutes. Compared to the total charge transferred by the AAA cell through the bulb, the total charge transferred by the C cell through the bulb is
   1. half as great
   2. twice as great
   3. the same
   4. four times as great

3. The current traveling from the cathode to the screen in a television picture tube is 5.0×10^{-5} ampere. How many electrons strike the screen in 5.0 seconds?
   1. 3.1×10^{24} electrons
   2. 6.3×10^{18} electrons
   3. 1.6×10^{15} electrons
   4. 1.0×10^{5} electrons

4. Charge flowing at the rate of 2.50×10^{16} elementary charges per second is equivalent to a current of
   1. 2.50×10^{13} A
   2. 6.25×10^{5} A
   3. 4.00×10^{-3} A
   4. 2.50×10^{-3} A

5. The current through a lightbulb is 2.0 amperes. How many coulombs of electric charge pass through the lightbulb in one minute?
   1. 60 C
   2. 2.0 C
   3. 120 C
   4. 240 C

6. If 10 coulombs of charge are transferred through an electric circuit in 5.0 seconds, then the current in the circuit is
   1. 0.50 A
   2. 2.0 A
   3. 15 A
   4. 50 A

7. A charge of 30 coulombs passes through a 24-ohm resistor in 6.0 seconds. What is the current through the resistor?
   1. 1.3 A
   2. 5.0 A
   3. 7.5 A
   4. 4.0 A

8. The diagram below shows two resistors, R_1 and R_2, connected in parallel in a circuit having a 120-volt power source. Resistor R_1 develops 150 watts and resistor R_2 develops an unknown power. Ammeter A in the circuit reads 0.50 ampere.

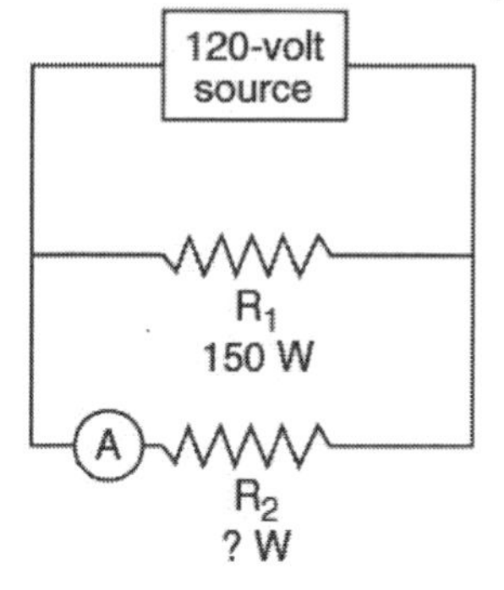

Calculate the amount of charge passing through resistor R_2 in 60 seconds. [Show all work, including the equation and substitution with units.]

9. What is the current in a wire if 3.4×10^{19} electrons pass by a point in this wire every 60 seconds?
   1. 1.8×10^{-18} A
   2. 3.1×10^{-11} A
   3. 9.1×10^{-2} A
   4. 11 A

Name: ____________________ Period: __________

Circuits-Current

10. The current in a wire is 4.0 amperes. The time required for 2.5×10^{19} electrons to pass a certain point in the wire is
    1. 1.0 s
    2. 0.25 s
    3. 0.50 s
    4. 4.0 s

11. An MP3 player draws a current of 0.120 ampere from a 3.00-volt battery. What is the total charge that passes through the player in 900 seconds?
    1. 324 C
    2. 108 C
    3. 5.40 C
    4. 1.80 C

12. A net charge of 5.0 coulombs passes a point on a conductor in 0.050 second. The average current is
    1. 8.0×10^{-8} A
    2. 1.0×10^{-2} A
    3. 2.5×10^{-1} A
    4. 1.0×10^{2} A

Name: ____________________ Period: __________

Circuits-Resistance

1. At 20°C, four conducting wires made of different materials have the same length and the same diameter. Which wire has the least resistance?
   1. aluminum
   2. gold
   3. nichrome
   4. tungsten

2. The graph below represents the relationship between the current in a metallic conductor and the potential difference across the conductor at constant temperature.

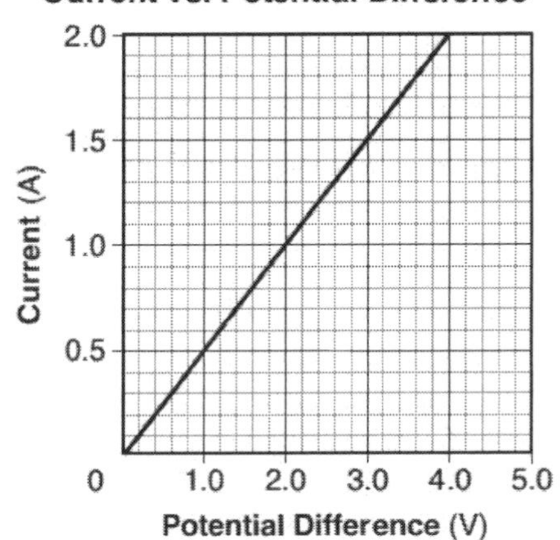

   The resistance of the conductor is
   1. 1.0 Ω
   2. 2.0 Ω
   3. 0.50 Ω
   4. 4.0 Ω

3. The diagram below represents a lamp, a 10-volt battery, and a length of nichrome wire connected in series.

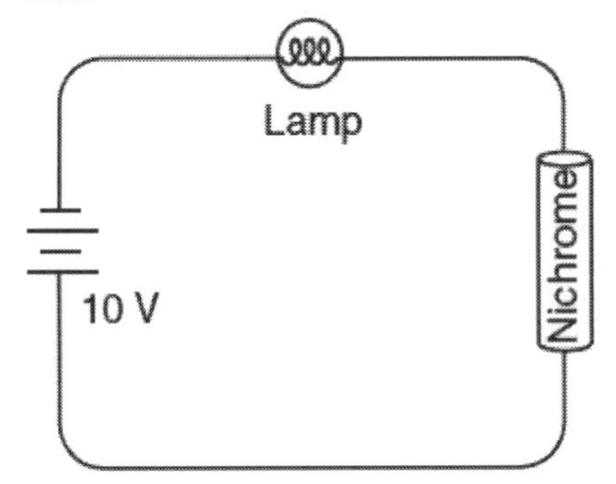

   As the temperature of the nichrome is decreased, the brightness of the lamp will
   1. decrease
   2. increase
   3. remain the same

Base your answers to questions 4 through 6 on the information and graph below.

A student conducted an experiment to determine the resistance of a lightbulb. As she applied various potential differences to the bulb, she recorded the voltages and corresponding currents and constructed the graph below.

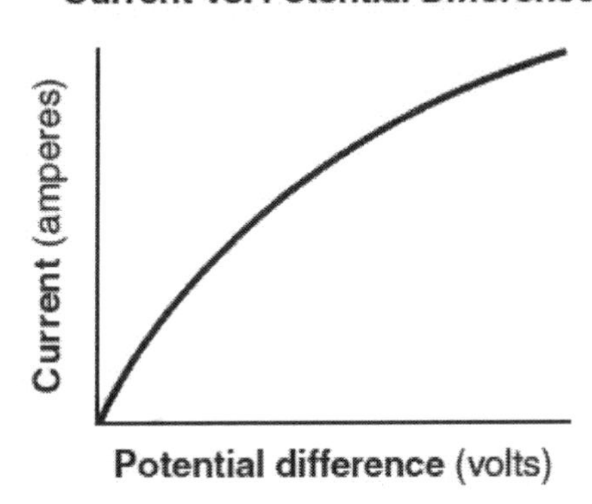

4. The student concluded that the resistance of the lightbulb was not constant. What evidence from the graph supports the student's conclusion?

5. According to the graph, as the potential difference increased, the resistance of the lightbulb
   1. decreased
   2. increased
   3. changed, but there is not enough information to know which way

6. While performing the experiment the student noticed that the lightbulb began to glow and became brighter as she increased the voltage. Of the factors affecting resistance, which factor caused the greatest change in the resistance of the bulb during her experiment?

Name:____________________ Period: ________

Circuits-Resistance

Base your answers to questions 7 through 10 on the information and data table below.

An experiment was performed using various lengths of a conductor of uniform cross-sectional area. The resistance of each length was measured and the data recorded in the data table.

Using the information in the data table, construct a graph on the grid below, following the directions provided.

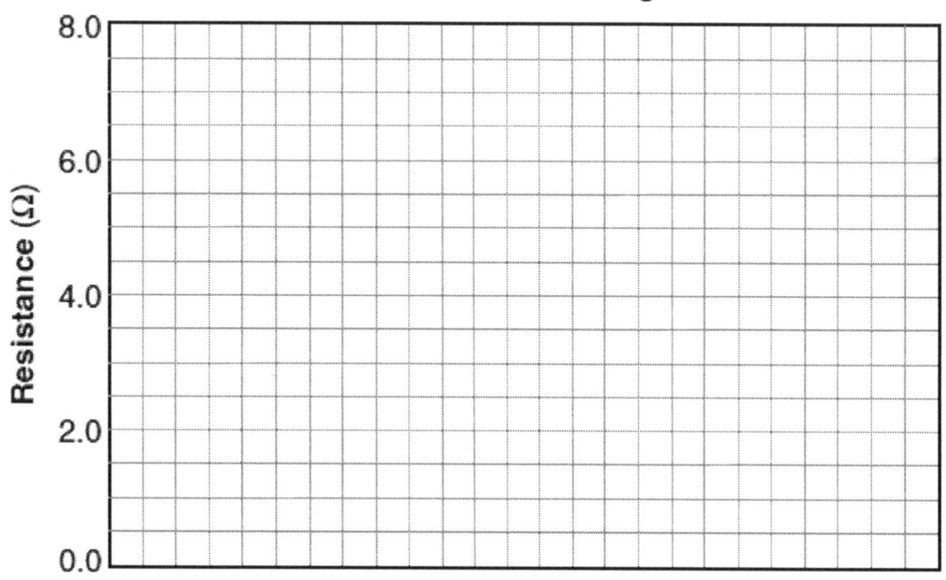

Length (meters)	Resistance (ohms)
5.1	1.6
11.0	3.8
16.0	4.6
18.0	5.9
23.0	7.5

7. Mark an appropriate scale on the axis labeled "Length (m)."

8. Plot the data points for resistance versus length.

9. Draw the best-fit line.

10. Calculate the slope of the best-fit line. [Show all work, including the equation and substitution with units.]

11. Which graph best represents the relationship between resistance and length of a copper wire of uniform cross-sectional area at constant temperature?

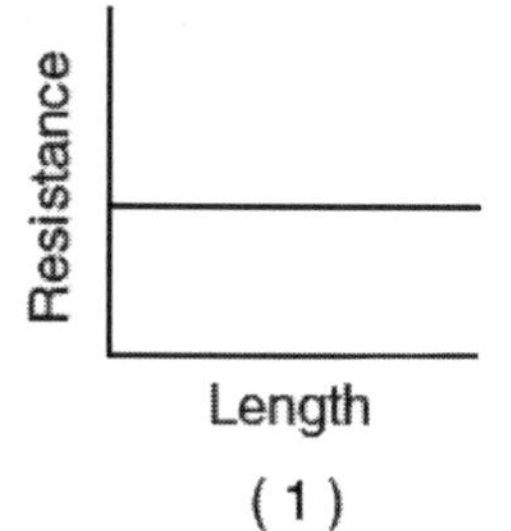

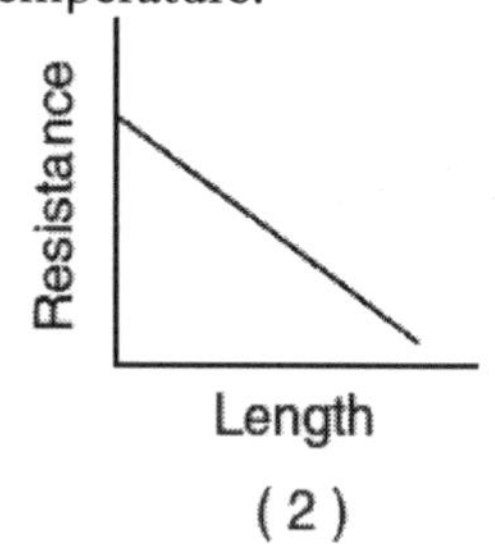

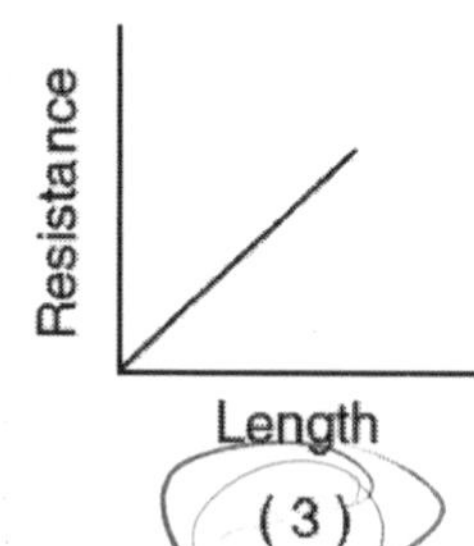

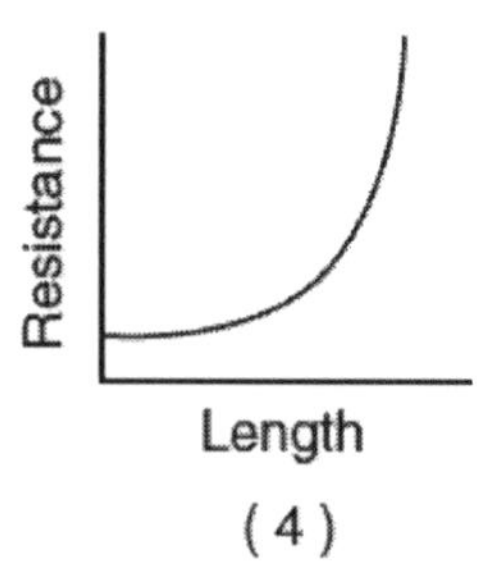

Name: ____________________ Period: ____________

Circuits-Resistance

12. The table below lists various characteristics of two metallic wires, A and B.

Wire	Material	Temperature (°C)	Length (m)	Cross-Sectional Area (m^2)	Resistance (Ω)
A	silver	20.	0.10	0.010	R
B	silver	20.	0.20	0.020	???

If wire A has resistance R, then wire B has resistance
1. R
2. 2R
3. R/2
4. 4R

Base your answers to questions 13 through 15 on the information and diagram below.

A circuit contains a 12.0-volt battery, an ammeter, a variable resistor, and connecting wires of negligible resistance, as shown below.

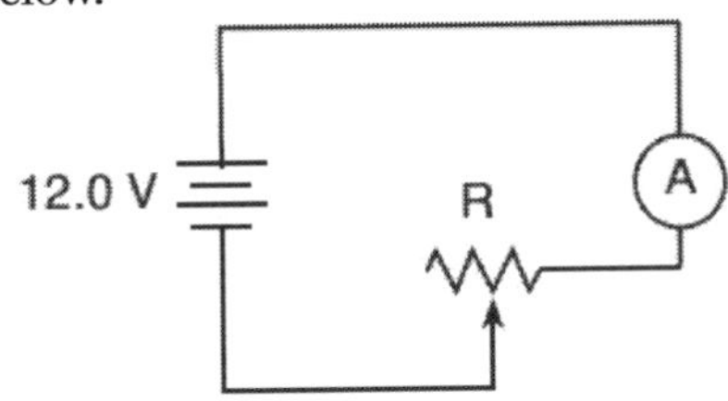

The variable resistor is a nichrome wire, maintained at 20°C. The length of the nichrome wire may be varied from 10 centimeters to 90 centimeters. The ammeter reads 2 amperes when the length of the wire is 10 centimeters.

13. Determine the resistance of the 10-centimeter length of nichrome wire.

14. Calculate the cross-sectional area of the nichrome wire. [Show all work, including the equation and substitution with units.]

15. What is the resistance at 20°C of a 1.50-meter-long aluminum conductor that has a cross-sectional area of 1.13×10^{-6} meter^2?
    1. 1.87×10^{-3} Ω
    2. 2.28×10^{-2} Ω
    3. 3.74×10^{-2} Ω
    4. 1.33×10^{6} Ω

16. A complete circuit is left on for several minutes, causing the connecting copper wire to become hot. As the temperature of the wire increases, the electrical resistance of the wire
    1. decreases
    2. increases
    3. remains the same

17. Which changes would cause the greatest increase in the rate of flow of charge through a conducting wire?
    1. increasing the applied potential difference and decreasing the length of wire
    2. increasing the applied potential difference and increasing the length of wire
    3. decreasing the applied potential difference and decreasing the length of wire
    4. decreasing the applied potential difference and increasing the length of wire

18. A length of copper wire and a 1.00-meter-long silver wire have the same cross-sectional area and resistance at 20°C. Calculate the length of the copper wire. [Show all work, including the equation and substitution with units.]

Name: ______________________ Period: __________

Circuits-Resistance

19. Several pieces of copper wire, all having the same length but different diameters, are kept at room temperature. Which graph best represents the resistance, R, of the wires as a function of their cross-sectional areas, A?

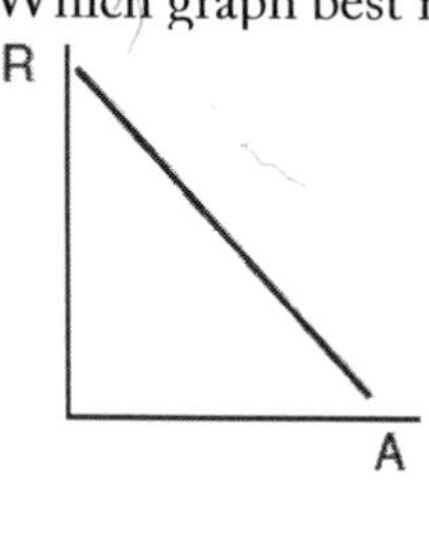

(1)

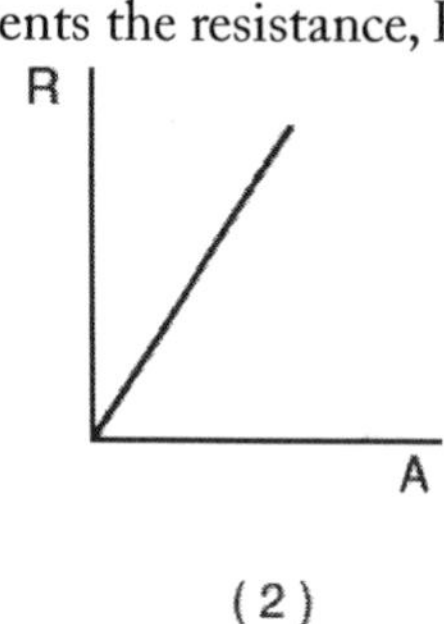

(2)

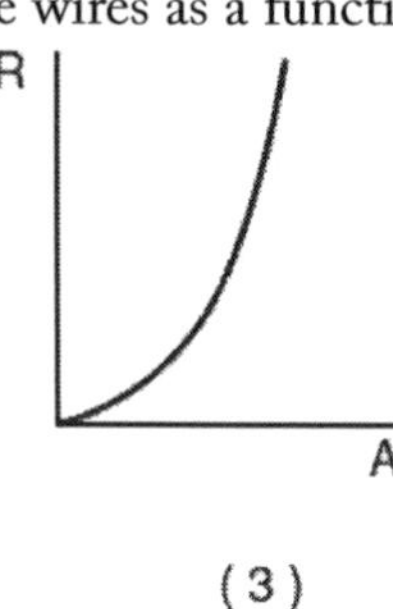

(3)

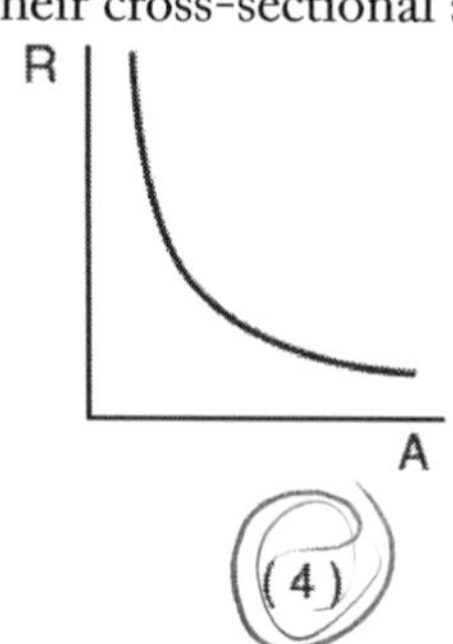

(4)

20. A copper wire at 20°C has a length of 10.0 meters and a cross-sectional area of 1.00×10^{-3} meter^2. The wire is stretched, becomes longer and thinner, and returns to 20°C. What effect does this stretching have on the wire's resistance?

21. A copper wire of length L and cross-sectional area A has resistance R. A second copper wire at the same temperature has a length of 2L and a cross-sectional area of 0.5A. What is the resistance of the second copper wire?
    1. R
    2. 2R
    3. 0.5R
    4. 4R

22. Pieces of aluminum, copper, gold, and silver wire each have the same length and the same cross-sectional area. Which wire has the *lowest* resistance at 20°C
    1. aluminum
    2. copper
    3. gold
    4. silver

23. Which quantity and unit are correctly paired?
    1. resistivity and Ω/m
    2. potential difference and eV
    3. current and C·s
    4. electric field strength and N/C

Base your answers to questions 24 and 25 on the information and diagram below.

A 10-meter length of copper wire is at 20°C. The radius of the wire is 1.0×10^{-3} meter.

Cross Section of Copper Wire

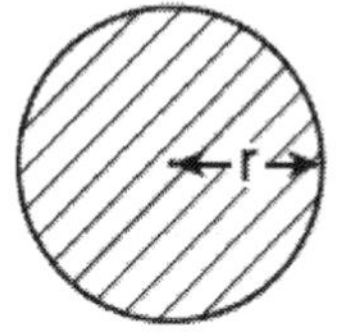

$r = 1.0 \times 10^{-3}$ m

24. Determine the cross-sectional area of the wire.

25. Calculate the resistance of the wire. [Show all work, including the equation and substitution with units.]

26. What is the resistance at 20°C of a 2.0-meter length of tungsten wire with a cross-sectional area of 7.9×10^{-7} meter^2?
    1. 5.7×10^{-1} Ω
    2. 1.4×10^{-1} Ω
    3. 7.1×10^{-2} Ω
    4. 4.0×10^{-2} Ω

Name: ______________________________ Period: ____________

Circuits-Resistance

Base your answers to questions 27 and 28 on the information below.

A 1.00-meter length of nichrome wire with a cross-sectional area of 7.85×10^{-7} meter^2 is connected to a 1.50-volt battery.

27. Calculate the resistance of the wire. [Show all work, including the equation and substitution with units.]

28. Determine the current in the wire.

29. The electrical resistance of a metallic conductor is inversely proportional to its
    1. temperature
    2. length
    3. cross-sectional area
    4. resistivity

30. A 0.686-meter-long wire has a cross-sectional area of 8.23×10^{-6} meter^2 and a resistance of 0.125 ohm at 20° Celsius. This wire could be made of
    1. aluminum
    2. copper
    3. nichrome
    4. tungsten

31. Calculate the resistance of 1.00-kilometer length of nichrome wire with a cross-sectional area of 3.50×10^{-6} meter^2 at 20°C. [Show all work, including the equation and substitution with units.]

Base your answers to questions 32 and 33 on the information below.

A 3.50-meter length of wire with a cross-sectional area of 3.14×10^{-6} meter^2 is at 20° Celsius. The current in the wire is 24.0 amperes when connected to a 1.50-volt source of potential difference.

32. Determine the resistance of the wire.

33. Calculate the resistivity of the wire. [Show all work, including the equation and substitution with units.]

34. Aluminum, copper, gold and nichrome wires of equal lengths of 0.1 meter and equal cross-sectional areas of 2.5×10^{-6} meter^2 are at 20°C. Which wire has the greatest electrical resistance?
    1. aluminum
    2. copper
    3. gold
    4. nichrome

35. A 10-meter length of wire with a cross-sectional area of 3.0×10^{-6} square meter has a resistance of 9.4×10^{-2} ohm at 20° Celsius. The wire is most likely made of
    1. silver
    2. copper
    3. aluminum
    4. tungsten

36. Which change decreases the resistance of a piece of copper wire?
    1. increasing the wire's length
    2. increasing the wire's resistivity
    3. decreasing the wire's temperature
    4. decreasing the wire's diameter

Name:_______________________________ Period: ____________

Circuits-Resistance

37. A 25.0-meter length of platinum wire with a cross-sectional area of 3.50×10^{-6} meter^2 has a resistance of 0.757 ohm at 20°C. Calculate the resistivity of the wire. [Show all work, including the equation and substitution with units.]

38. What is the resistance of a 20.0-meter-long tungsten rod with a cross-sectional area of 1.00×10^{-4} meter^2 at 20° Celsius.
    1. $2.80 \times 10^{-5}\ \Omega$
    2. $1.12 \times 10^{-2}\ \Omega$
    3. $89.3\ \Omega$
    4. $112\ \Omega$

39. During a laboratory experiment, a student finds that at 20° Celsius, a 6.0-meter length of copper wire has a resistance of 1.3 ohms. The cross-sectional area of this wire is
    1. $7.9 \times 10^{-8}\ m^2$
    2. $1.1 \times 10^{-7}\ m^2$
    3. $4.6 \times 10^{0}\ m^2$
    4. $1.3 \times 10^{7}\ m^2$

Name: ______________________________ Period: __________

Circuits-Ohm's Law

1. Which graph best represents the relationship between the electrical power and the current in a resistor that obeys Ohm's Law?

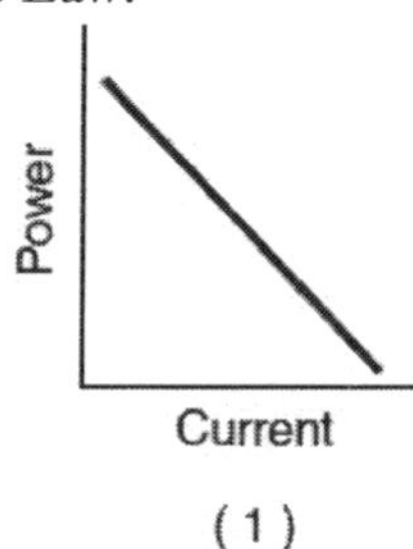

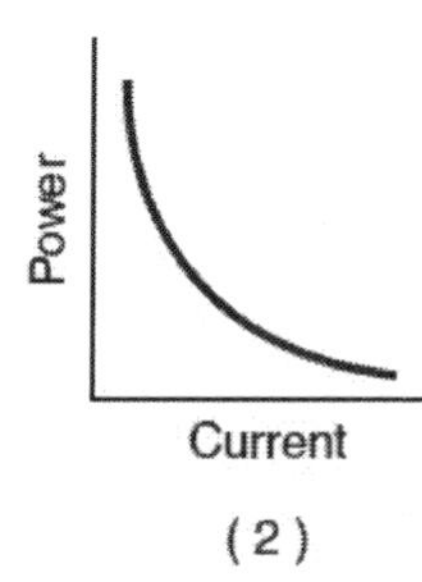

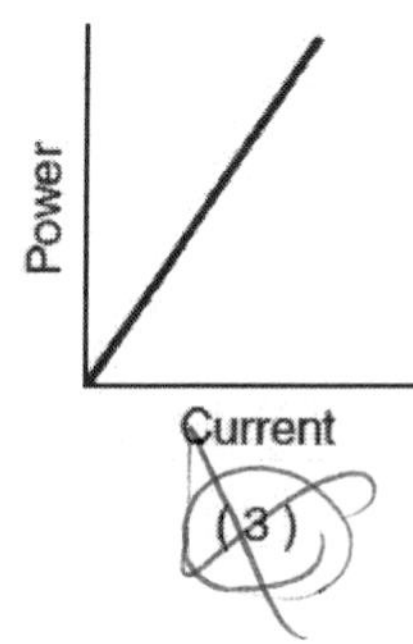

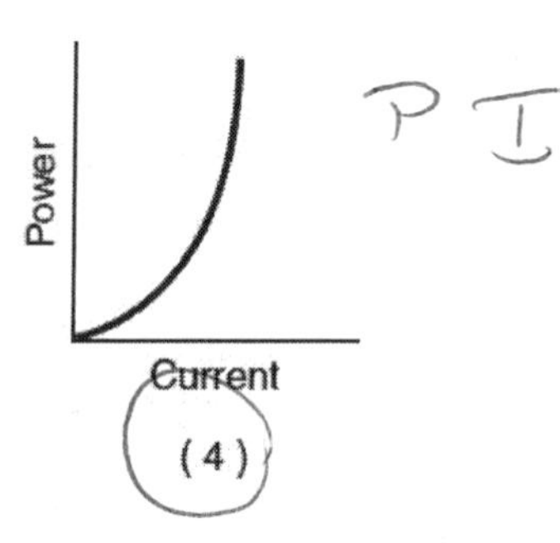

2. A potential drop of 50 volts is measured across a 250-ohm resistor. What is the power developed in the resistor?
   1. 0.20 W
   2. 5.0 W
   3. 10 W
   4. 50 W

3. How much electrical energy is required to move a 4.00-microcoulomb charge through a potential difference of 36 volts?
   1. 9.00×10^{6} J
   2. 144 J
   3. 1.44×10^{-4} J
   4. 1.11×10^{-7} J

4. A circuit consists of a resistor and a battery. Increasing the voltage of the battery while keeping the temperature of the circuit constant would result in an increase in
   1. current, only
   2. resistance, only
   3. both current and resistance
   4. neither current nor resistance

5. A generator produces a 115-volt potential difference and a maximum of 20 amperes of current. Calculate the total electrical energy the generator produces operating at maximum capacity for 60 seconds. [Show all work, including the equation and substitution with units.]

6. An electric circuit contains a variable resistor connected to a source of constant voltage. As the resistance of the variable resistor is increased, the power dissipated in the circuit
   1. decreases
   2. increases
   3. remains the same

7. An electric circuit contains a variable resistor connected to a source of constant potential difference. Which graph best represents the relationship between current and resistance in this circuit?

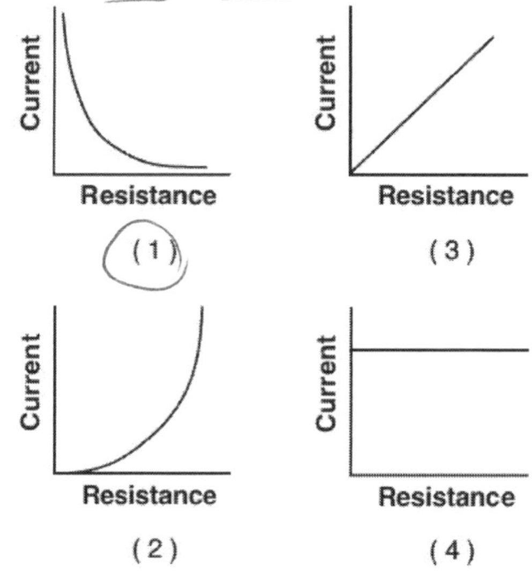

8. In a simple electric circuit, a 24-ohm resistor is connected across a 6-volt battery. What is the current in the circuit?
   1. 1.0 A
   2. 0.25 A
   3. 140 A
   4. 4.0 A

Name:______________________________ Period: ____________

Circuits-Ohm's Law

9. Which graph best represents the relationship between the power expended by a resistor that obeys Ohm's Law and the potential difference applied to the resistor?

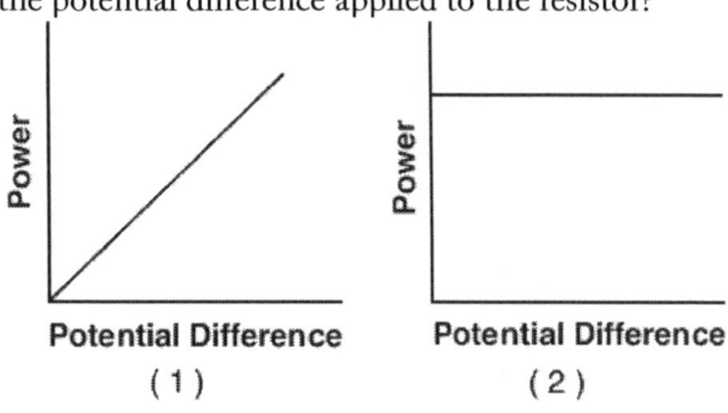

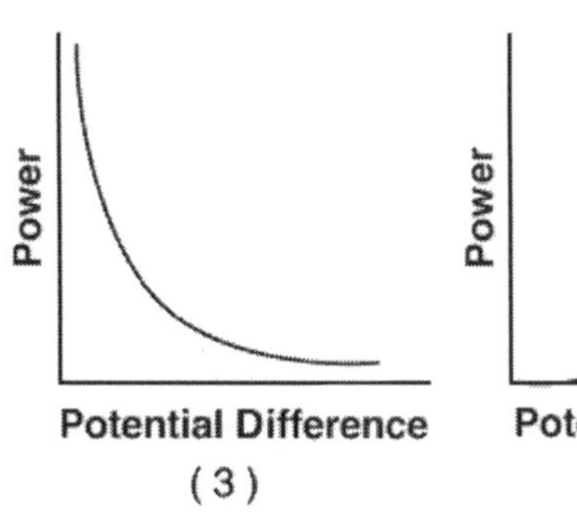

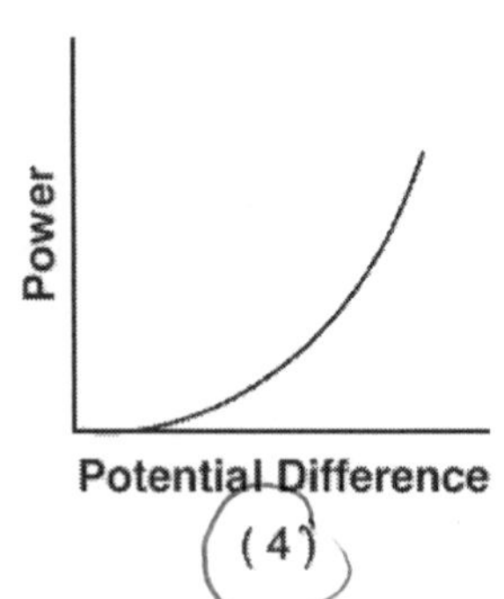

10. The current through a 10-ohm resistor is 1.2 amperes. What is the potential difference across the resistor?
    1. 8.3 V
    2. 12 V
    3. 14 V
    4. 120 V

11. The graph below represents the relationship between the current in a metallic conductor and the potential difference across the conductor at constant temperature.

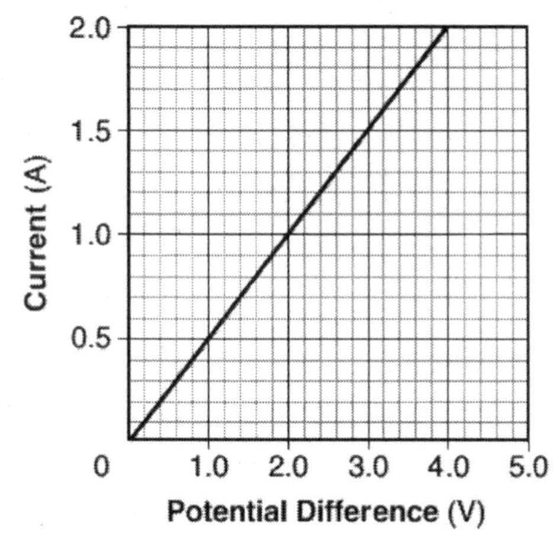

The resistance of the conductor is
1. 1.0 ohms
2. 2.0 ohms
3. 0.5 ohms
4. 4.0 ohms

12. A 330-ohm resistor is connected to a 5-volt battery. The current through the resistor is
    1. 0.152 mA
    2. 15.2 mA
    3. 335 mA
    4. 1650 mA

13. An immersion heater has a resistance of 5 ohms while drawing a current of 3 amperes. How much electrical energy is delivered to the heater during 200 seconds of operation?
    1. 3.0×10^3 J
    2. 6.0×10^3 J
    3. 9.0×10^3 J
    4. 1.5×10^4 J

14. An operating 100-watt lamp is connected to a 120-volt outlet. What is the total electrical energy used by the lamp in 60 seconds?
    1. 0.60 J
    2. 1.7 J
    3. 6.0×10^3 J
    4. 7.2×10^3 J

15. A 150-watt lightbulb is brighter than a 60-watt lightbulb when both are operating at a potential difference of 110 volts. Compared to the resistance of and the current drawn by the 150-watt lightbulb, the 60-watt lightbulb has
    1. less resistance and draws more current
    2. less resistance and draws less current
    3. more resistance and draws more current
    4. more resistance and draws less current

16. A light bulb attached to a 120-volt source of potential difference draws a current of 1.25 amperes for 35 seconds. Calculate how much electrical energy is used by the bulb. [Show all work, including the equation and substitution with units.]

Name:________________________ Period: __________

Circuits-Ohm's Law

Base your answers to questions 17 through 20 on the information, circuit diagram, and data table below.

In a physics lab, a student used the circuit shown to measure the current through and the potential drop across a resistor of unknown resistance, R. The instructor told the student to use the switch to operate the circuit only long enough to take each reading. The student's measurements are recorded in the data table.

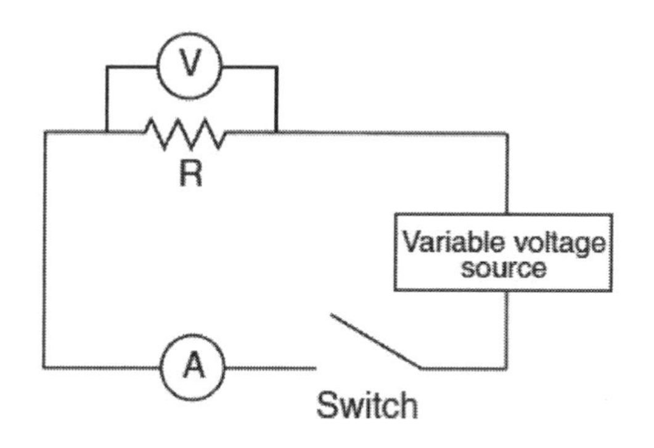

Data Table

Current (A)	Potential Drop (V)
0.80	21.4
1.20	35.8
1.90	56.0
2.30	72.4
3.20	98.4

Using the information in the data table, construct a graph on the grid following the directions below.

17. Mark an appropriate scale on the axis labeled "Potential Drop (V)."

18. Plot the data points for potential drop versus current.

19. Draw the line or curve of best fit.

20. Calculate the slope of the line or curve of best fit. [Show all work, including the equation and substitution with units.]

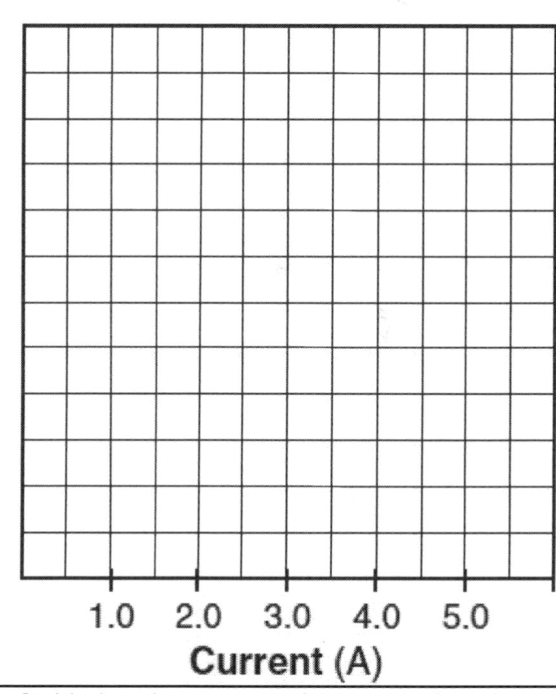

21. A 100-ohm resistor and an unknown resistor are connected in series to a 10-volt battery. If the potential drop across the 100-ohm resistor is 4 volts, the resistance of the unknown resistor is
    1. 50 ohms
    2. 100 ohms
    3. 150 ohms
    4. 200 ohms

22. In a flashlight, a battery provides a total of 3 volts to a bulb. If the flashlight bulb has an operating resistance of 5 ohms, the current through the bulb is
    1. 0.30 A
    2. 0.60 A
    3. 1.5 A
    4. 1.7 A

Name: ______________________ Period: __________

Circuits-Ohm's Law

23. The graph at right shows the relationship between the potential difference across a metallic conductor and the electric current through the conductor at constant temperature T_1.

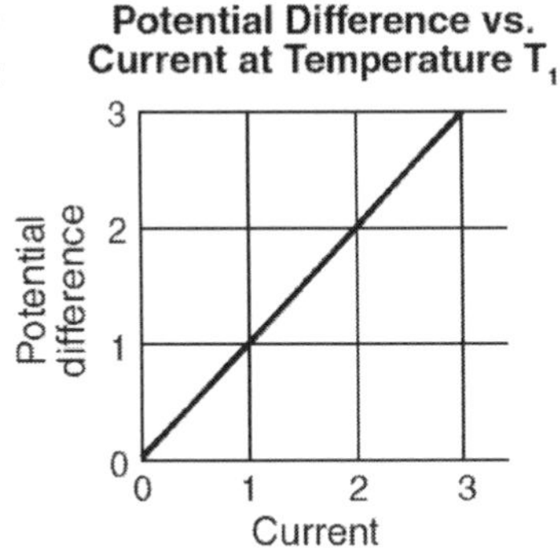

Which graph below best represents the relationship between potential difference and current for the same conductor maintained at a higher constant temperature, T_2?

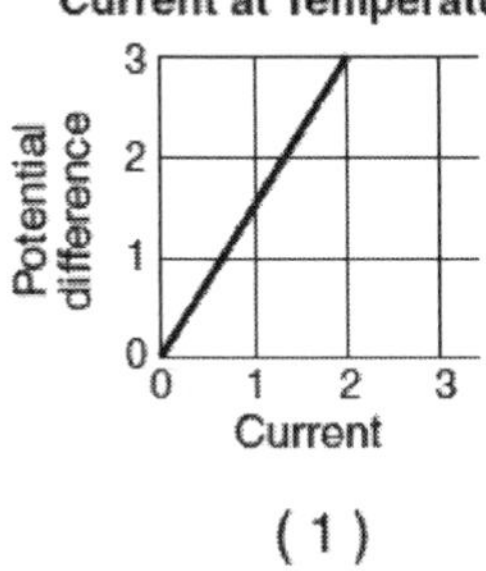

(1)

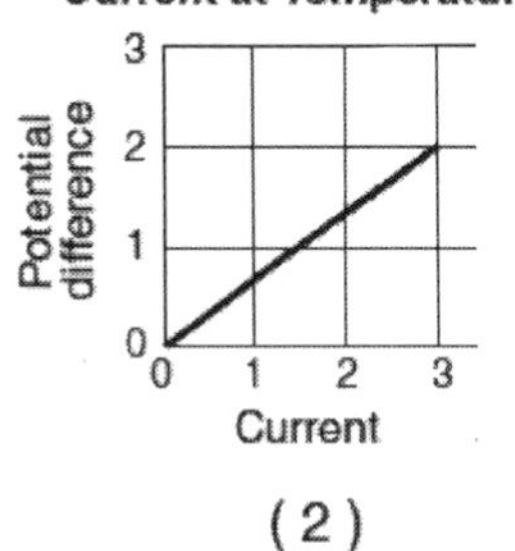

(2)

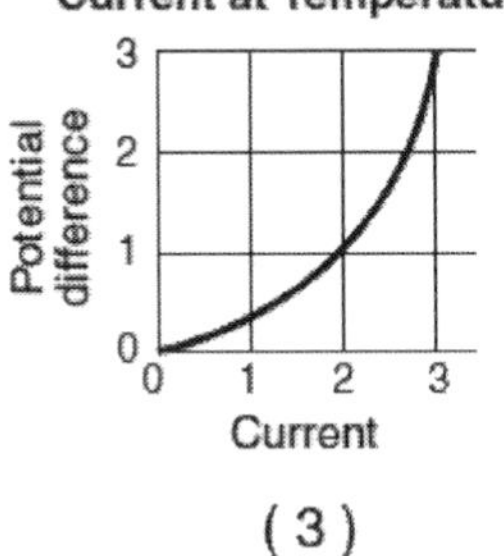

(3)

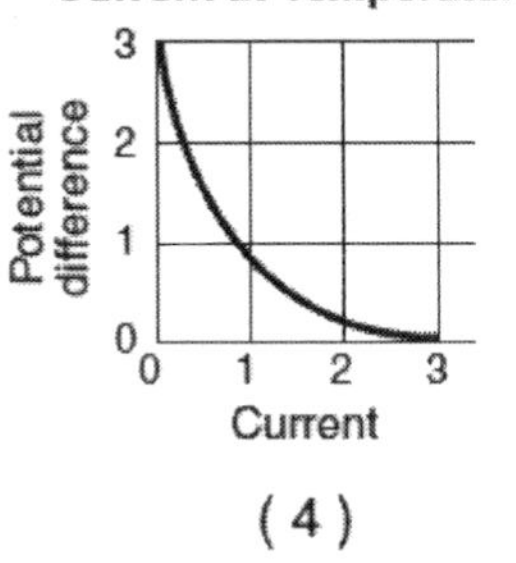

(4)

24. A long copper wire was connected to a voltage source. The voltage was varied and the current through the wire measured, while temperature was held constant. The collected data are represented by the graph below.

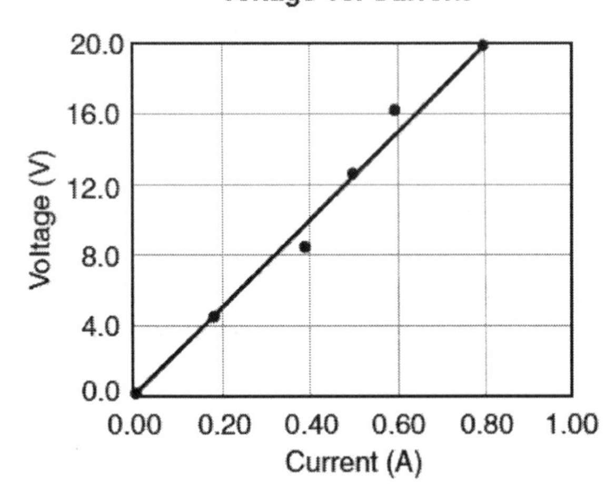

Using the graph, determine the resistance of the copper wire.

25. A constant potential difference is applied across a variable resistor held at constant temperature. Which graph best represents the relationship between the resistance of the variable resistor and the current through it?

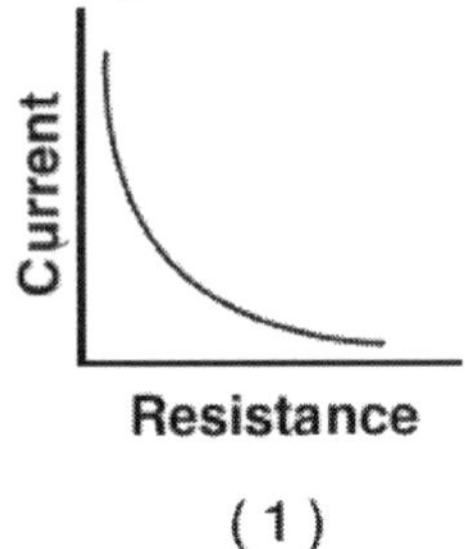

(1)

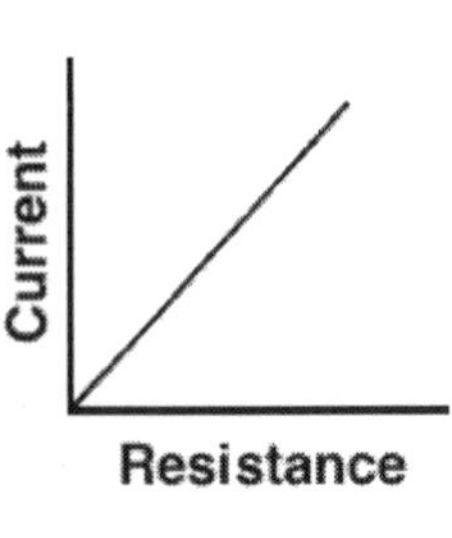
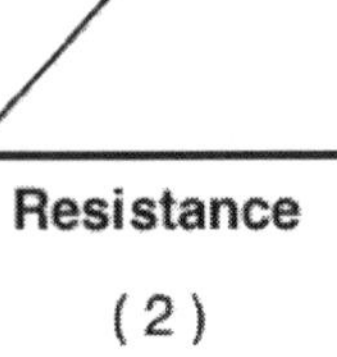

(2)

Current

Resistance

(3)

Current

Resistance

(4)

Name: ______________________ Period: __________

Circuits-Ohm's Law

26. An electrical appliance draws 9.0 amperes of current when connected to a 120-volt source of potential difference. What is the total amount of power dissipated by this appliance?
    1. 13 W
    2. 110 W
    3. 130 W
    4. 1100 W

27. In a simple electric circuit, a 110-volt electric heater draws 2.0 amperes of current. The resistance of the heater is
    1. 0.018 ohms
    2. 28 ohms
    3. 55 ohms
    4. 220 ohms

28. If the potential difference applied to a fixed resistance is doubled, the power dissipated by that resistance
    1. remains the same
    2. doubles
    3. halves
    4. quadruples

29. A 4.50-volt personal stereo uses 1950 joules of electrical energy in an hour. What is the electrical resistance of the personal stereo?
    1. 433 ohms
    2. 96.3 ohms
    3. 37.4 ohms
    4. 0.623 ohms

30. If 20 joules of work is used to transfer 20 coulombs of charge through a 20-ohm resistor, the potential difference across the resistor is
    1. 1 V
    2. 20 V
    3. 0.05 V
    4. 400 V

31. A 50-watt lightbulb and a 100-watt lightbulb are each operated at 110 volts. Compared to the resistance of the 50-watt bulb, the resistance of the 100-watt bulb is
    1. half as great
    2. twice as great
    3. one-fourth as great
    4. four times as great

32. The graph below represents the relationship between the potential difference (V) across a resistor and the current (I) through the resistor.

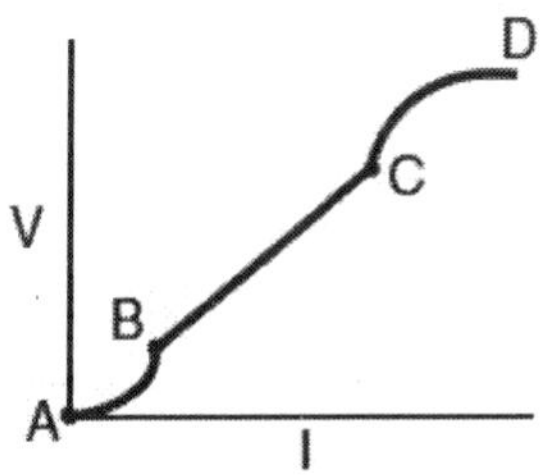

Through which entire interval does the resistor obey Ohm's law?
    1. AB
    2. BC
    3. CD
    4. AD

33. The resistance of a 60-watt lightbulb operated at 120 volts is approximately
    1. 720 ohms
    2. 240 ohms
    3. 120 ohms
    4. 60 ohms

34. An electric drill operating at 120 volts draws a current of 3 amperes. What is the total amount of electrical energy used by the drill during one minute of operation?
    1. 2.16×10^4 J
    2. 2.40×10^3 J
    3. 3.60×10^2 J
    4. 4.00×10^1 J

35. An electric iron operating at 120 volts draws 10 amperes of current. How much heat energy is delivered by the iron in 30 seconds?
    1. 3.0×10^2 J
    2. 1.2×10^3 J
    3. 3.6×10^3 J
    4. 3.6×10^4 J

36. In a series circuit containing two lamps, the battery supplies a potential difference of 1.5 volts. If the current in the circuit is 0.10 ampere, at what rate does the circuit use energy?
    1. 0.015 W
    2. 0.15 W
    3. 1.5 W
    4. 15 W

Name: ______________________ Period: __________

Circuits-Ohm's Law

37. An electric circuit consists of a variable resistor connected to a source of constant potential difference. If the resistance of the resistor is doubled, the current through the resistor is
    1. halved
    2. doubled
    3. quartered
    4. quadruples

38. Which physical quantity is correctly paired with its unit?
    1. power and watt·seconds
    2. energy and newton·seconds
    3. electric current and amperes/coulomb
    4. electrical potential difference and joules/coulomb

39. A 6-ohm resistor that obeys Ohm's Law is connected to a source of variable potential difference. When the applied voltage is decreased from 12 V to 6 V, the current passing through the resistor
    1. remains the same
    2. is doubled
    3. is halved
    4. is quadruples

40. An electric heater operating at 120 volts draws 8 amperes of current through its 15 ohms of resistance. The total amount of heat energy produced by the heater in 60 seconds is
    1. 7.20×10^3 J
    2. 5.76×10^4 J
    3. 8.64×10^4 J
    4. 6.91×10^6 J

41. A device operating at a potential difference of 1.5 volts draws a current of 0.20 ampere. How much energy is used by the device in 60 seconds?
    1. 4.5 J
    2. 8.0 J
    3. 12 J
    4. 18 J

42. Calculate the resistance of a 900-watt toaster operating at 120 volts. [Show all work, including the equation and substitution with units.]

43. What is the current in a 100-ohm resistor connected to a 0.40-volt source of potential difference?
    1. 250 mA
    2. 40 mA
    3. 2.5 mA
    4. 4.0 mA

44. How much total energy is dissipated in 10 seconds in a 4-ohm resistor with a current of 0.50 ampere?
    1. 2.5 J
    2. 5.0 J
    3. 10 J
    4. 20 J

45. If a 1.5-volt cell is to be completely recharged, each electron must be supplied with a minimum energy of
    1. 1.5 eV
    2. 1.5 J
    3. 9.5×10^{18} eV
    4. 9.5×10^{18} J

46. As the potential difference across a given resistor is increased, the power expended in moving charge through the resistor
    1. decreases
    2. increases
    3. remains the same

47. The heating element in an automobile window has a resistance of 1.2 ohms when operated at 12 volts. Calculate the power dissipated in the heating element. [Show all work, including the equation and substitution with units.]

Name:_______________________________ Period: ____________

Circuits-Ohm's Law

48. The resistance of a circuit remains constant. Which graph best represents the relationship between the current in the circuit and the potential difference provided by the battery?

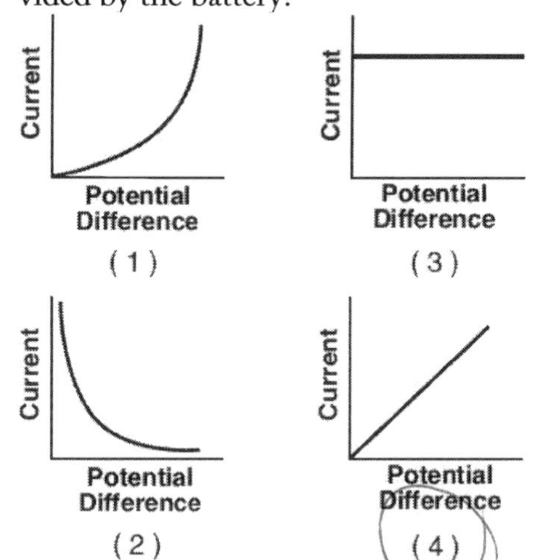

49. Moving 4.0 coulombs of charge through a circuit requires 48 joules of electric energy. What is the potential difference across this circuit?
    1. 190 V
    2. 48 V
    3. 12 V
    4. 4.0 V

50. An electric dryer consumes 6.0×10^6 joules of electrical energy when operating at 220 volts for 1.8×10^3 seconds. During operation, the dryer draws a current of
    1. 10 A
    2. 15 A
    3. 9.0×10^2 A
    4. 3.3×10^3 A

51. The total amount of electrical energy used by a 315-watt television during 30.0 minutes of operation is
    1. 5.67×10^5 J
    2. 9.45×10^3 J
    3. 1.05×10^1 J
    4. 1.75×10^{-1} J

52. A radio operating at 3.0 volts and a constant temperature draws a current of 1.8×10^{-4} ampere. What is the resistance of the radio circuit?
    1. $1.7 \times 10^4\ \Omega$
    2. $3.0 \times 10^1\ \Omega$
    3. $5.4 \times 10^{-4}\ \Omega$
    4. $6.0 \times 10^{-5}\ \Omega$

Name: ______________________________ Period: __________

Circuits-Circuit Analysis

Base your answers to questions 1 through 3 on the information and diagram below.

A 3.0-ohm resistor, an unknown resistor, R, and two ammeters, A_1 and A_2, are connected as shown with a 12-volt source. Ammeter A_2 reads a current of 5.0 amperes.

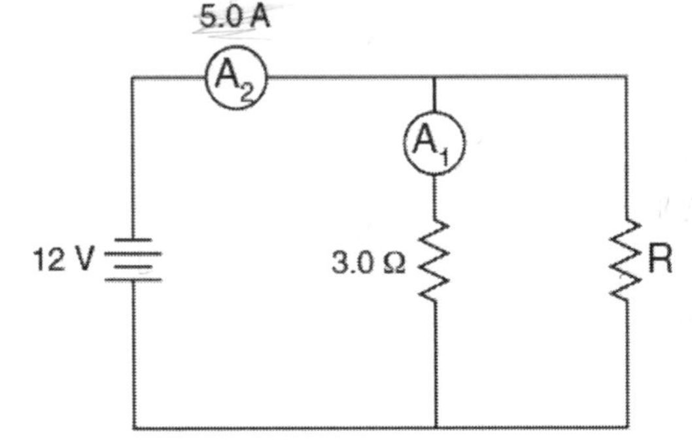

1. Determine the equivalent resistance of the circuit.

2. Calculate the current measured by ammeter A_1. [Show all work, including the equation and substitution with units.

3. Calculate the resistance of the unknown resistor, R. [Show all work, including the equation and substitution with units.]

4. A 9-volt battery is connected to a 4-ohm resistor and a 5-ohm resistor as shown in the diagram below.

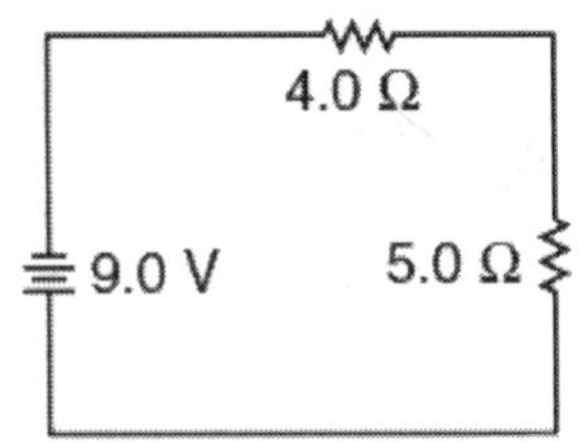

What is the current in the 5-ohm resistor?
1. 1.0 A
2. 1.8 A
3. 2.3 A
4. 4.0 A

Base your answers to questions 5 through 7 on the information below.

An 18-ohm resistor and a 36-ohm resistor are connected in parallel with a 24-volt battery. A single ammeter is placed in the circuit to read its total current.

5. Draw a diagram of this circuit.

6. Calculate the equivalent resistance of the circuit.

7. Calculate the total power dissipated in the circuit.

Name:____________________________ Period: __________

Circuits-Circuit Analysis

8. In which circuit would current flow through resistor R1 but not through resistor R2 while switch S is open?

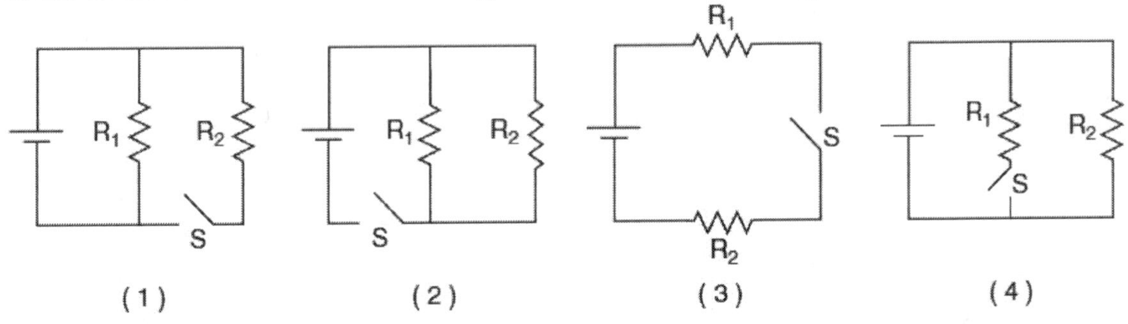

9. Which circuit diagram below correctly shows the connection of ammeter A and voltmeter V to measure the current through and potential difference across resistor R?

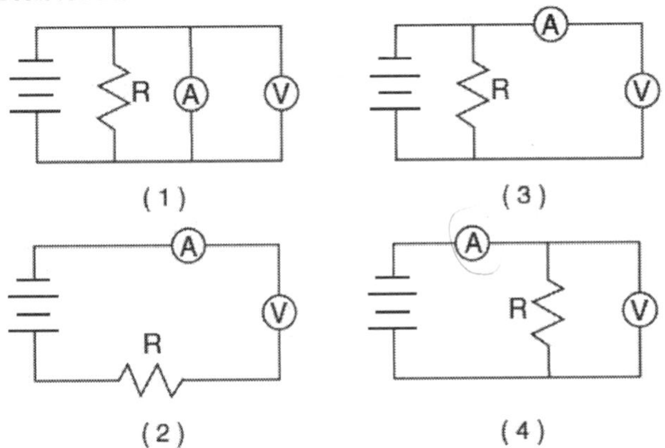

10. In the space below, draw a diagram of an operating circuit that includes:
   - a battery as a source of potential difference
   - two resistors in parallel with each other
   - an ammeter that reads the total current in the circuit

Base your answers to questions 11 through 13 on the information and diagram below.

A 15-ohm resistor, R_1, and a 30-ohm resistor, R_2, are to be connected in parallel between points A and B in a circuit containing a 90-volt battery.

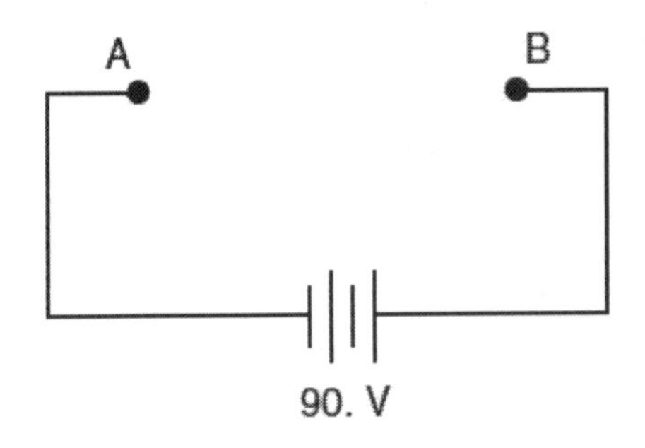

11. Complete the diagram above to show the two resistors in parallel between points A and B.

12. Determine the potential difference across resistor R_1.

13. Calculate the current in resistor R_1.

Name:______________________________ Period: __________

Circuits-Circuit Analysis

Base your answers to questions 14 through 16 on the information and diagram below, showing all work including the equation and substitution with units.

A 50-ohm resistor, an unknown resistor R, a 120-volt source, and an ammeter are connected in a complete circuit. The ammeter reads 0.50 ampere.

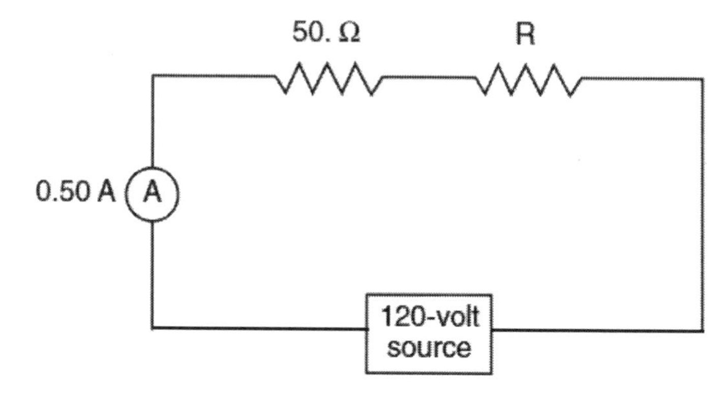

14. Calculate the equivalent resistance of the circuit.

15. Determine the resistance of resistor R.

16. Calculate the power dissipated by the 50-ohm resistor.

---

17. In which circuit would an ammeter show the greatest total current?

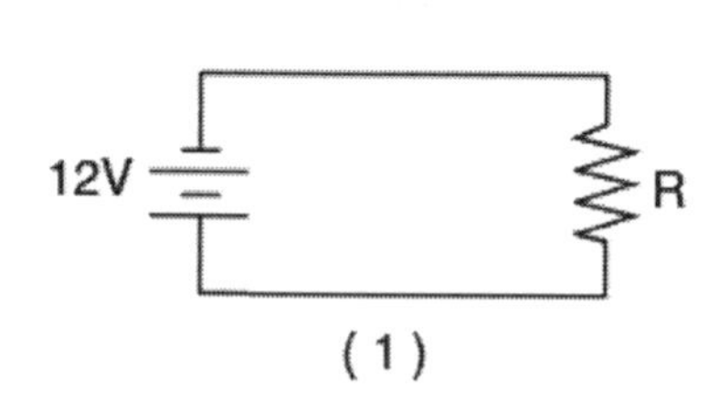

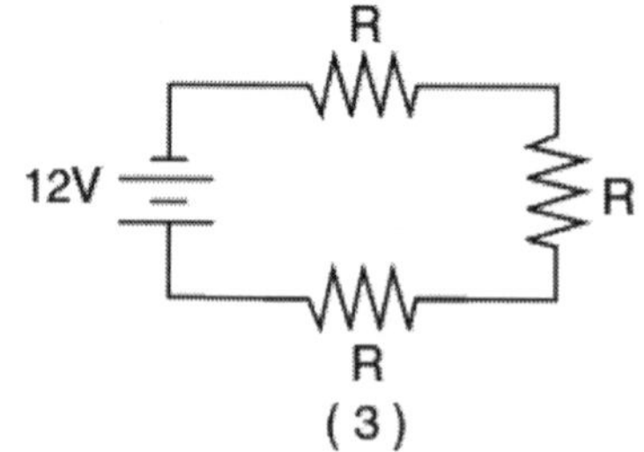

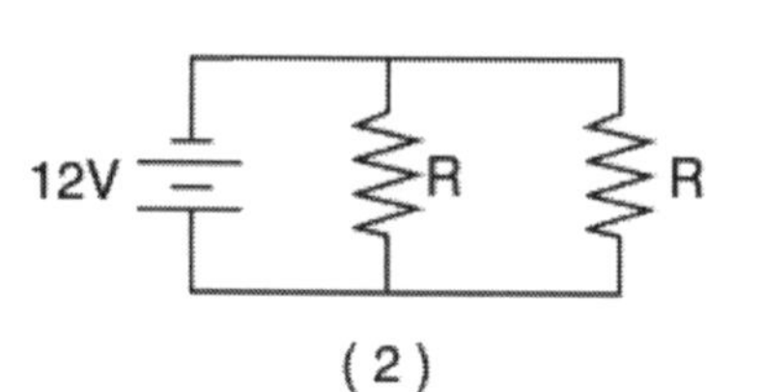

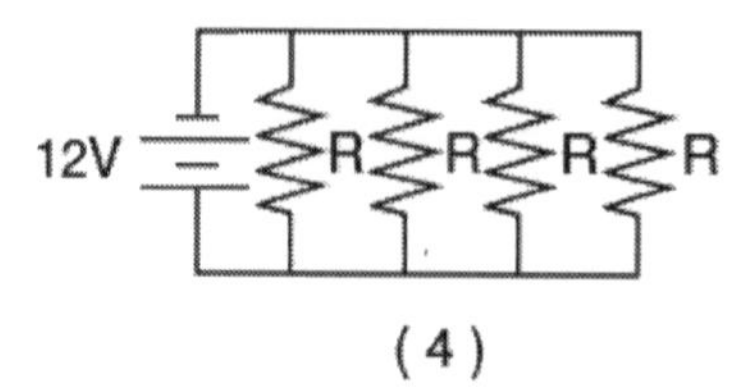

Name: ____________________ Period: __________

Circuits-Circuit Analysis

Base your answers to questions 18 through 22 on the information below and data table at right.

Three lamps were connected in a circuit with a battery of constant potential. The current, potential difference, and resistance for each lamp are listed in the data table. [There is negligible resistance in the wires and battery.]

	Current (A)	Potential Difference (V)	Resistance (Ω)
lamp 1	0.45	40.1	89
lamp 2	0.11	40.1	365
lamp 3	0.28	40.1	143

18. Using standard circuit symbols, draw a circuit showing how the lamps and battery are connected.

19. What is the potential difference supplied by the battery?

20. Calculate the equivalent resistance of the circuit.

21. If lamp 3 is removed from the circuit, what would be the value of the potential difference across lamp 1 after lamp 3 is removed?

22. If lamp 3 is removed from the circuit, what would be the value of the current in lamp 2 after lamp 3 is removed?

---

23. In which circuit would ammeter A show the greatest current?

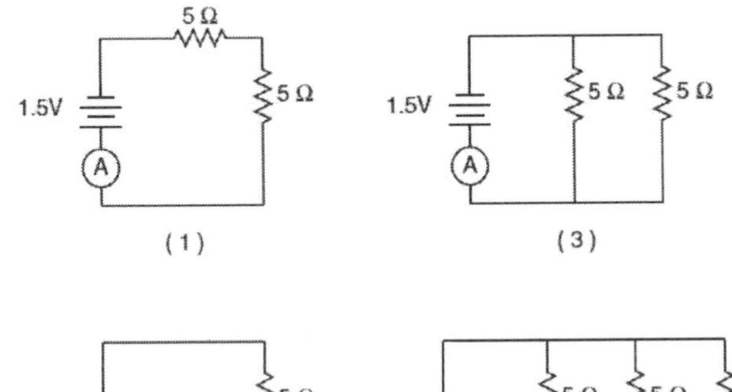

1.5V, A, 5 Ω
(2)

1.5V, A, 5 Ω, 5 Ω, 5 Ω
(4)

Name:______________________________ Period: __________

Circuits-Circuit Analysis

24. A 6-ohm resistor and a 4-ohm resistor are connected in series with a 6-volt battery in an operating electric circuit. A voltmeter is connected to measure the potential difference across the 6-ohm resistor. Draw a diagram of this circuit including the battery, resistors, and voltmeter. Label each resistor with its value.

25. What is the total current in a circuit consisting of six operating 100-watt lamps connected in parallel to a 120-volt source?
    1. 5 A
    2. 20 A
    3. 600 A
    4. 12,000 A

26. The circuit diagram below represents four resistors connected to a 12-volt source.

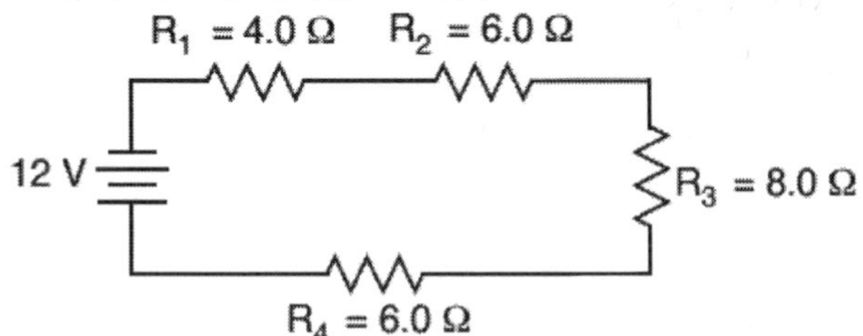

What is the total current in the circuit?
1. 0.50 A
2. 2.0 A
3. 8.6 A
4. 24 A

27. As the number of resistors in a parallel circuit is increased, what happens to the equivalent resistance of the circuit and total current in the circuit?
    1. Both equivalent resistance and total current decrease.
    2. Both equivalent resistance and total current increase.
    3. Equivalent resistance decreases and total current increases.
    4. Equivalent resistance increases and total current decreases.

Base your answers to questions 28 and 29 on the circuit diagram below.

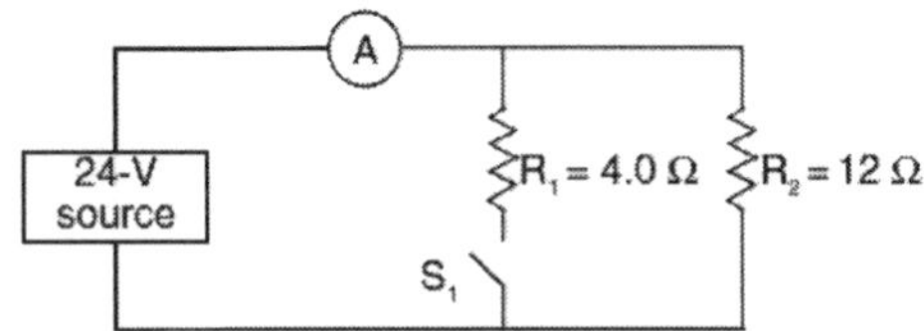

28. If switch S_1 is open, the reading of ammeter A is
    1. 0.50 A
    2. 2.0 A
    3. 1.5 A
    4. 6.0 A

29. If switch S_1 is closed, the equivalent resistance of the circuit is
    1. 8 ohms
    2. 2 ohms
    3. 3 ohms
    4. 16 ohms

30. Which circuit has the smallest equivalent resistance?

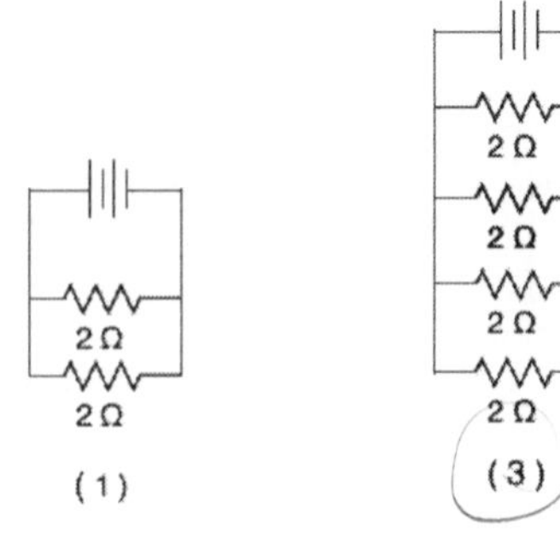

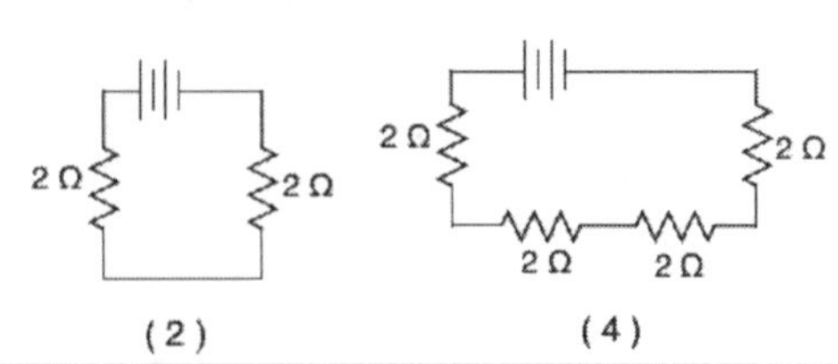

Name: ______________________________ Period: ____________

Circuits-Circuit Analysis

Base your answers to questions 31 through 33 on the information below.

A 5-ohm resistor, a 10-ohm resistor, and a 15-ohm resistor are connected in parallel with a battery. The current through the 5-ohm resistor is 2.4 amperes.

31. Using standard circuit symbols, draw a diagram of this electric circuit. in the space at right.

32. Calculate the amount of electrical energy expended in the 5-ohm resistor in 2 minutes.

33. A 20-ohm resistor is added to the circuit in parallel with the other resistors. Describe the effect the addition of this resistor has on the amount of electrical energy expended in the 5-ohm resistor in 2 minutes.

34. In the circuit diagram below, two 4-ohm resistors are connected to a 16-volt battery as shown.

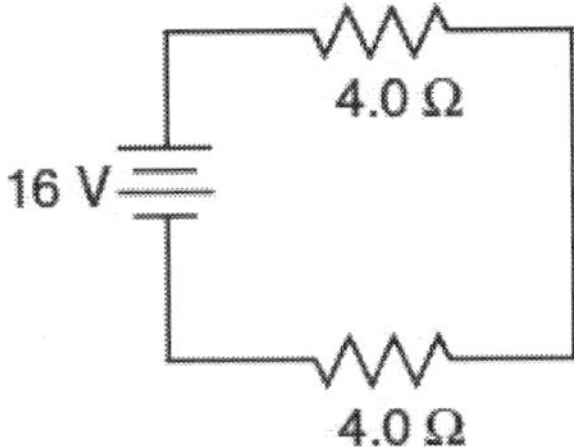

The rate at which electrical energy is expended in this circuit is
1. 8.0 W
2. 16 W
3. 32 W
4. 64 W

35. Two identical resistors connected in series have an equivalent resistance of 4 ohms. The same two resistors, when connected in parallel, have an equivalent resistance of
1. 1 ohm
2. 2 ohms
3. 8 ohms
4. 4 ohms

36. An electric circuit contains a source of potential difference and 5-ohm resistors that combine to give the circuit an equivalent resistance of 15 ohms. In the space below, draw a diagram of this circuit using standard circuit symbols. [Assume the availability of any number of 5-ohm resistors and wires of negligible resistance.]

Name:________________________ Period: ____________

Circuits-Circuit Analysis

Base your answers to questions 37 through 39 on the diagram below, which represents an electrical circuit consisting of four resistors and a 12-volt battery.

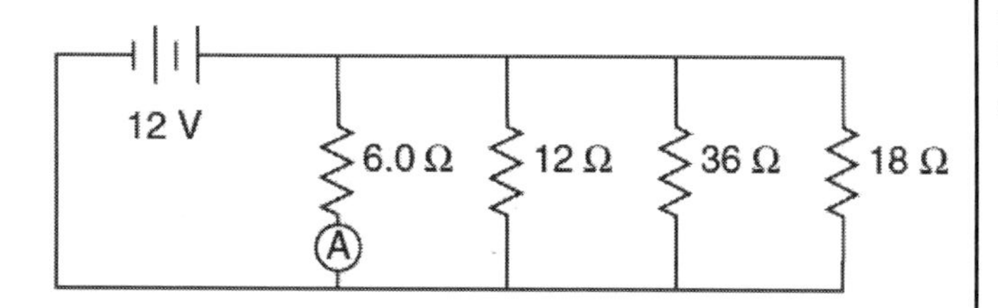

37. What is the current measured by ammeter A?
    1. 0.50 A
    2. 2.0 A
    3. 72 A
    4. 4.0 A

38. What is the equivalent resistance of this circuit?
    1. 72 ohms
    2. 18 ohms
    3. 3.0 ohms
    4. 0.33 ohms

39. How much power is dissipated in the 36-ohm resistor?
    1. 110 W
    2. 48 W
    3. 3.0 W
    4. 4.0 W

---

40. Three resistors, 4 ohms, 6 ohms, and 8 ohms, are connected in parallel in an electric circuit. The equivalent resistance of the circuit is
    1. less than 4 ohms
    2. between 4 ohms and 8 ohms
    3. between 10 ohms and 18 ohms
    4. 18 ohms

41. A simple circuit consists of a 100-ohm resistor connected to a battery. A 25-ohm resistor is to be connected in the circuit. Determine the smallest equivalent resistance possible when both resistors are connected to the battery.

Base your answers to questions 42 through 44 on the information and diagram below.

A 20-ohm resistor and a 30-ohm resistor are connected in parallel to a 12-volt battery as shown. An ammeter is connected as shown.

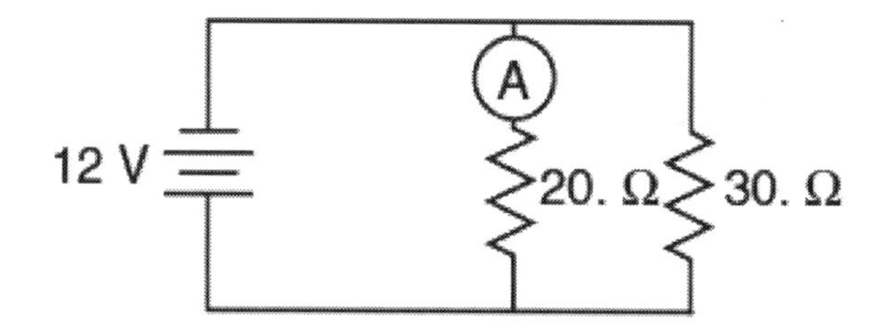

42. What is the equivalent resistance of the circuit?
    1. 10 Ω
    2. 12 Ω
    3. 25 Ω
    4. 50 Ω

43. What is the current reading of the ammeter?
    1. 1.0 A
    2. 0.60 A
    3. 0.40 A
    4. 0.20 A

44. What is the power of the 30-ohm resistor?
    1. 4.8 W
    2. 12 W
    3. 30 W
    4. 75 W

---

45. The diagram below shows a circuit with two resistors.

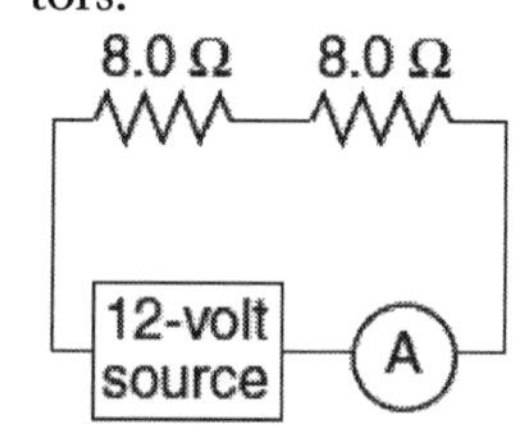

What is the reading on ammeter A?
1. 1.3 A
2. 1.5 A
3. 3.0 A
4. 0.75 A

Name: ______________________________ Period: __________

Circuits-Circuit Analysis

Base your answers to questions 46 and 47 on the circuit diagram below, which shows two resistors connected to a 24-volt source of potential difference.

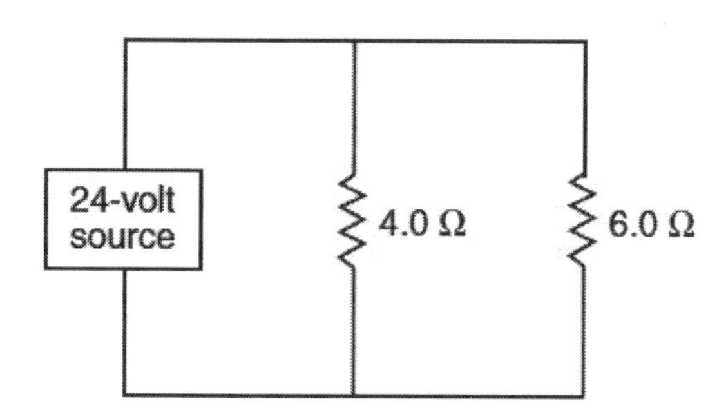

46. On the diagram above, use the appropriate circuit symbol to indicate a correct placement of a voltmeter to determine the potential difference across the circuit.

47. What is the total resistance of the circuit?
    1. 0.42 Ω
    2. 2.4 Ω
    3. 5.0 Ω
    4. 10 Ω

48. The diagram below represents an electric circuit consisting of a 12-volt battery, a 3-ohm resistor, R_1, and a variable resistor, R_2.

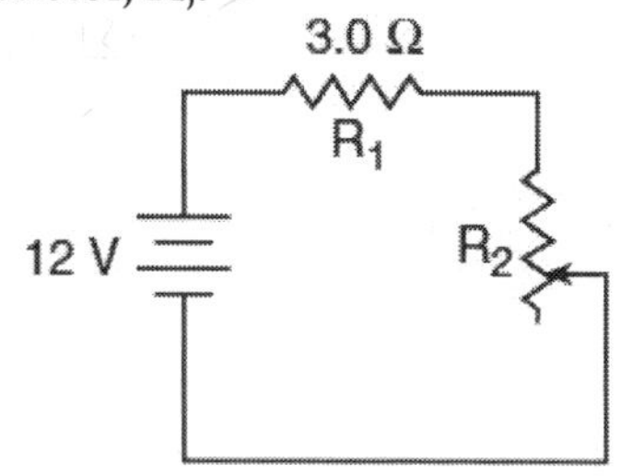

At what value must the variable resistor be set to produce a current of 1.0 ampere through R_1?
    1. 6.0 Ω
    2. 9.0 Ω
    3. 3.0 Ω
    4. 12 Ω

49. Two identical resistors connected in parallel have an equivalent resistance of 40 ohms. What is the resistance of each resistor?
    1. 20 Ω
    2. 40 Ω
    3. 80 Ω
    4. 160 Ω

50. A 6-ohm lamp requires 0.25 ampere of current to operate. In which circuit below would the lamp operate correctly when switch S is closed?

(1) 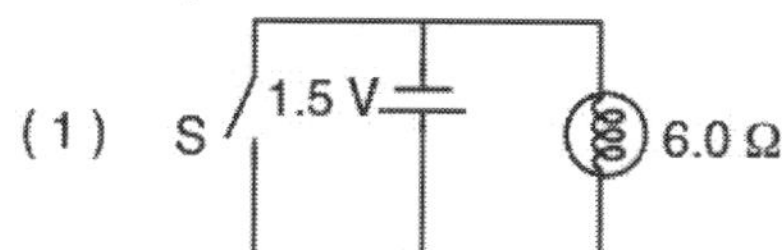

(2) 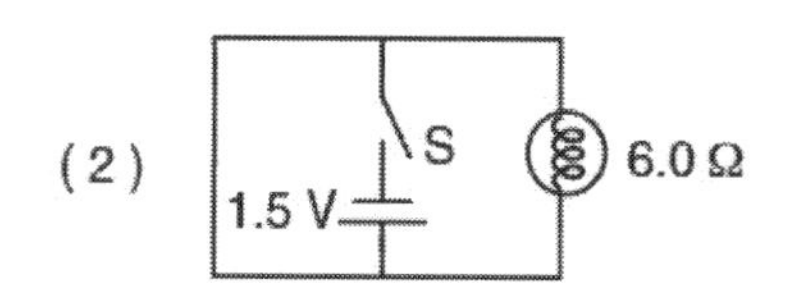

(3) 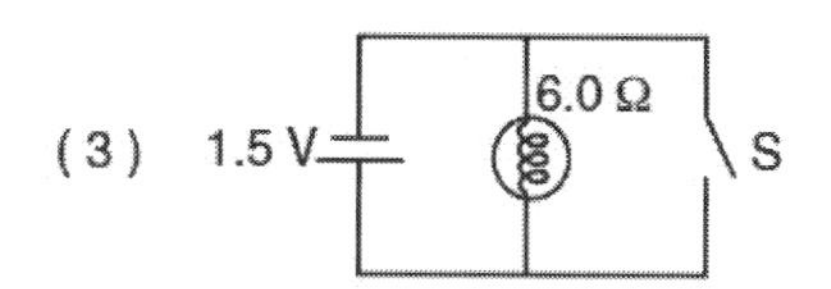

(4) 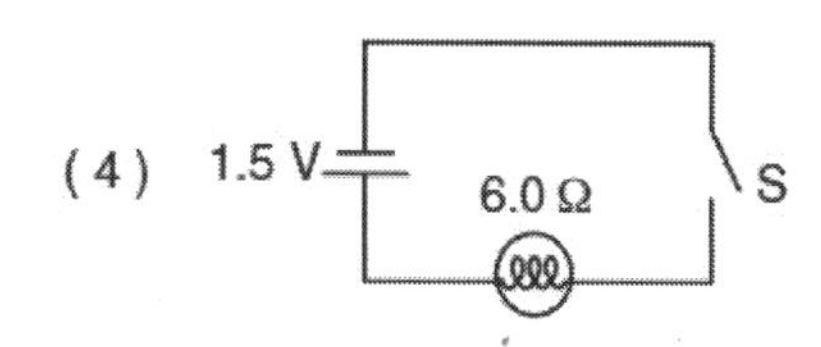

51. In which circuit represented below are meters properly connected to measure the current through resistor R1 and the potential difference across resistor R2?

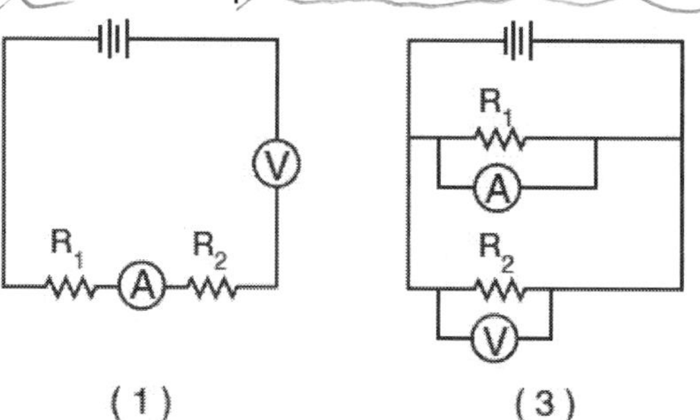

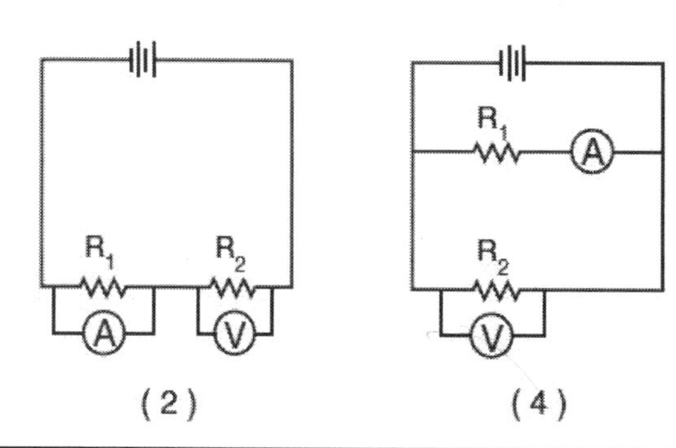

Name: ______________________ Period: ____________

Circuits-Circuit Analysis

52. Which combination of resistors has the smallest equivalent resistance?

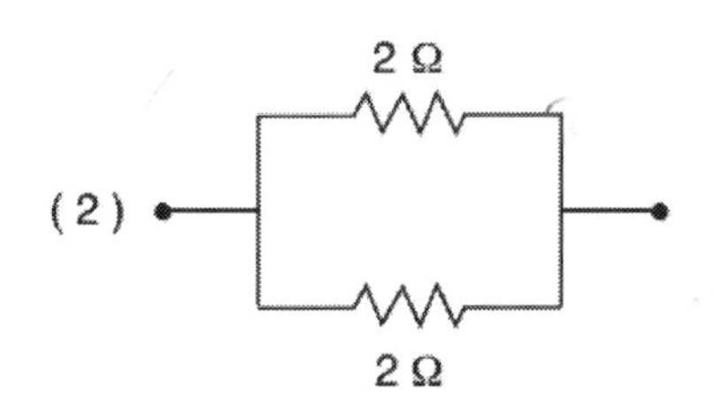

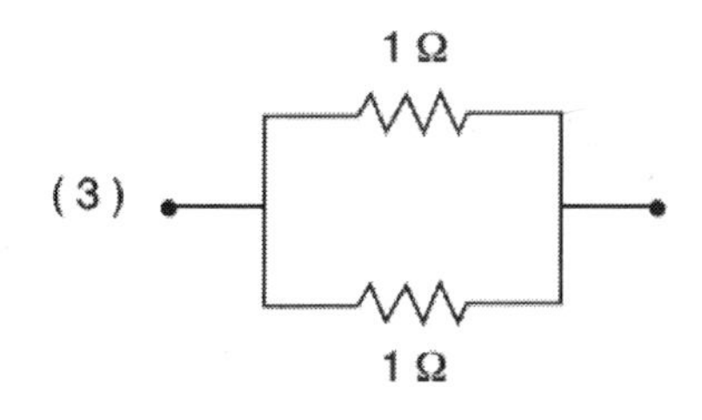

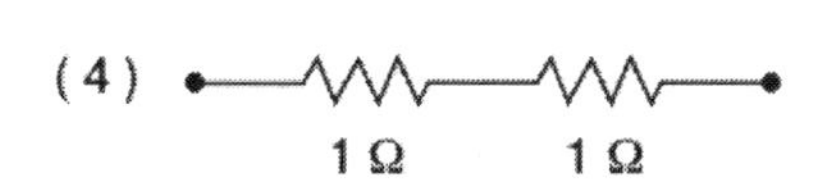

53. The diagram below represents currents in a segment of an electric circuit.

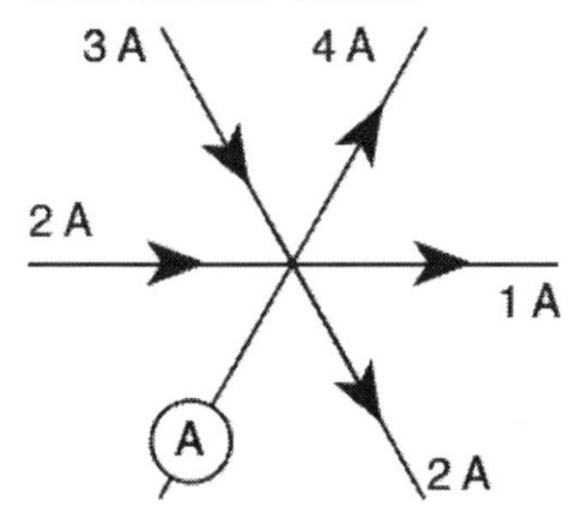

What is the reading of ammeter A?
1. 1 A
2. 2 A
3. 3 A
4. 4 A

54. What is the minimum equipment needed to determine the power dissipated in a resistor of unknown value?
1. a voltmeter, only
2. an ammeter, only
3. a voltmeter and an ammeter, only
4. a voltmeter, an ammeter, and a stopwatch

55. The diagram below represents a circuit consisting of two resistors connected to a source of potential difference.

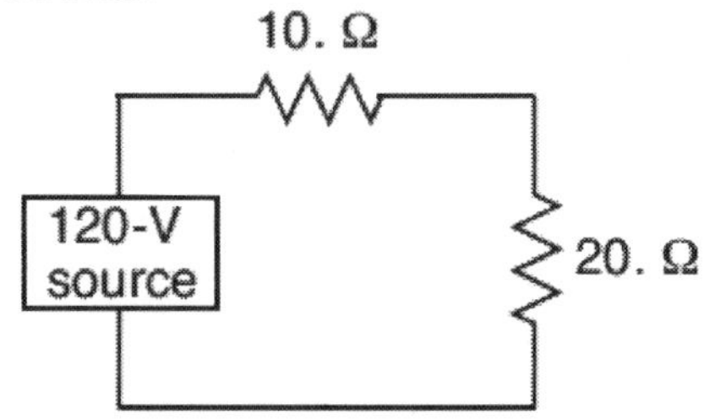

What is the current through the 20-ohm resistor?
1. 0.25 A
2. 6.0 A
3. 12 A
4. 4.0 A

56. In the circuit diagram shown below, ammeter A_1 reads 10 amperes.

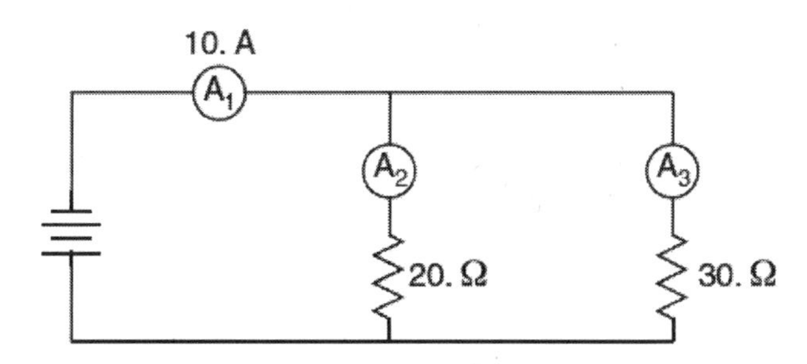

What is the reading of ammeter A_2?
1. 6.0 A
2. 10 A
3. 20 A
4. 4.0 A

57. In the circuit represented by the diagram below, what is the reading of voltmeter V?

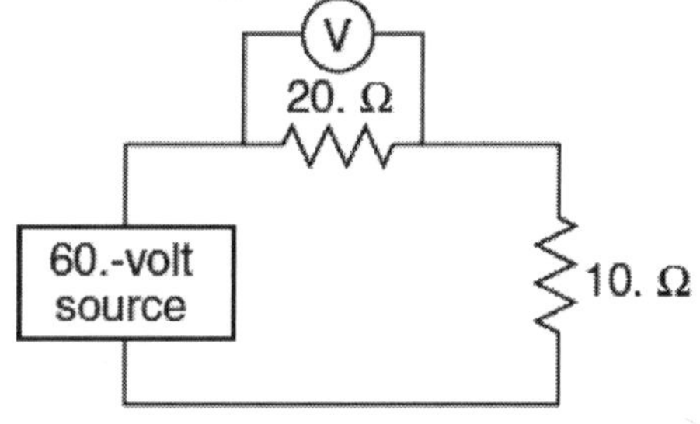

1. 20 V
2. 2.0 V
3. 30 V
4. 40 V

Name: ______________________________ Period: ____________

Circuits-Circuit Analysis

58. In the electric circuit diagram below, possible locations of an ammeter and a voltmeter are indicated by circles 1,2, 3, and 4.

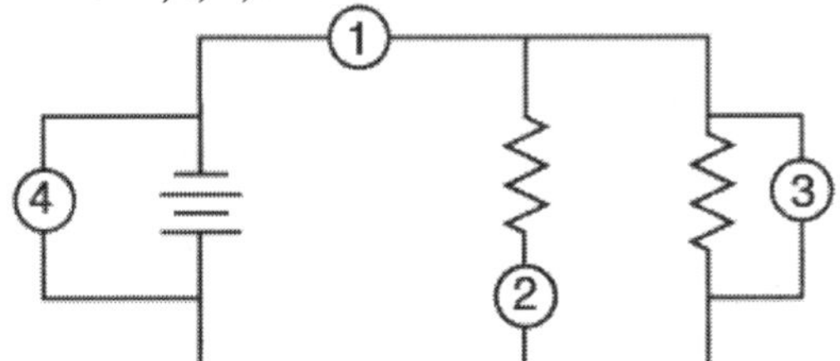

Where should an ammeter be located to correctly measure the total current and where should a voltmeter be located to correctly measure the total voltage?
1. ammeter at 1 and voltmeter at 4
2. ammeter at 2 and voltmeter at 3
3. ammeter at 3 and voltmeter at 4
4. ammeter at 1 and voltmeter at 2

59. What must be inserted between points A and B to establish a steady electric current in the incomplete circuit represented in the diagram below?

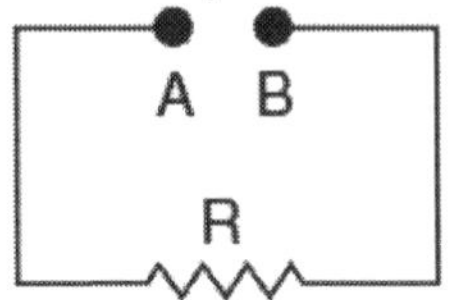

1. switch
2. voltmeter
3. magnetic field source
4. source of potential difference

60. The diagram below represents part of an electric circuit containing three resistors.

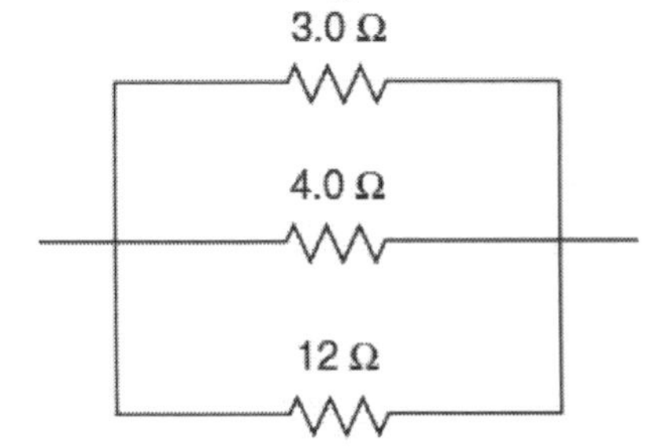

What is the equivalent resistance of this part of the circuit?
1. 0.67 Ω
2. 1.5 Ω
3. 6.3 Ω
4. 19 Ω

61. The diagram below represents a simple circuit consisting of a variable resistor, a battery, an ammeter, and a voltmeter

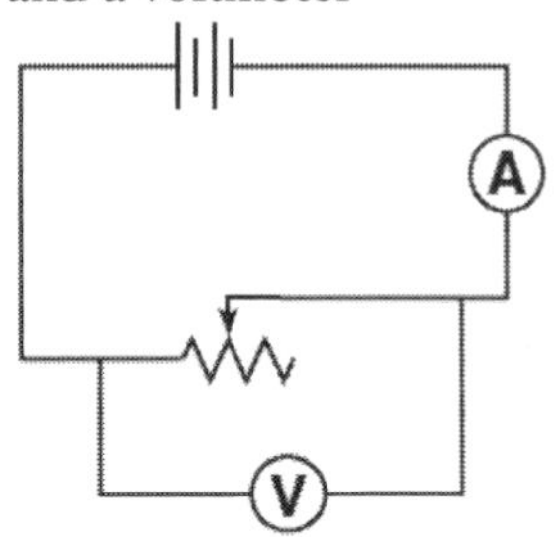

What is the effect of increasing the resistance of the variable resistor from 1000 Ω to 10000 Ω? [Assume constant temperature.]
1. The ammeter reading decreases.
2. The ammeter reading increases.
3. The voltmeter reading decreases.
4. The voltmeter reading increases.

62. Three identical lamps are connected in parallel with each other. If the resistance of each lamp is X ohms, what is the equivalent resistance of this parallel combination?
1. X Ω
2. X/3 Ω
3. 3X Ω
4. 3/X Ω

63. A 3-ohm resistor and a 6-ohm resistor are connected in series in an operating electric circuit. If the current through the 3-ohm resistor is 4 amperes, what is the potential difference across the 6-ohm resistor?
1. 8.0 V
2. 2.0 V
3. 12 V
4. 24 V

64. Circuit A has four 3-ohm resistors connected in series with a 24-volt battery, and circuit B has two 3-ohm resistors connected in series with a 24-volt battery. Compared to the total potential drop across circuit A, the total potential drop across circuit B is
1. one-half as great
2. twice as great
3. the same
4. four times as great

Name:__________ Period: __________

Circuits-Circuit Analysis

65. A circuit consists of a 10-ohm resistor, a 15-ohm resistor, and a 20-ohm resistor connected in parallel across a 9-volt battery. What is the equivalent resistance of this circuit?
    1. 0.200 Ω
    2. 1.95 Ω
    3. 4.62 Ω
    4. 45.0 Ω

66. A 2-ohm resistor and a 4-ohm resistor are connected in series with a 12-volt battery. If the current through the 2-ohm resistor is 2.0 amperes, the current through the 4-ohm resistor is
    1. 1.0 A
    2. 2.0 A
    3. 3.0 A
    4. 4.0 A

67. A 3-ohm resistor and a 6-ohm resistor are connected in parallel across a 9-volt battery. Which statement best compares the potential difference across each resistor?
    1. The potential difference across the 6-ohm resistor is the same as the potential difference across the 3-ohm resistor.
    2. The potential difference across the 6-ohm resistor is twice as great as the potential difference across the 3-ohm resistor.
    3. The potential difference across the 6-ohm resistor is half as great as the potential difference across the 3-ohm resistor.
    4. The potential difference across the 6-ohm resistor is four times as great as the potential difference across the 3-ohm resistor.

68. A 3.6-volt battery is used to operate a cell phone for 5 minutes. If the cell phone dissipates 0.064 watt of power during its operation, current that passes through the phone is
    1. 0.018 A
    2. 5.3 A
    3. 19 A
    4. 56 A

69. To increase the brightness of a desk lamp, a student replaces a 50-watt incandescent lightbulb with a 100-watt incandescent lightbulb. Compared to the 50-watt lightbulb, the 100-watt lightbulb has
    1. less resistance and draws more current
    2. less resistance and draws less current
    3. more resistance and draws more current
    4. more resistance and draws less current

Base your answers to questions 70 and 71 on the information below.

A 15-ohm resistor and a 20-ohm resistor are connected in parallel with a 9-volt battery. A single ammeter is connected to measure the total current of the circuit.

70. Draw a diagram of this circuit using standard circuit schematic symbols.

71. Calculate the equivalent resistance of the circuit. [Show all work including the equation and substitution with units.]

Name: ______________________ Period: __________

Circuits-Circuit Analysis

72. The diagram below shows currents in a segment of an electric circuit.

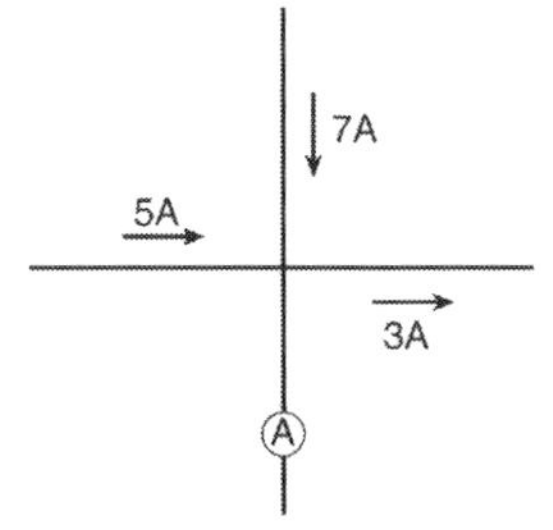

What is the reading of ammeter A?
1. 1 A
2. 5 A
3. 9 A
4. 15 A

Base your answers to questions 73 and 74 on the information below.

A 20-ohm resistor, R_1, and a resistor of unknown resistance, R_2, are connected in parallel to a 30-volt source, as shown in the circuit diagram below. An ammeter in the circuit reads 2.0 amperes.

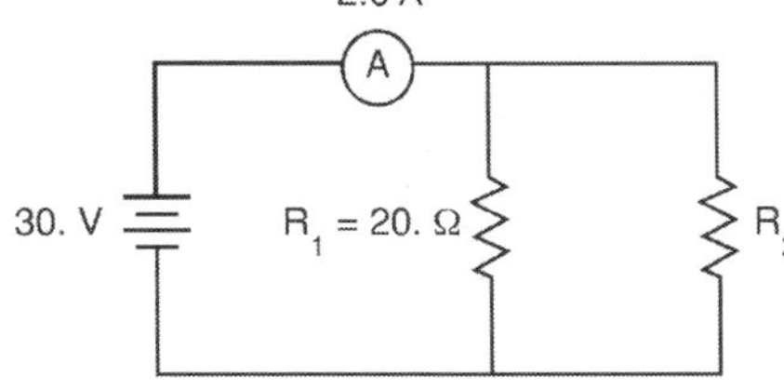

73. Determine the equivalent resistance of the circuit.

74. Calculate the resistance of resistor R_2. [Show all work including the equation and substitution with units.]

Base your answers to questions 75 through 78 on the information below.

A student constructed a series circuit consisting of a 12.0-volt battery, a 10.0-ohm lamp, and a resistor. The circuit does not contain a voltmeter or an ammeter. When the circuit is operating, the total current through the circuit is 0.50 ampere.

75. In the space below, draw a diagram of the series circuit constructed to operate the lamp, using symbols from the Reference Tables for Physical Setting/Physics.

76. Determine the equivalent resistance of the circuit.

77. Determine the resistance of the resistor.

78. Calculate the power consumed by the lamp.

79. If several resistors are connected in series in an electric circuit, the potential difference across each resistor
    1. varies directly with its resistance
    2. varies inversely with its resistance
    3. varies inversely with the square of its resistance
    4. is independent of its resistance

Name:______________________________ Period: __________

Circuits-Circuit Analysis

Base your answers to questions 80 through 83 on the information and circuit diagram below and on your knowledge of physics. Three lamps are connected in parallel to a 120-volt source of potential difference, as represented below.

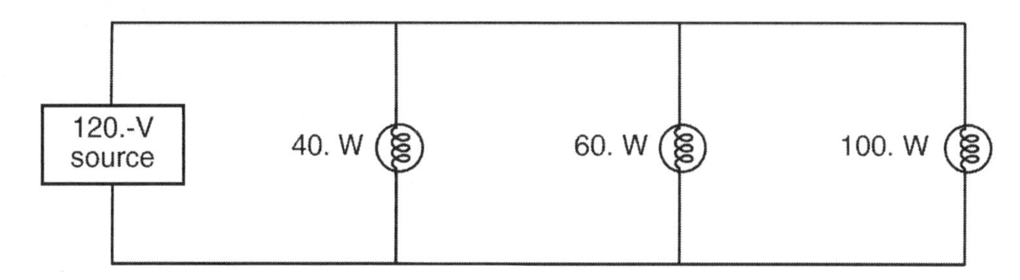

80. Calculate the resistance of the 40-watt lamp. [Show all work, including the equation, substitution with units, and answer with units.]

81. Describe what change, if any, would occur in the power dissipated by the 100-watt lamp if the 60-watt lamp were to burn out.

82. Describe what change, if any, would occur in the equivalent resistance of the circuit if the 60-watt lamp were to burn out.

83. The circuit is disassembled. The same three lamps are then connected in series with each other and the source. Compare the equivalent resistance of this series circuit to the equivalent resistance of the parallel circuit.

Name: ____________________ Period: __________

Magnetism

1. In order to produce a magnetic field, an electric charge must be
   1. stationary
   2. moving
   3. positive
   4. negative

2. The diagram below shows a bar magnet.

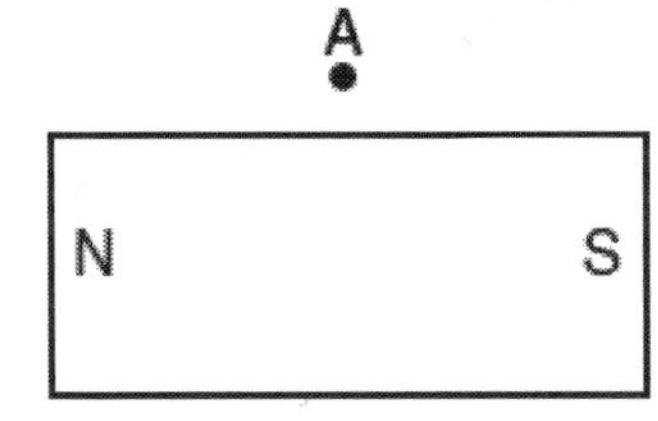

What is the direction of a compass needle placed at point A?
   1. up
   2. down
   3. right
   4. left

3. The diagram below represents the magnetic field near point P.

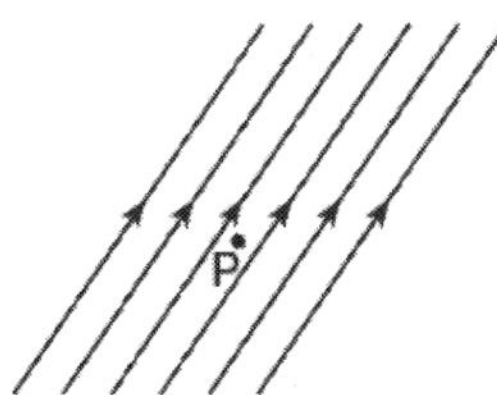

If a compass is placed at point P in the same plane as the magnetic field, which arrow represents the direction the north end of the compass needle will point?

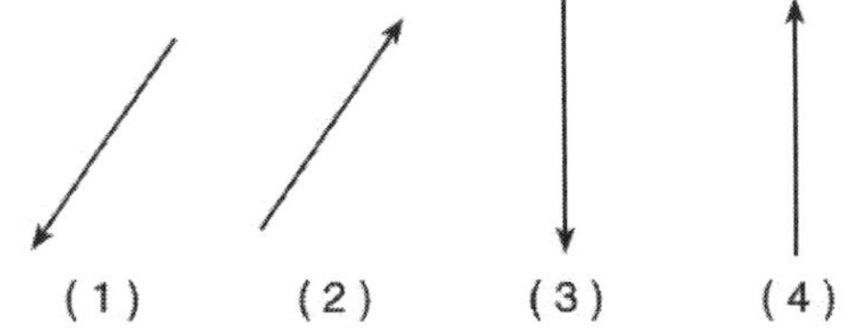

4. Which type of field is present near a moving electric charge?
   1. an electric field, only
   2. a magnetic field, only
   3. both an electric field and a magnetic field
   4. neither an electric field nor a magnetic field

5. A student is given two pieces of iron and told to determine if one or both of the pieces are magnets. First, the student touches an end of one piece to one end of the other. The two pieces of iron attract. Next, the student reverses one of the pieces and again touches the ends together. The two pieces attract again. What does the student definitely know about the initial magnetic properties of the two pieces of iron?

6. The diagram below represents a wire conductor, RS, positioned perpendicular to a uniform magnetic field directed into the page.

R
Magnetic field directed into the page
S

Describe the direction in which the wire could be moved to produce the maximum potential difference across its ends, R and S.

7. The diagram below shows the lines of magnetic force between two north magnetic poles.

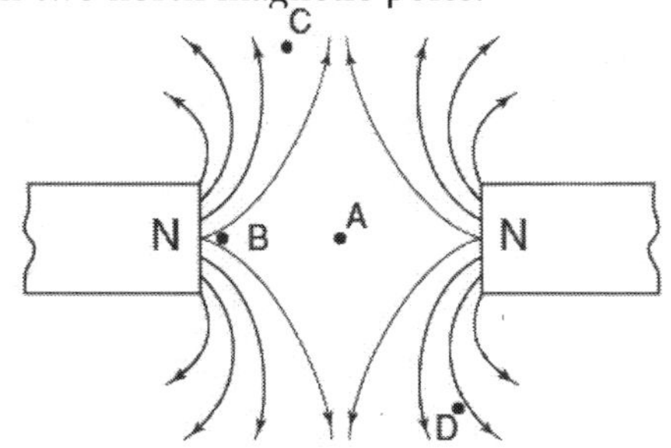

At which point is the magnetic field strength greatest?
   1. A
   2. B
   3. C
   4. D

Name:______________________ Period: __________

Magnetism

8. The diagram below shows a wire moving to the right at speed v through a uniform magnetic field that is directed into the page.

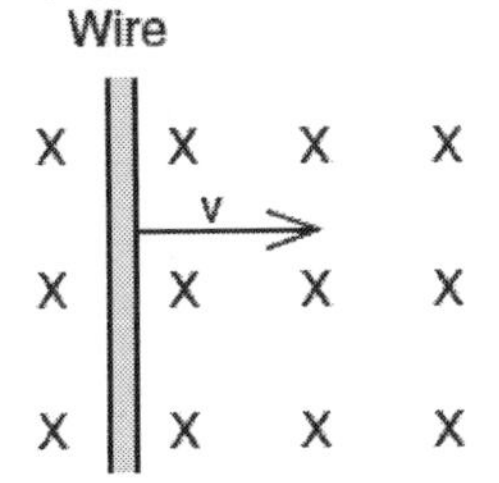

As the speed of the wire is increased, the induced potential difference will
1. decrease
2. increase
3. remain the same

9. Which is *not* a vector quantity?
1. electric charge
2. magnetic field strength
3. velocity
4. displacement

10. When two ring magnets are placed on a pencil, magnet A remains suspended above magnet B, as shown below.

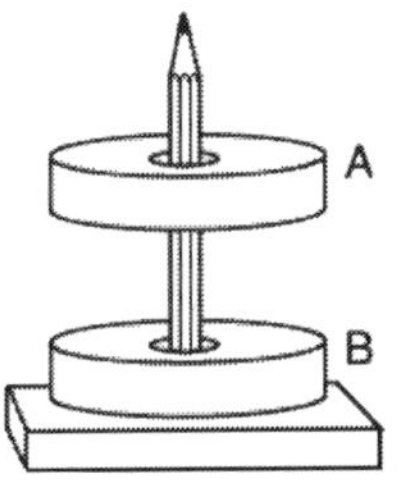

Which statement describes the gravitational force and the magnetic force acting on magnet A due to magnet B?
1. The gravitational force is attractive and the magnetic force is repulsive.
2. The gravitational force is ~~repulsive~~ and the magnetic force is attractive.
3. Both the gravitational force and the magnetic force are attractive.
4. Both the gravitational force and the magnetic force are ~~repulsive.~~

11. Moving a length of copper wire through a magnetic field may cause the wire to have a
1. potential difference across it
2. lower temperature
3. lower resistivity
4. higher resistance

12. The diagram below shows the magnetic field lines between two magnetic poles, A and B.

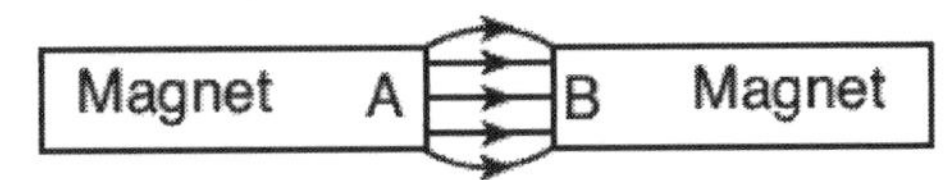

Which statement describes the polarity of magnetic poles A and B?
1. A is a north pole and B is a south pole.
2. A is a south pole and B is a north pole.
3. Both A and B are north poles.
4. Both A and B are south poles.

13. Magnetic fields are produced by particles that are
1. moving and charged
2. moving and neutral
3. stationary and charged
4. stationary and neutral

14. The diagram below represents a 0.5-kilogram bar magnet and a 0.7-kilogram bar magnet with a distance of 0.2 meter between their centers.

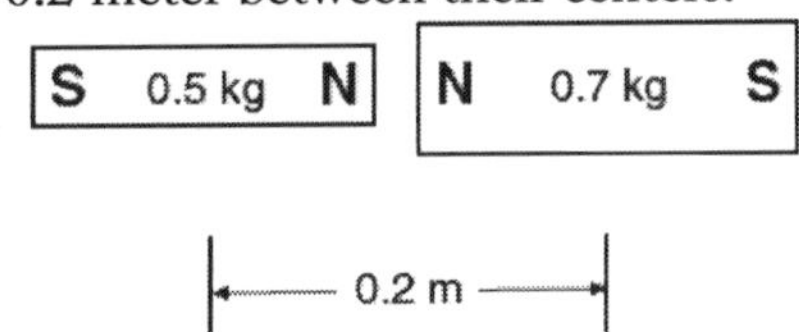

Which statement best describes the forces between the bar magnets?
1. Gravitational force and magnetic force are both repulsive
2. Gravitational force is repulsive and magnetic force is attractive.
3. Gravitational force is attractive and magnetic force is repulsive.
4. Gravitational force and magnetic force are both attractive.

15. Draw a diagram of a bar magnet, with a minimum of four field lines to show the magnitude and direction of the magnetic field in the region surrounding the bar magnet.

Name: ______________________________ Period: __________

Magnetism

16. A small object is dropped through a loop of wire connected to a sensitive ammeter on the edge of a table, as shown in the diagram below.

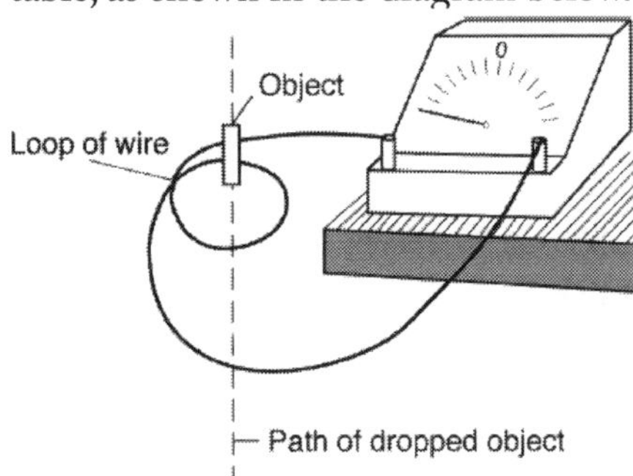

A reading on the ammeter is most likely produced when the object falling through the loop of wire is a
1. flashlight battery
2. bar magnet
3. brass mass
4. plastic ruler

17. Which particle would produce a magnetic field?
1. a neutral particle moving in a straight line
2. a neutral particle moving in a circle
3. a stationary charged particle
4. a moving charged particle

18. The diagram below shows the north pole of one bar magnet located near the south pole of another bar magnet. On the diagram, draw three magnetic field lines in the region between the magnets.

19. An electron moving at constant speed produces
1. a magnetic field, only
2. an electric field, only
3. both a magnetic and an electric field
4. neither a magnetic nor an electric field

Name: ____________________ Period: __________

Waves-Wave Basics

1. Which type of wave requires a material medium through which to travel?
   1. sound
   2. television
   3. radio
   4. x ray

2. A single vibratory disturbance moving through a medium is called
   1. a node
   2. an antinode
   3. a standing wave
   4. a pulse

3. The diagram below represents a transverse wave traveling to the right through a medium. Point A represents a particle of the medium.

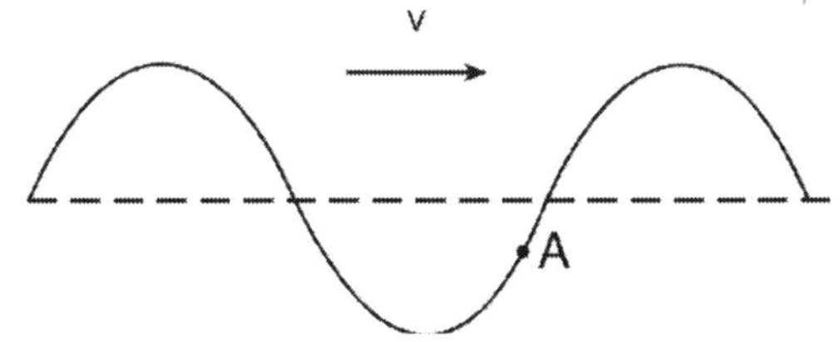

   In which direction will particle A move in the next instant of time?
   1. up
   2. down
   3. left
   4. right

4. As a transverse wave travels through a medium, the individual particles of the medium move
   1. perpendicular to the direction of wave travel
   2. parallel to the direction of wave travel
   3. in circles
   4. in ellipses

5. A periodic wave transfers
   1. energy, only
   2. mass, only
   3. both energy and mass
   4. neither energy nor mass

6. Which type of wave requires a material medium through which to travel?
   1. radio wave
   2. microwave
   3. light wave
   4. mechanical wave

7. A ringing bell is located in a chamber. When the air is removed from the chamber, why can the bell be seen vibrating but not be heard?
   1. Light waves can travel through a vacuum, but sound waves cannot.
   2. Sound waves have greater amplitude than light waves.
   3. Light waves travel slower than sound waves.
   4. Sound waves have higher frequencies than light waves.

8. Which statement correctly describes one characteristic of a sound wave?
   1. A sound wave can travel through a vacuum
   2. A sound wave is a transverse wave
   3. The amount of energy a sound wave transmits is directly related to the wave's amplitude.
   4. The amount of energy a sound wave transmits is inversely related to the wave's frequency

9. A television remote control is used to direct pulses of electromagnetic radiation to a receiver on a television. This communication from the remote control to the television illustrates that electromagnetic radiation
   1. is a longitudinal wave
   2. possesses energy inversely proportional to its frequency
   3. diffracts and accelerates in air
   4. transfers energy without transferring mass

10. The diagram below represents a transverse water wave propagating toward the left. A cork is floating on the water's surface at point P.

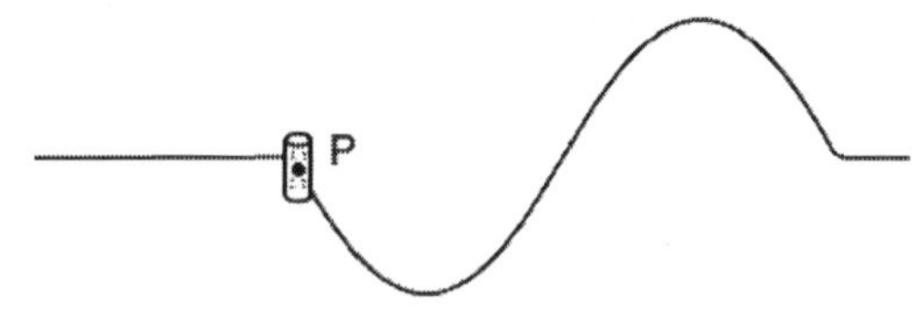

   In which direction will the cork move as the wave passes point P?
   1. up, then down, then up
   2. down, then up, then down
   3. left, then right, then left
   4. right, then left, then right

Name: ____________________ Period: __________

Waves-Wave Basics

11. A pulse traveled the length of a stretched spring. The pulse transferred
    1. energy, only
    2. mass, only
    3. both energy and mass
    4. neither energy nor mass

12. A transverse wave passes through a uniform material medium from left to right, as shown in the diagram below.

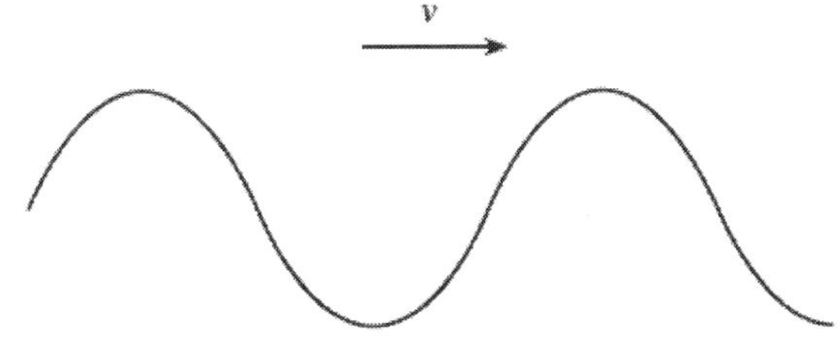

Which diagram best represents the direction of vibration of the particles of the medium?

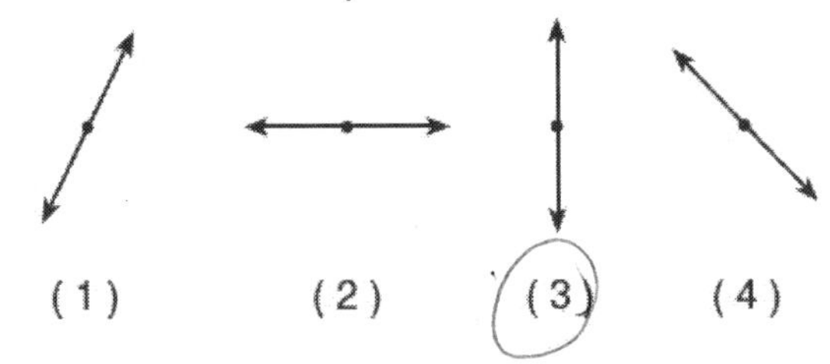

13. A tuning fork vibrating in air produces sound waves. These waves are best classified as
    1. transverse, because the air molecules are vibrating parallel to the direction of wave motion
    2. transverse, because the air molecules are vibrating perpendicular to the direction of wave motion
    3. longitudinal, because the air molecules are vibrating parallel to the direction of wave motion
    4. longitudinal, because the air molecules are vibrating perpendicular to the direction of wave motion

14. Which form(s) of energy can be transmitted through a vacuum?
    1. light, only
    2. sound, only
    3. both light and sound
    4. neither light nor sound

15. How are electromagnetic waves that are produced by oscillating charges and sound waves that are produced by oscillating tuning forks similar?
    1. Both have the same frequency as their respective sources.
    2. Both require a matter medium for propagation.
    3. Both are longitudinal waves.
    4. Both are transverse waves.

16. A student strikes the top rope of a volleyball net, sending a single vibratory disturbance along the length of the net, as shown in the diagram below.

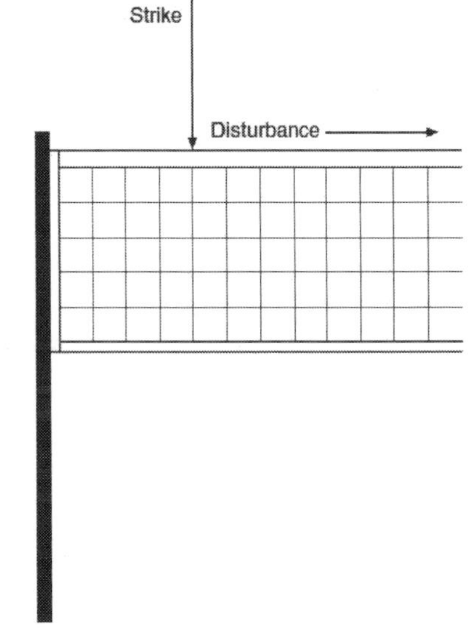

This disturbance is best described as
    1. a pulse
    2. a periodic wave
    3. a longitudinal wave
    4. an electromagnetic wave

17. Which diagram below does not represent a periodic wave?

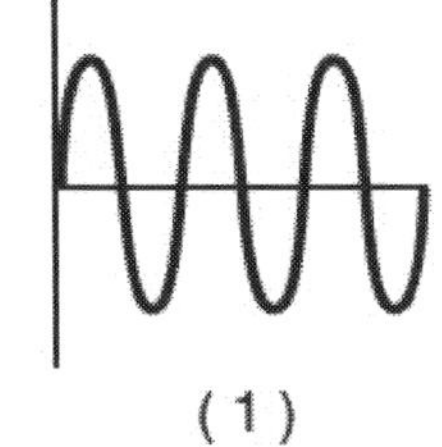
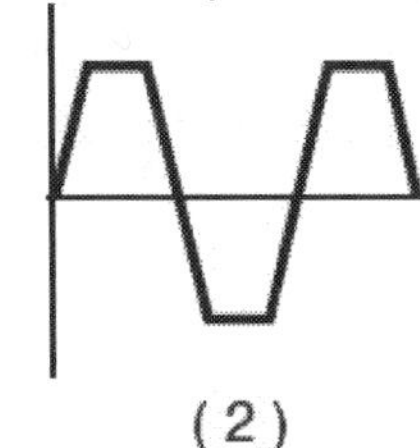
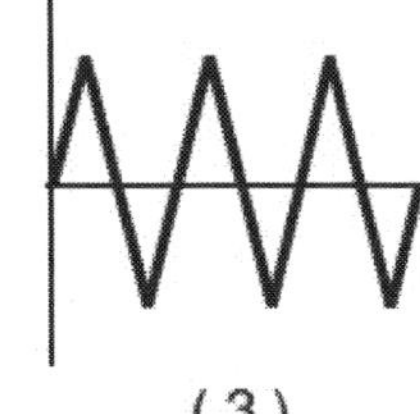
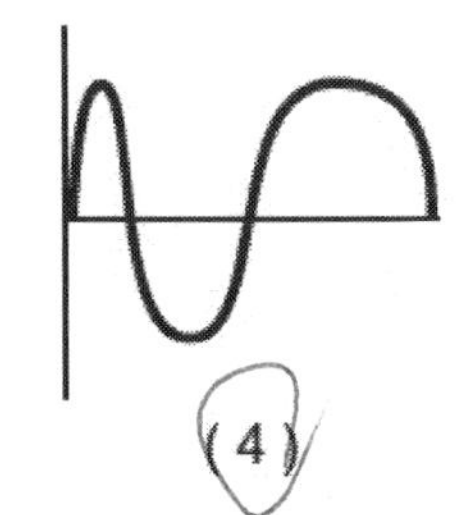

(1) (2) (3) (4)

Name:______________________________ Period: __________

Waves-Wave Basics

18. An earthquake wave is traveling from west to east through rock. If the particles of the rock are vibrating in a north-south direction, the wave must be classified as
    1. transverse
    2. longitudinal
    3. a microwave
    4. a radio wave

19. As a transverse wave travels through a medium, the individual particles of the medium move
    1. perpendicular to the direction of wave travel
    2. parallel to the direction of wave travel
    3. in circles
    4. in ellipses

20. A tuning fork oscillates with a frequency of 256 hertz after being struck by a rubber hammer. Which phrase best describes the sound waves produced by this oscillating tuning fork?
    1. electromagnetic waves that require no medium for transmission
    2. electromagnetic waves that require a medium for transmission
    3. mechanical waves that require no medium for transmission
    4. mechanical waves that require a medium for transmission

A longitudinal wave moves to the right through a uniform medium, as shown below. Points A, B, C, D, and E represent the positions of particles of the medium.

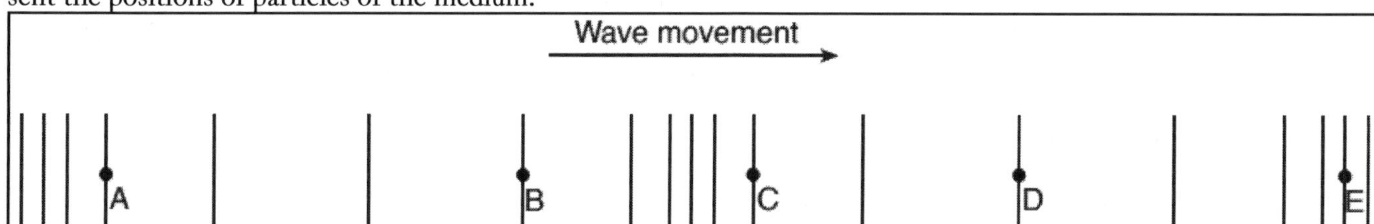

21. Which diagram best represents the motion of the particle at position C as the wave moves to the right?

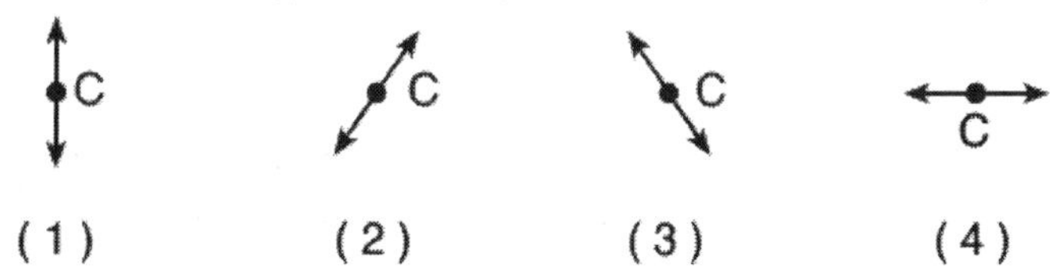

22. The wavelength of this wave is equal to the distance between points
    1. A and B
    2. A and C
    3. B and C
    4. B and E

23. The energy of this wave is related to its
    1. amplitude
    2. period
    3. speed
    4. wavelength

Name: ______________________________ Period: ____________

Waves-Wave Basics

Base your answers to questions 24 and 25 on the diagram at right, which shows a wave in a rope.

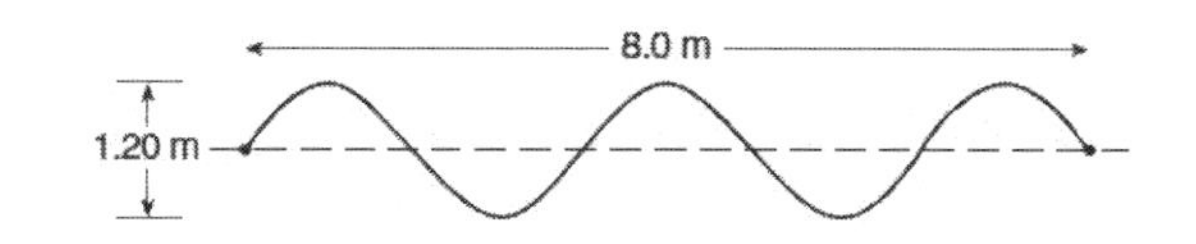

24. Determine the wavelength of the wave.

25. Determine the amplitude of the wave.

26. The energy of a sound wave is most closely related to the wave's
    1. frequency
    2. amplitude
    3. wavelength
    4. speed

27. Which statement describes a characteristic common to all electromagnetic waves and mechanical waves?
    1. Both types of waves travel at the same speed.
    2. Both types of waves require a material medium for propagation.
    3. Both types of waves propagate in a vacuum.
    4. Both types of waves transfer energy.

28. The diagram below represents a periodic wave moving along a rope.

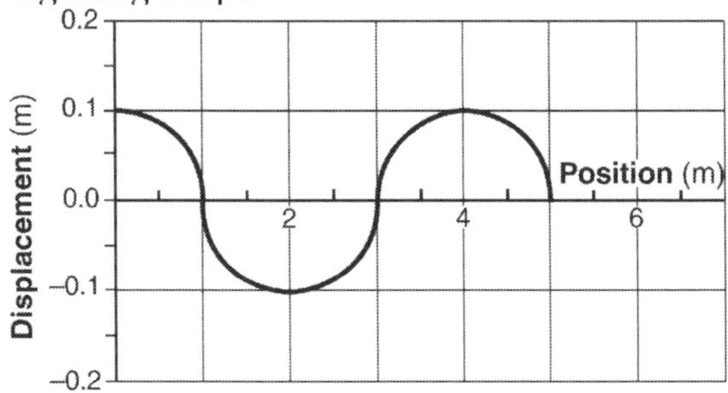

On the grid below, draw at least one full wave with the same amplitude and half the wavelength of the given wave.

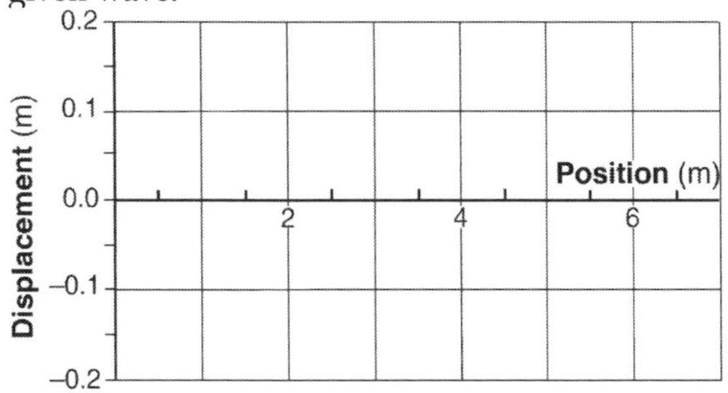

29. Transverse waves are to radio waves as longitudinal waves are to
    1. light waves
    2. microwaves
    3. ultraviolet waves
    4. sound waves

30. The diagram below shows a periodic wave.

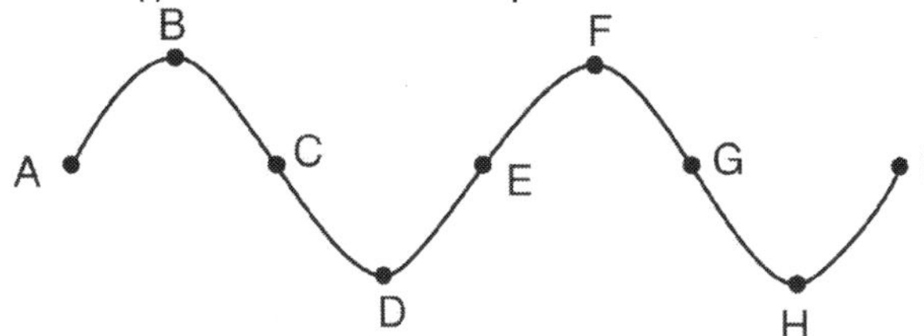

Which two points on the wave are 180° out of phase?
1. A and C
2. B and E
3. F and G
4. D and H

31. The diagram below shows a mechanical transverse wave traveling to the right in a medium. Point A represents a particle in the medium. Draw an arrow originating at point A to indicate the initial direction that the particle will move as the wave continues to travel to the right in the medium.

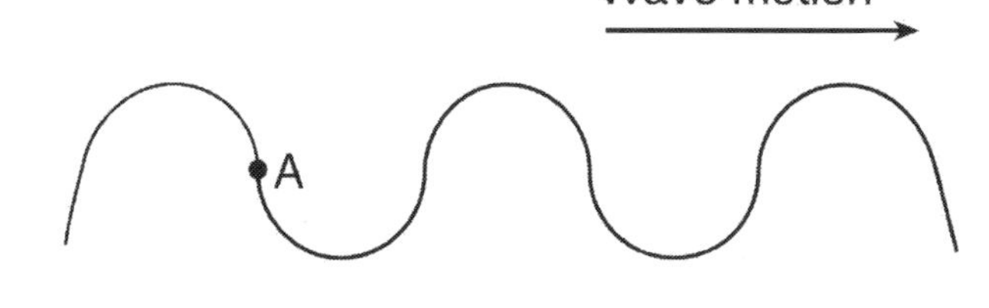

Name: ______________________ Period: __________

Waves-Wave Basics

32. The amplitude of a sound wave is most closely related to the sound's
    1. speed
    2. wavelength
    3. loudness
    4. pitch

33. As a longitudinal wave moves through a medium, the particles of the medium
    1. vibrate parallel to the direction of the wave's propagation
    2. vibrate perpendicular to the direction of the wave's propagation
    3. are transferred in the direction of the wave's motion, only
    4. are stationary

Name:______________________________ Period: __________

Waves-Wave Characteristics

1. What is the wavelength of a 256-hertz sound wave in air at STP?
   1. 1.17×10^6 m
   2. 1.29 m
   3. 0.773 m
   4. 8.53×10^{-7} m

2. The graph below represents the relationship between wavelength and frequency of waves created by two students shaking the ends of a loose spring.

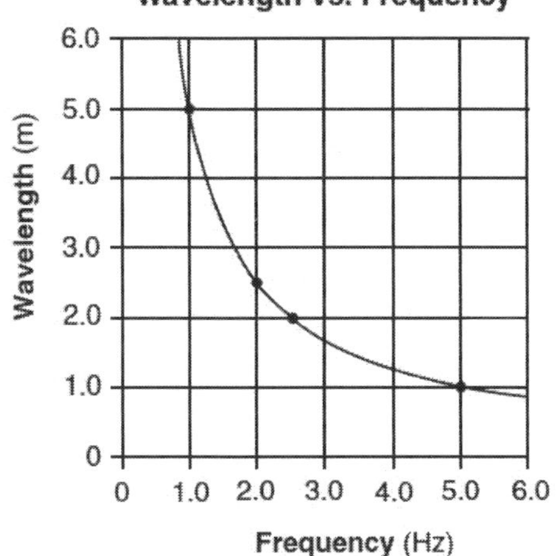

Calculate the speed of the waves generated in the spring. [Show all work, including the equation and substitution with units.]

3. What is the period of a water wave if 4 complete waves pass a fixed point in 10 seconds?
   1. 0.25 s
   2. 0.40 s
   3. 2.5 s
   4. 4.0 s

4. If the frequency of a periodic wave is doubled, the period of the wave will be
   1. halved
   2. doubled
   3. quartered
   4. quadrupled

5. A 512-hertz sound wave travels 100 meters to an observer through air at STP. What is the wavelength of this sound wave?

6. The diagram below represents a periodic wave.

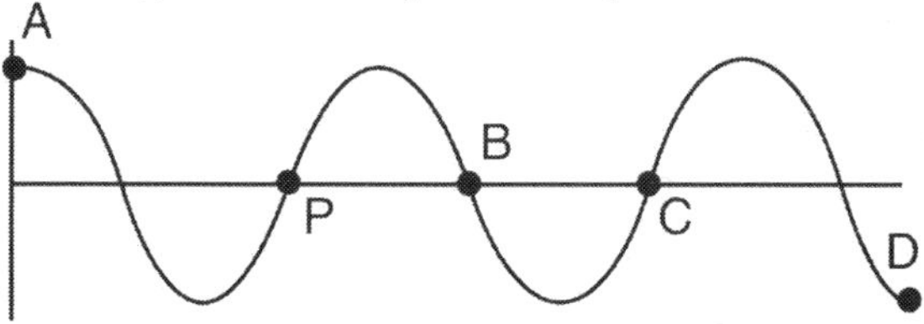

Which point on the wave is in phase with point P?
   1. A
   2. B
   3. C
   4. D

7. A periodic wave having a frequency of 5.0 hertz and a speed of 10 meters per second has a wavelength of
   1. 0.50 m
   2. 2.0 m
   3. 5.0 m
   4. 50 m

8. A ringing bell is located in a chamber. When the air is removed from the chamber, why can the bell be seen vibrating but not be heard?
   1. Light waves can travel through a vacuum, but sound waves cannot.
   2. Sound waves have greater amplitude than light waves.
   3. Light waves travel slower than sound waves.
   4. Sound waves have higher frequencies than light waves.

9. The diagram below represents a transverse wave.

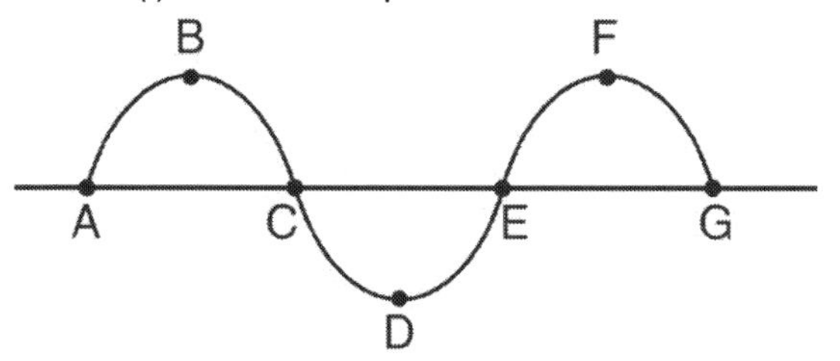

The wavelength of the wave is equal to the distance between points
   1. A and G
   2. B and F
   3. C and E
   4. D and F

Name:________________________________ Period: ____________

Waves-Wave Characteristics

10. The diagram below represents a periodic transverse wave traveling in a uniform medium.

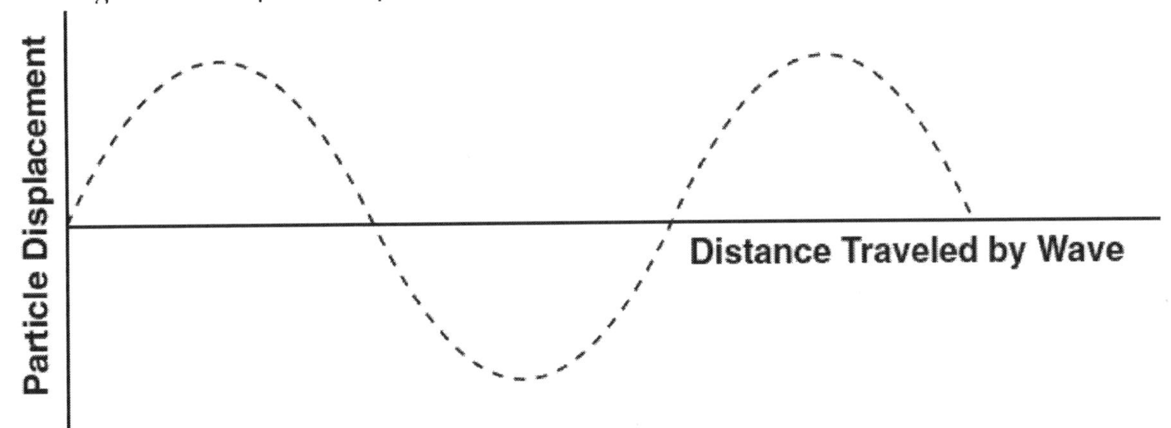

On the diagram above, draw a wave having both a smaller amplitude and the same wavelength as the given wave.

Base your answers to questions 11 through 13 on the information and diagram below.

A longitudinal wave moves to the right through a uniform medium, as shown below. Points A, B, C, D, and E represent the positions of particles of the medium.

Wave movement

11. Which diagram best represents the motion of the particle at position C as the wave moves to the right?

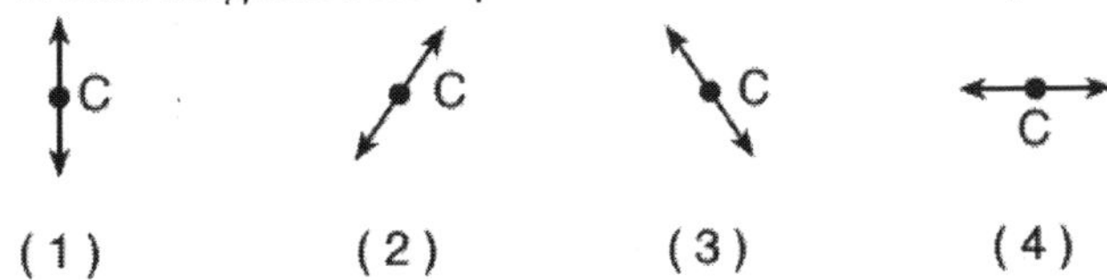

12. The wavelength of this wave is equal to the distance between points
    1. A and B
    2. B and C
    3. A and C
    4. B and E

13. The energy of this wave is related to its
    1. amplitude
    2. period
    3. speed
    4. wavelength

Name: ______________________ Period: ________

Waves-Wave Characteristics

14. Which wave diagram has *both* wavelength (λ) and amplitude (A) labeled correctly?

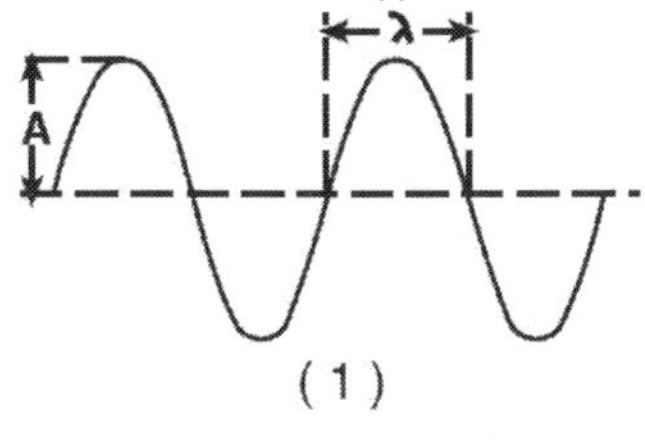

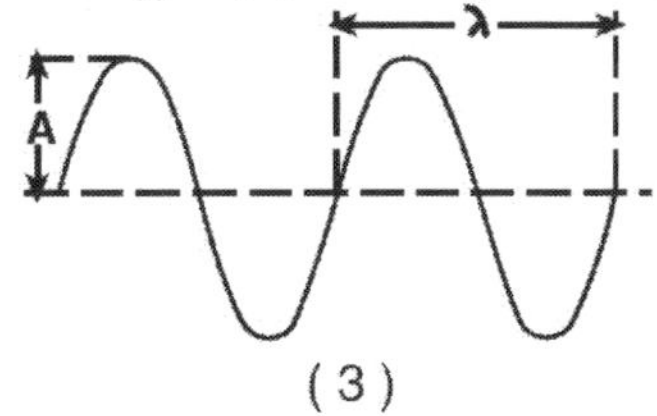

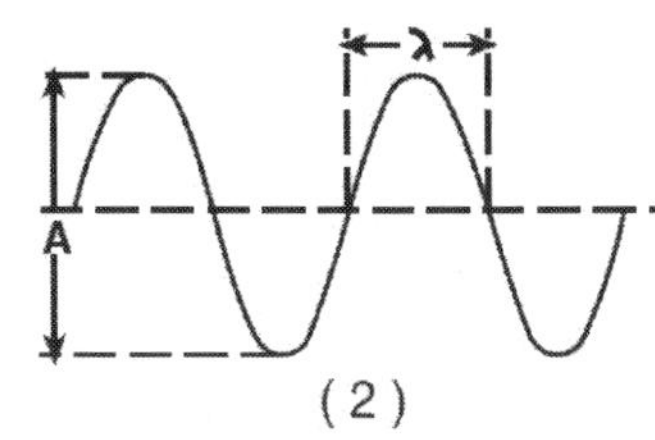

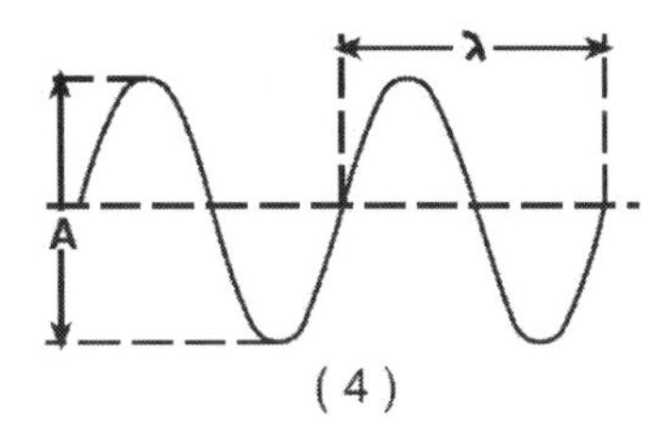

15. The diagram below represents a transverse wave moving on a uniform rope with point A labeled as shown. On the diagram, mark an X at the point on the wave that is 180° out of phase with point A.

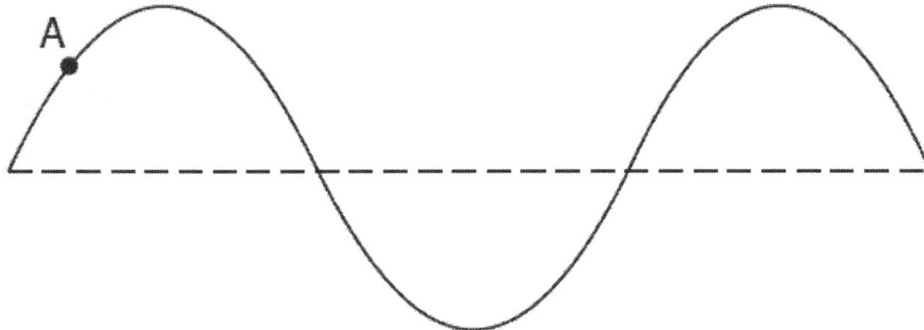

Base your answers to questions 16 through 18 on the information and diagram below.

Three waves, A, B, and C, travel 12 meters in 2.0 seconds through the same medium as shown in the diagram below.

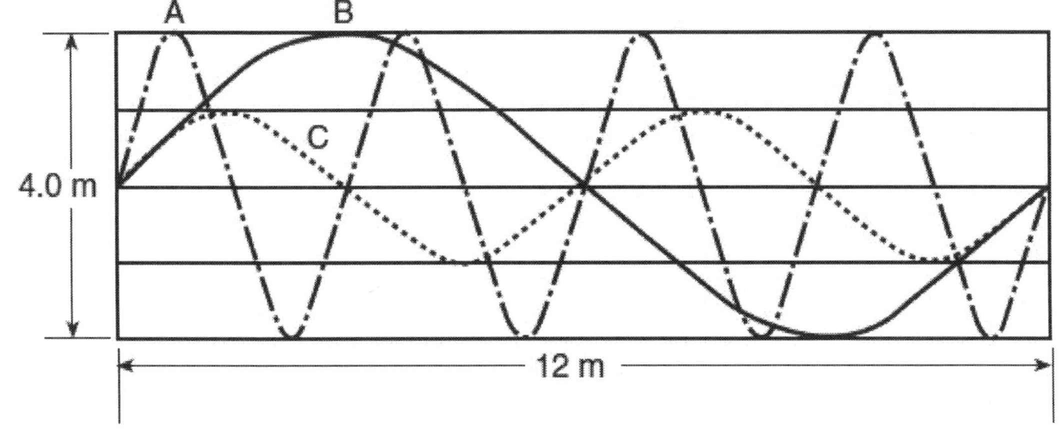

16. What is the amplitude of wave C?

17. What is the period of wave A?

18. What is the speed of wave B?

Name:______________________________ Period: ____________

Waves-Wave Characteristics

Base your answers to questions 19 through 21 on the information below. [Show all work, including the equation and substitution with units.]

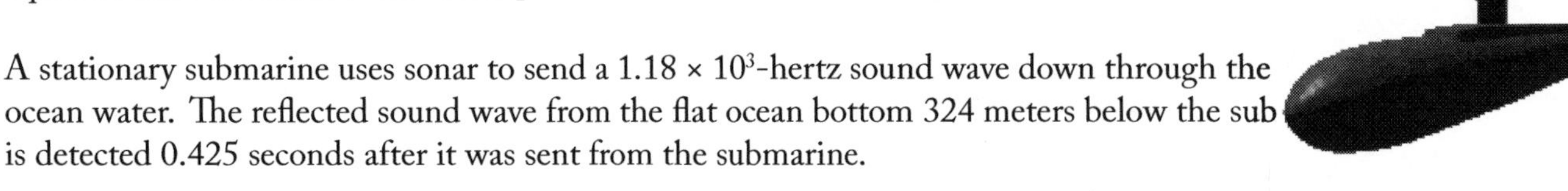

A stationary submarine uses sonar to send a 1.18×10^3-hertz sound wave down through the ocean water. The reflected sound wave from the flat ocean bottom 324 meters below the sub is detected 0.425 seconds after it was sent from the submarine.

19. Calculate the speed of the sound wave in the ocean water.

20. Calculate the wavelength of the sound wave in the ocean water.

21. Determine the period of the sound wave in the ocean water.

---

22. A motor is used to produce 4.0 waves each second in a string. What is the frequency of the waves?
    1. 0.25 Hz
    2. 15 Hz
    3. 25 Hz
    4. 4.0 Hz

23. If the amplitude of a wave is increased, the frequency of the wave will
    1. decrease
    2. increase
    3. remain the same

24. The time required for a wave to complete one full cycle is called the wave's
    1. frequency
    2. period
    3. velocity
    4. wavelength

25. The graph below represents the displacement of a particle in a medium over a period of time.

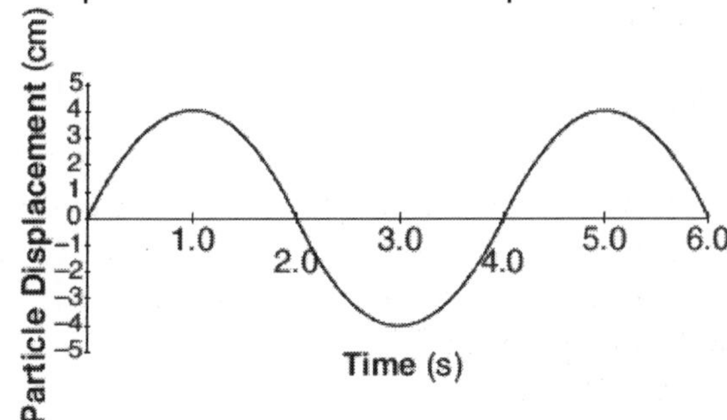

The amplitude of the wave is
1. 4.0 s
2. 6.0 s
3. 8 cm
4. 4 cm

26. A periodic wave is produced by a vibrating tuning fork. The amplitude of the wave would be greater if the tuning fork were
    1. struck more softly
    2. struck harder
    3. replaced by a lower frequency tuning fork
    4. replaced by a higher frequency tuning fork

27. The diagram below represents a periodic wave.

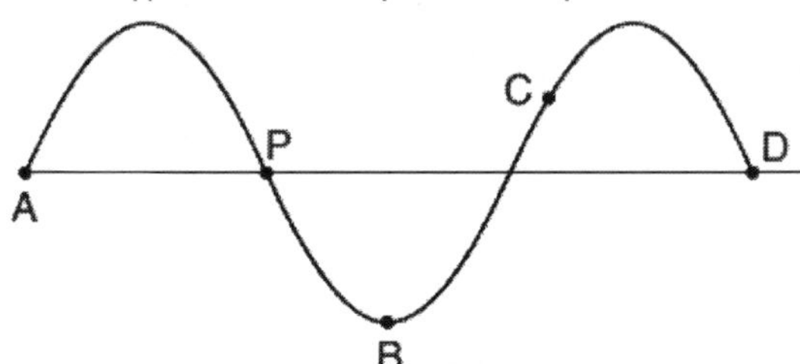

Which point on the wave is 90° out of phase with point P?
1. A
2. B
3. C
4. D

Name: ______________________ Period: ________

Waves-Wave Characteristics

28. A periodic wave travels at speed v through medium A. The wave passes with all its energy into medium B. The speed of the wave through medium B is v/2. On the diagram below, draw the wave as it travels through medium B. Show at least one full wave.

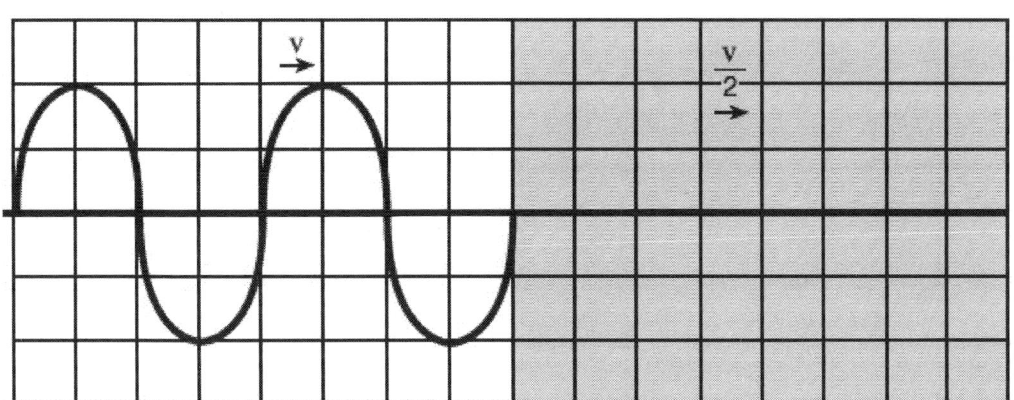

29. The energy of a water wave is closely related to its
    1. frequency
    2. wavelength
    3. period
    4. amplitude

30. The diagram below represents a transverse wave.

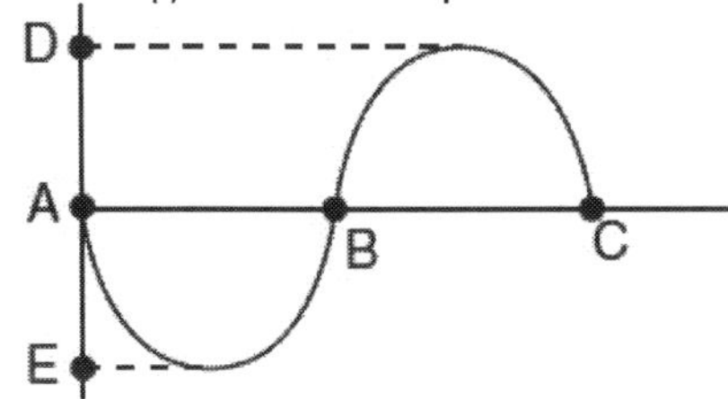

The distance between which two points identifies the amplitude of the wave?
    1. A and B
    2. A and C
    3. A and E
    4. D and E

31. If the amplitude of a wave traveling in a rope is doubled, the speed of the wave in the rope will
    1. decrease
    2. increase
    3. remain the same

32. Increasing the amplitude of a sound wave produces a sound with
    1. lower speed
    2. higher pitch
    3. shorter wavelength
    4. greater loudness

33. The energy of a sound wave is closely related to its
    1. period
    2. amplitude
    3. frequency
    4. wavelength

34. The diagram below shows a periodic wave.

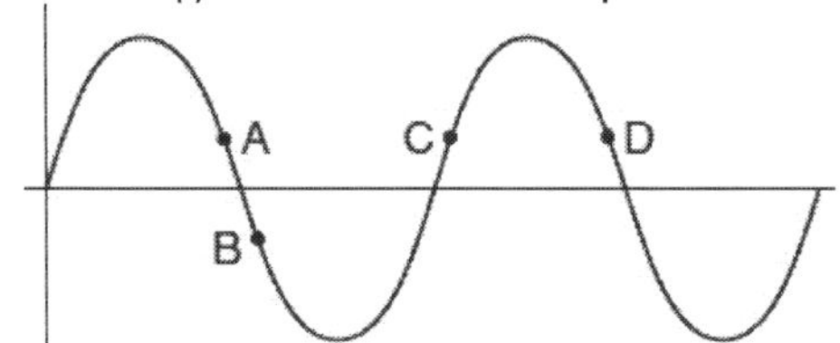

Which points are in phase with each other?
    1. A and C
    2. A and D
    3. B and C
    4. C and D

35. Which unit is equivalent to meters per second?
    1. Hz·s
    2. Hz·m
    3. s/Hz
    4. m/Hz

36. The sound wave produced by a trumpet has a frequency of 440 hertz. What is the distance between successive compressions in this sound wave as it travels through air at STP?
    1. 1.5×10^{-6} m
    2. 0.75 m
    3. 1.3 m
    4. 6.8×10^{5} m

Name: ________________________________ Period: ____________

Waves-Wave Characteristics

Base your answers to questions 37 and 38 on the information and diagram below.

A student standing on a dock observes a piece of wood floating on the water as shown below. As a water wave passes, the wood moves up and down, rising to the top of a wave crest every 5.0 seconds.

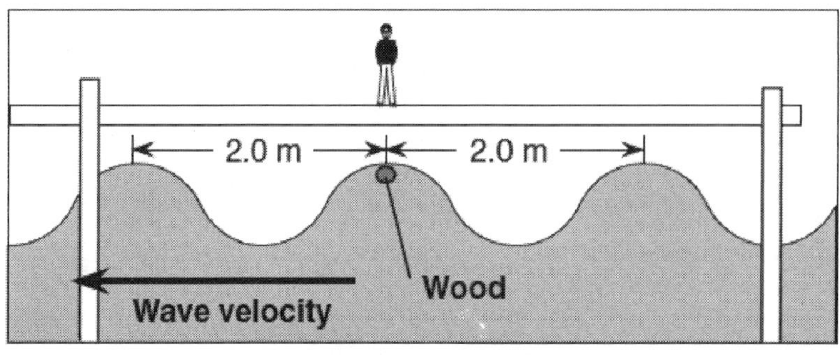

(Not drawn to scale)

37. Calculate the frequency of the passing water waves. [Show all work, including the equation and substitution with units.]

38. Calculate the speed of the water waves. [Show all work, including the equation and substitution with units.]

39. The diagram below represents a periodic wave traveling through a uniform medium.

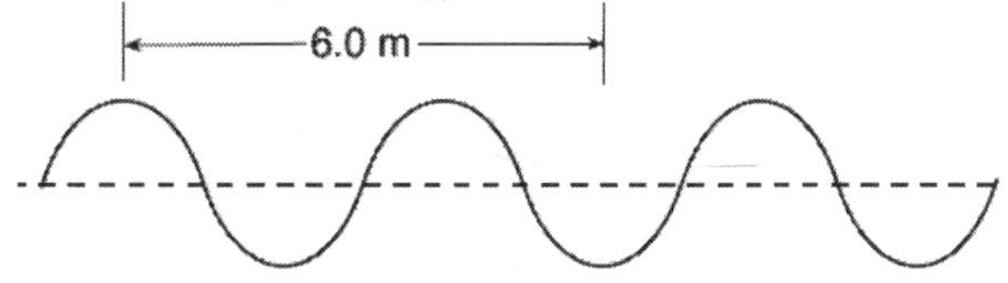

If the frequency of the wave is 2.0 hertz, the speed of the wave is
1. 6.0 m/s
2. 2.0 m/s
3. 8.0 m/s
4. 4.0 m/s

40. Two waves having the same frequency and amplitude are traveling in the same medium. Maximum constructive interference occurs at points where the phase difference between the two superimposed waves is
1. 0°
2. 90°
3. 180°
4. 270°

41. A surfacing whale in an aquarium produces water wave crests every 0.40 seconds. If the water wave travels at 4.5 meters per second, the wavelength of the wave is
1. 1.8 m
2. 2.4 m
3. 3.0 m
4. 11 m

42. The diagram below represents a transverse wave moving along a string.

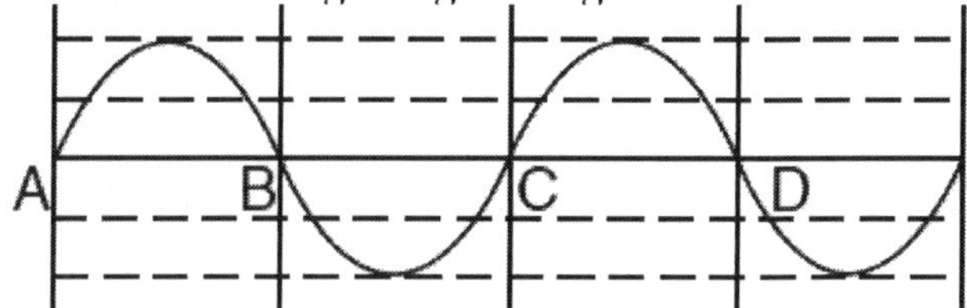

On the diagram above, draw a transverse wave that would produce complete destructive interference when superimposed with the original wave.

43. The product of a wave's frequency and its period is
1. one
2. its velocity
3. its wavelength
4. Planck's constant

Name: ________________________________ Period: ____________

Waves-Wave Characteristics

Base your answers to questions 44 and 45 on the information below.

A student plucks a guitar string and the vibrations produce a sound wave with a frequency of 650 hertz.

44. The sound wave produced can best be described as a
    1. transverse wave of constant amplitude
    2. longitudinal wave of constant frequency
    3. mechanical wave of varying frequency
    4. electromagnetic wave of varying wavelengths

45. Calculate the wavelength of the sound wave in air at STP. [Show all work, including the equation and substitution with units.]

Base your answers to questions 46 and 47 on the information below.

A transverse wave with an amplitude of 0.20 meters and wavelength of 3.0 meters travels toward the right in a medium with a speed of 4.0 meters per second.

46. On the diagram below, place an X at each of *two* points that are in phase with each other.

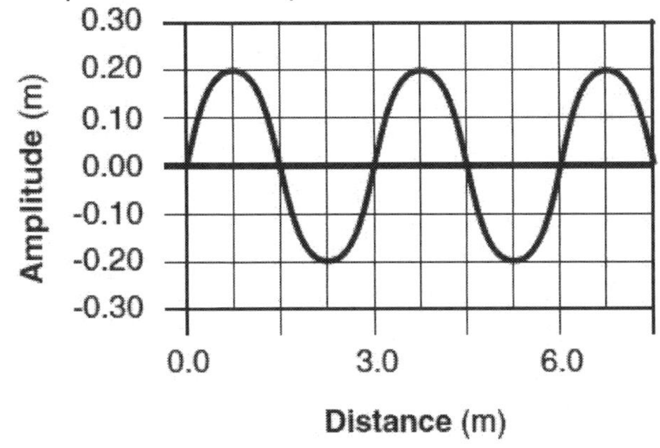

47. Calculate the period of the wave. [Show all work, including the equation and substitution with units.]

48. A tuning fork vibrates at a frequency of 512 hertz when struck with a rubber hammer. The sound produced by the tuning fork will travel through the air as a
    1. longitudinal wave with air molecules vibrating parallel to the direction of travel
    2. transverse wave with air molecules vibrating parallel to the direction of travel
    3. longitudinal wave with air molecules vibrating perpendicular to the direction of travel
    4. transverse wave with air molecules vibrating perpendicular to the direction of travel

49. What is characteristic of both sound waves and electromagnetic waves?
    1. They require a medium.
    2. They transfer energy.
    3. They are mechanical waves.
    4. They are longitudinal waves.

50. What is the wavelength of a 2.50-kilohertz sound wave traveling at 326 meters per second through air?
    1. 0.130 m
    2. 1.30 m
    3. 7.67 m
    4. 130 m

51. While sitting in a boat, a fisherman observes that two complete waves pass by his position every 4 seconds. What is the period of these waves?
    1. 0.5 s
    2. 2 s
    3. 8 s
    4. 4 s

52. A sound wave traveling eastward through air causes the air molecules to
    1. vibrate east and west
    2. vibrate north and south
    3. move eastward, only
    4. move northward, only

Name: ______________________________ Period: __________

Waves-Wave Characteristics

53. The diagram below represents a periodic wave.

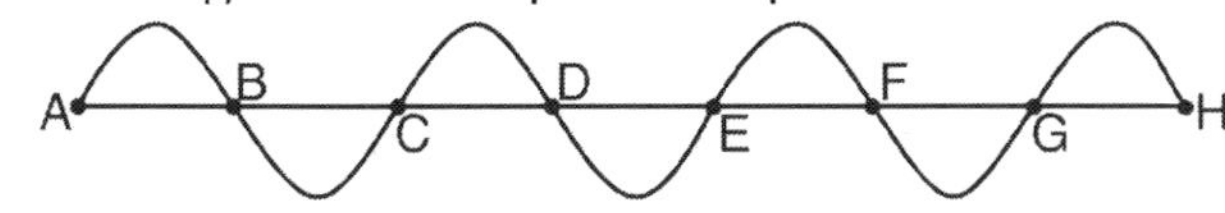

Which two points on the wave are out of phase?
1. A and C
2. B and F
3. C and E
4. D and G

54. A distance of 1.0×10^{-2} meter separates successive crests of a periodic wave produced in a shallow tank of water. If a crest passes a point in the tank every 4.0×10^{-1} second, what is the speed of this wave?
1. 2.5×10^{-4} m/s
2. 4.0×10^{-3} m/s
3. 2.5×10^{-2} m/s
4. 4.0×10^{-1} m/s

55. A nurse takes the pulse of a heart and determines the heart beats periodically 60 times in 60 seconds. The period of the heartbeat is
1. 1 Hz

2. 60 Hz
3. 1 s
4. 60 s

56. What is the period of a sound wave having a frequency of 340 hertz?
1. 3.40×10^{2} s
2. 1.02×10^{0} s
3. 9.73×10^{-1} s
4. 2.94×10^{-3} s

57. A beam of light has a wavelength of 4.5×10^{-7} meter in a vacuum. The frequency of this light is
1. 1.5×10^{-15} Hz
2. 4.5×10^{-7} Hz
3. 1.4×10^{2} Hz
4. 6.7×10^{14} Hz

58. The diagram below represents two waves, A and B, traveling through the same uniform medium.

Which characteristic is the same for both waves?
1. amplitude
2. frequency
3. period
4. wavelength

59. A duck floating in a pool oscillates up and down 5.0 times during a 10.-second interval as a periodic wave passes by. What is the frequency of the duck's oscillations?

1. 0.10 Hz
2. 0.50 Hz
3. 2.0 Hz
4. 50. Hz

60. A student produces a wave in a long spring by vibrating its end. As the frequency of the vibration is doubled, the wavelength in the spring is
1. quartered
2. halved
3. unchanged
4. doubled

61. The diagram below shows waves A and B in the same medium.

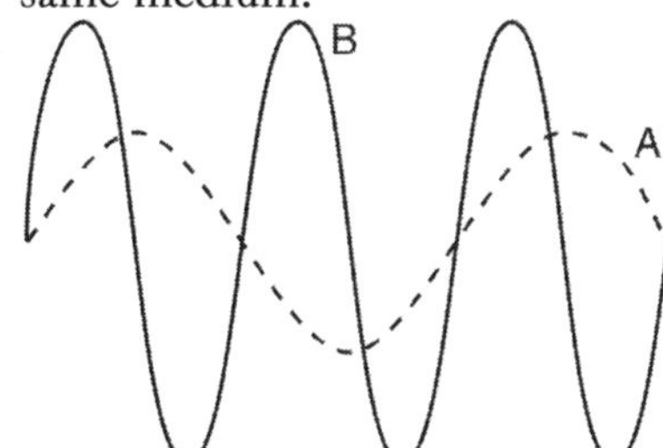

Compared to wave A, wave B has
1. twice the amplitude and twice the wavelength
2. twice the amplitude and half the wavelength
3. the same amplitude and half the wavelength
4. half the amplitude and the same wavelength

Name: ______________________ Period: ____________

Waves-Wave Characteristics

62. Which two points on the wave shown in the diagram below are in phase with each other?

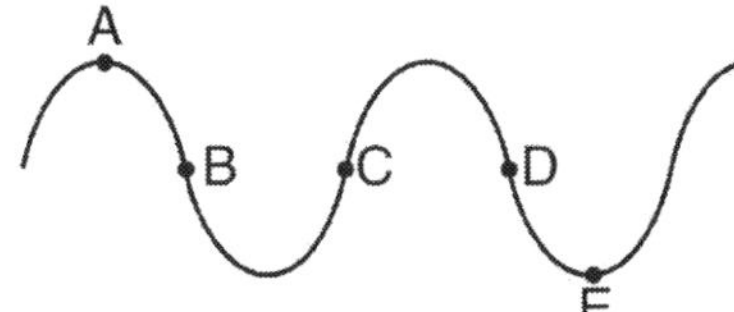

1. A and B
2. A and E
3. B and C
4. B and D

63. A microwave oven emits a microwave with a wavelength of 2.00×10^{-2} meter in air. Calculate the frequency of the microwave. [Show all work, including the equation and substitution with units.]

Name: ______________________ Period: __________

Waves-Wave Behaviors

1. While playing, two children create a standing wave in a rope, as shown in the diagram below. A third child participates by jumping the rope.

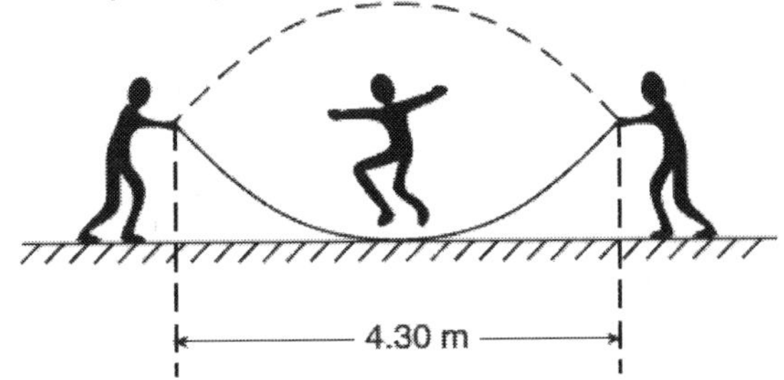

What is the wavelength of this standing wave?
   1. 2.15 m
   2. 4.30 m
   3. 6.45 m
   4. 8.60 m

2. The diagram below shows two pulses approaching each other in a uniform medium.

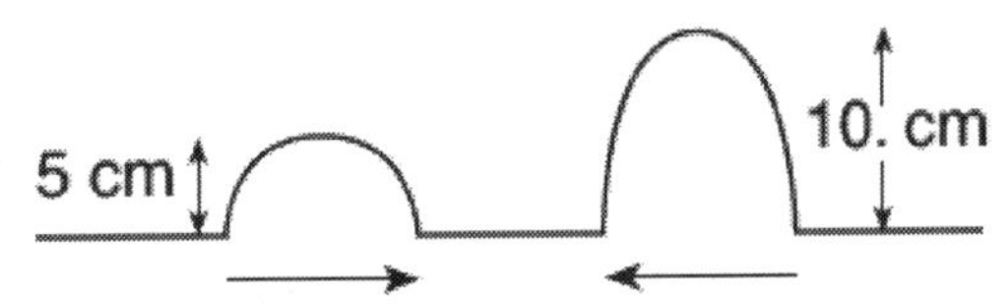

Which diagram best represents the superposition of the two pulses?

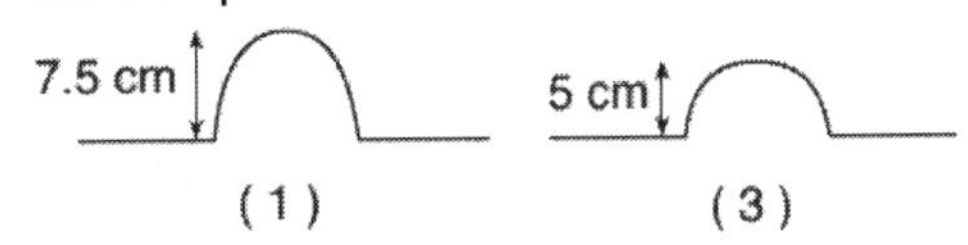

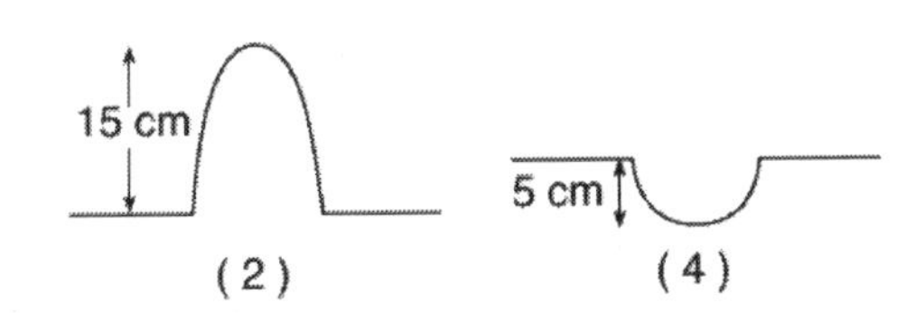

3. Sound waves strike a glass and cause it to shatter. This phenomenon illustrates
   1. resonance
   2. refraction
   3. reflection
   4. diffraction

4. A sound of constant frequency is produced by the siren on top of a firehouse. Compared to the frequency produced by the siren, the frequency observed by a firefighter approaching the firehouse is
   1. lower
   2. higher
   3. the same

5. The superposition of two waves traveling in the same medium produces a standing wave pattern if the two waves have
   1. the same frequency, the same amplitude, and travel in the same direction
   2. the same frequency, the same amplitude, and travel in opposite directions
   3. the same frequency, different amplitudes, and travel in the same direction
   4. the same frequency, different amplitudes, and travel in opposite directions

6. The diagram below represents a standing wave.

The number of nodes and antinodes shown in the diagram is
   1. 4 nodes and 5 antinodes
   2. 5 nodes and 6 antinodes
   3. 6 nodes and 5 antinodes
   4. 6 nodes and 10 antinodes

7. A car's horn is producing a sound wave having a constant frequency of 350 hertz. If the car moves toward a stationary observer at constant speed, the frequency of the car's horn detected by this observer may be
   1. 320 Hz
   2. 330 Hz
   3. 350 Hz
   4. 380 Hz

8. Standing waves in water are produced most often by periodic water waves
   1. being absorbed at the boundary with a new medium
   2. refracting at a boundary with a new medium
   3. diffracting around a barrier
   4. reflecting from a barrier

Name: ______________________________ Period: __________

Waves-Wave Behaviors

9. Two pulses, A and B, travel toward each other along the same rope, as shown below.

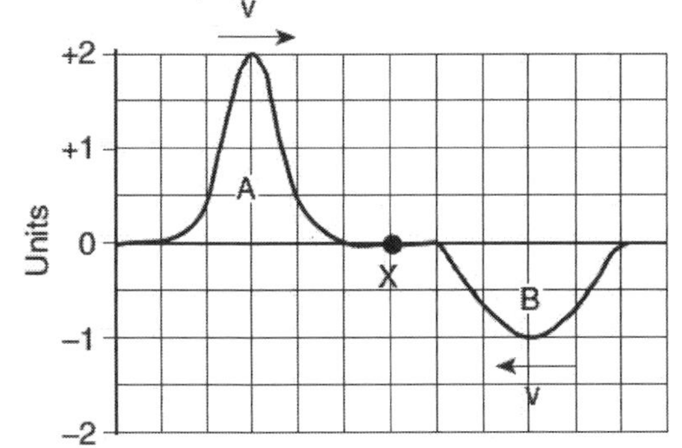

When the centers of the two pulses meet at point X, the amplitude at the center of the resultant pulse will be

1. +1 unit
2. +2 units
3. 0
4. -1 unit

10. A car's horn produces a sound wave of constant frequency. As the car speeds up going away from a stationary spectator, the sound wave deteced by the spectator

1. decreases in amplitude and decreases in frequency
2. decreases in amplitude and increases in frequency
3. increases in amplitude and decreases in frequency
4. increases in amplitude and increases in frequency

11. Playing a certain musical note on a trumpet causes the spring on the bottom of a nearby snare drum to vibrate. This phenomenon is an example of

1. resonance
2. refraction
3. reflection
4. diffraction

A system consists of an oscillator and a speaker that emits a 1000-hertz sound wave. A microphone detects the sound wave 1.00 meter from the speaker.

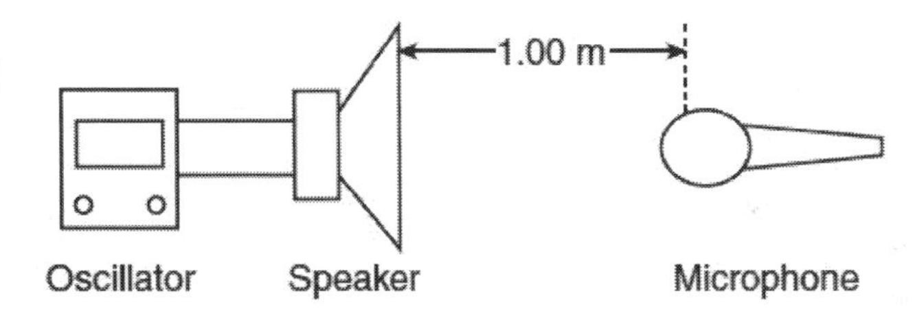

12. Which type of wave is emitted by the speaker?

1. transverse
2. longitudinal
3. circular
4. electromagnetic

13. The microphone is moved to a new fixed location 0.50 meter in front of the speaker. Compared to the sound waves detected at the 1.00-meter position, the sound waves detected at the 0.50-meter position have a different

1. wave speed
2. frequency
3. wavelength
4. amplitude

14. The microphone is moved at constant speed from the 0.50-meter position back to its original position 1.00 meter from the speaker. Compared to the 1000-hertz frequency emitted by the speaker, the frequency detected by the moving microphone is

1. lower
2. higher
3. the same

Name: ______________________ Period: __________

Waves-Wave Behaviors

15. Two pulses traveling in the same uniform medium approach each other, as shown in the diagram below.

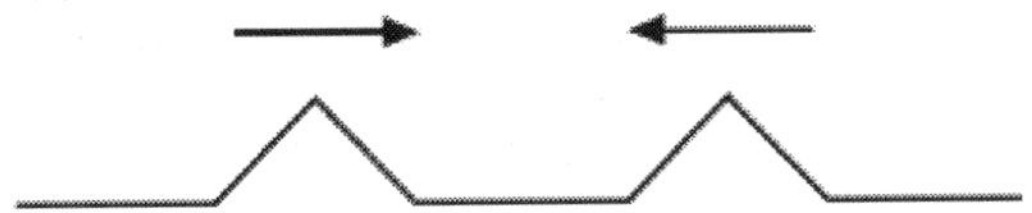

Which diagram best represents the superposition of the two pulses?

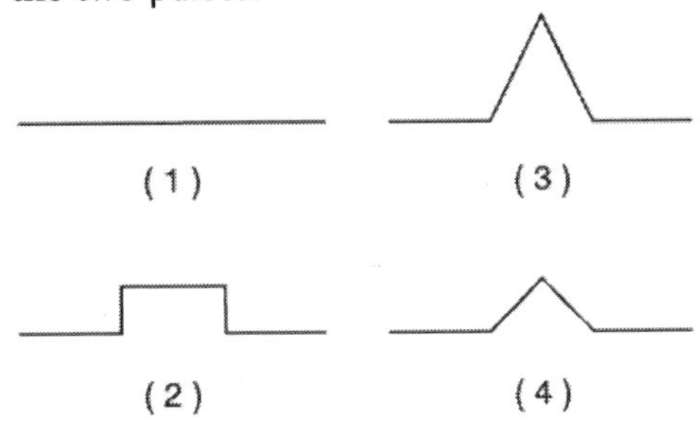

16. The diagram below shows a standing wave.

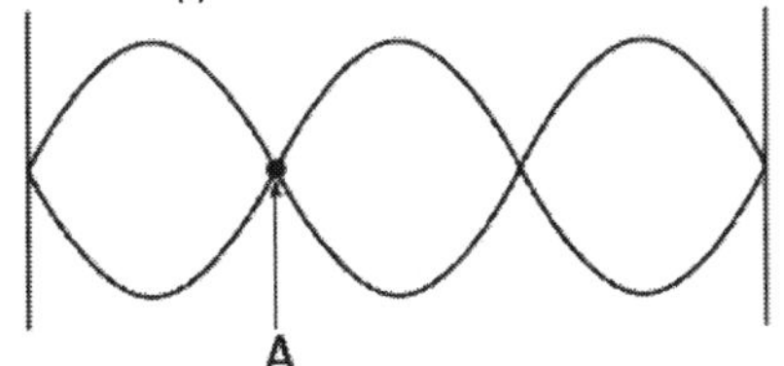

Point A on the standing wave is
1. a node resulting from constructive interference
2. a node resulting from destructive interference
3. an antinode resulting from constructive interference
4. an antinode resulting from destructive interference

17. A source of waves and an observer are moving relative to each other. The observer will detect a steadily increasing frequency if
1. he moves toward the source at a constant speed
2. the source moves away from him at a constant speed
3. he accelerates toward the source
4. the source accelerates away from him

18. The diagram below shows two pulses traveling toward each other in a uniform medium.

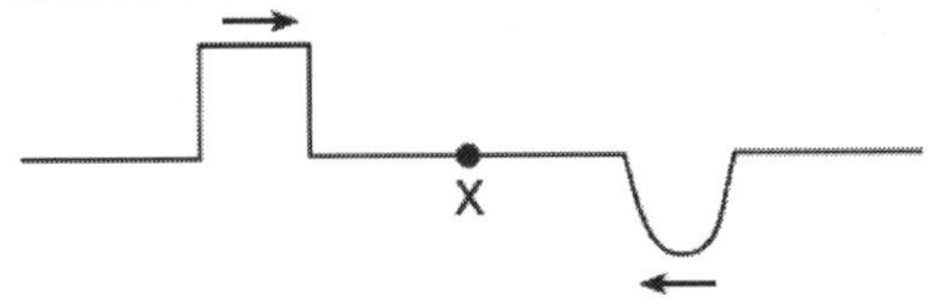

Which diagram best represents the medium when the pulses meet at point X?

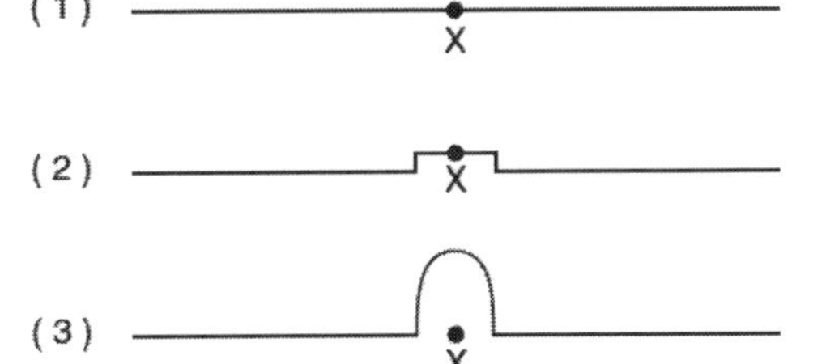

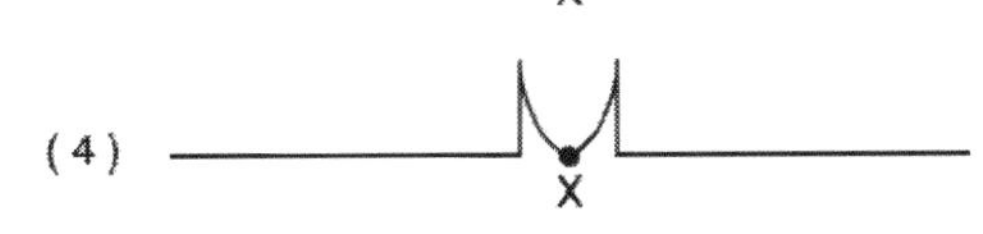

19. A dampened fingertip rubbed around the rim of a crystal stemware glass causes the glass to vibrate and produce a musical note. This effect is due to
1. resonance
2. refraction
3. reflection
4. rarefaction

20. The diagram below shows a standing wave in a string clamped at each end.

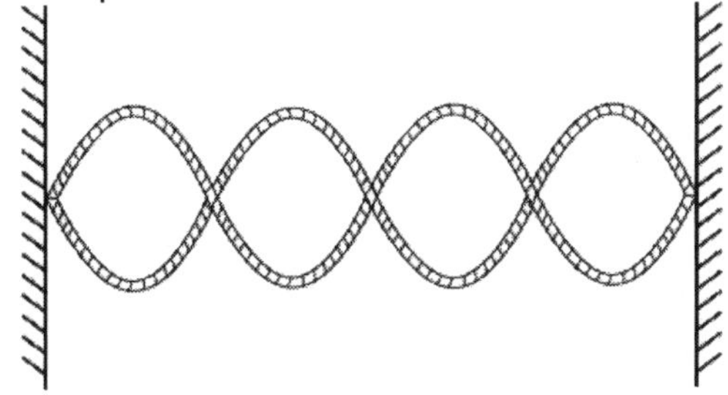

What is the total number of nodes and antinodes in the standing wave?
1. 3 nodes and 2 antinodes
2. 2 nodes and 3 antinodes
3. 5 nodes and 4 antinodes
4. 4 nodes and 5 antinodes

Name: ______________________________ Period: ____________

Waves-Wave Behaviors

21. A radar gun can determine the speed of a moving automobile by measuring the difference in frequency between emitted and reflected radar waves. This process illustrates
    1. resonance
    2. the Doppler effect
    3. diffraction
    4. refraction

22. A 256-hertz vibrating tuning fork is brought near a nonvibrating 256-hertz tuning fork. The second tuning fork begins to vibrate. Which phenomenon causes the nonvibrating tuning fork to begin to vibrate?
    1. resistance
    2. resonance
    3. refraction
    4. reflection

23. The diagram below represents two pulses approaching each other from opposite directions in the same medium.

Which diagram best represents the medium after the pulses have passed through each other?

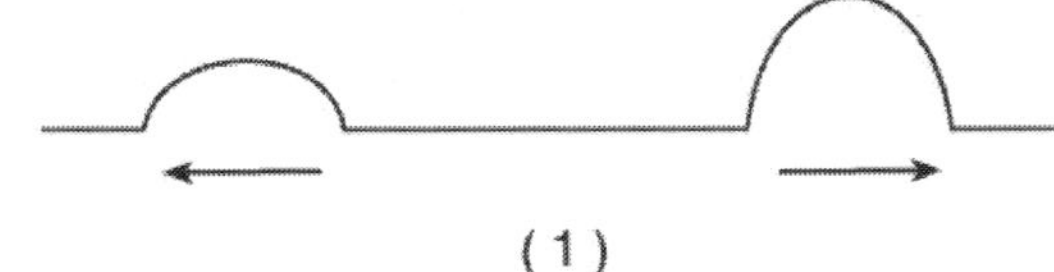

(1)

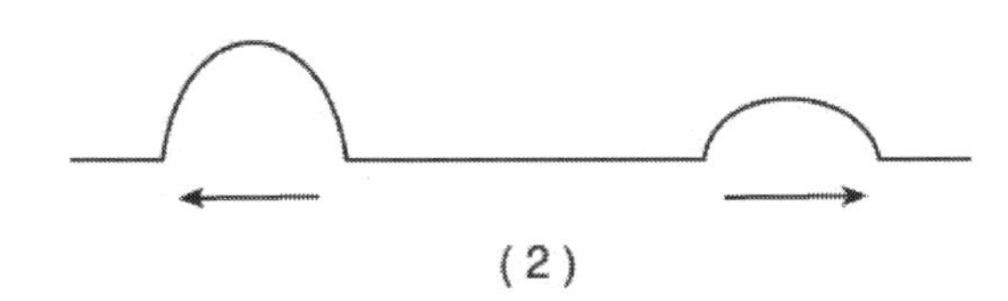

(2)

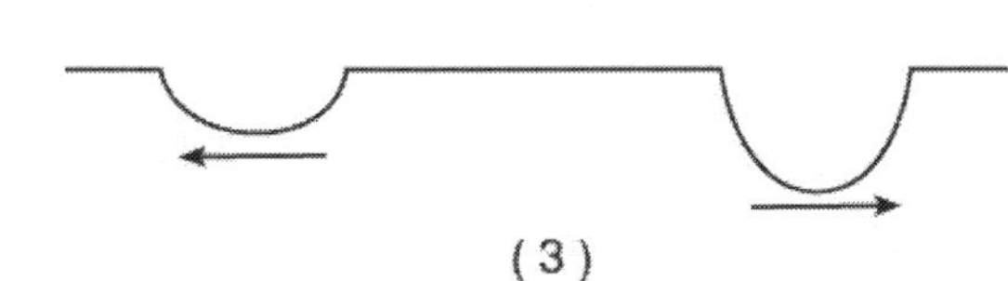

(3)

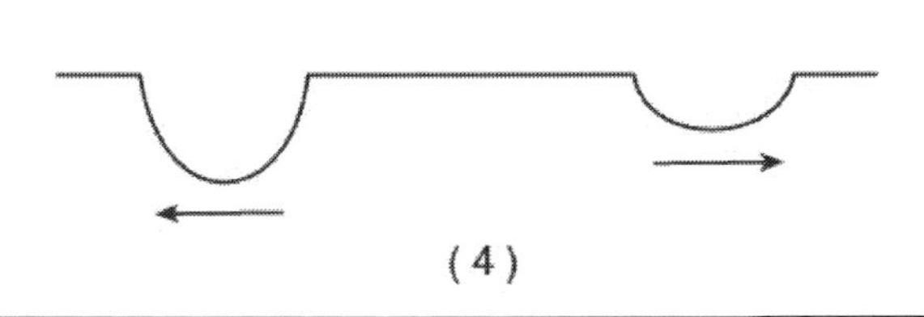

(4)

24. The diagram below shows two waves traveling in the same medium. Points A, B, C, and D are located along the rest position of the medium. The waves interfere to produce a resultant wave.

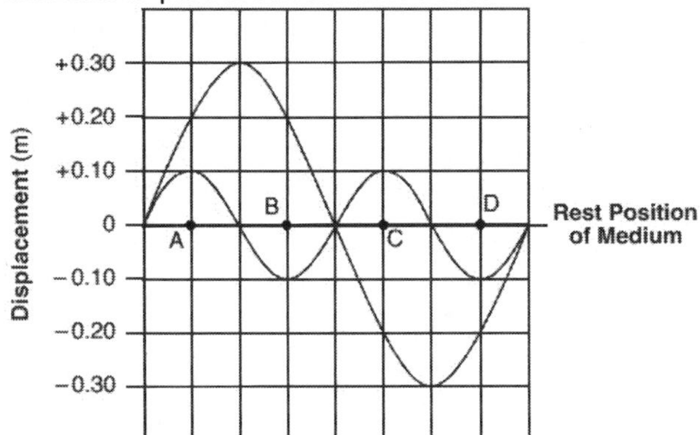

The superposition of the waves produces the greatest positive displacement of the medium from its rest position at point
    1. A
    2. B
    3. C
    4. D

25. The diagram below represents a wave moving toward the right side of this page.

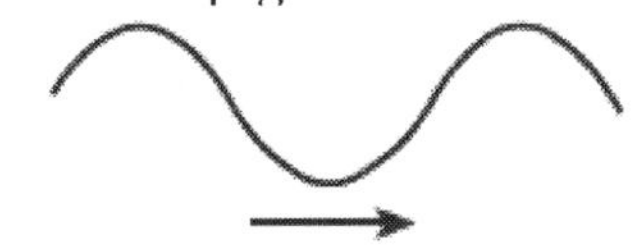

Which wave shown below could produce a standing wave with the original wave?

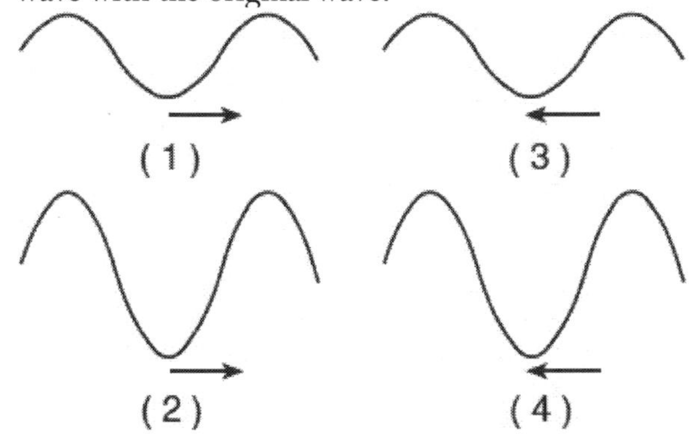

26. Which wave phenomenon occurs when vibrations in one object cause vibrations in a second object?
    1. reflection
    2. resonance
    3. intensity
    4. tuning

Name:____________________________ Period: __________

Waves-Wave Behaviors

Base your answers to questions 27 and 28 on the information below.

One end of a rope is attached to a variable-speed drill and the other end is attached to a 5.0-kilogram mass. The rope is draped over a hook on a wall opposite the drill. When the drill rotates at a frequency of 20.0 Hz, standing waves of the same frequency are set up in the rope. The diagram below shows such a wave pattern.

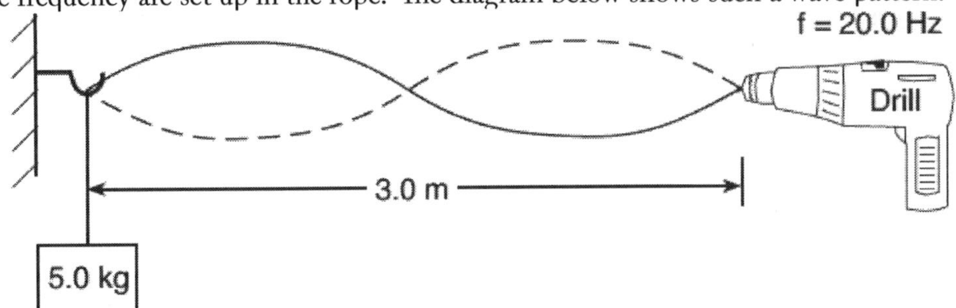

27. Determine the wavelength of the waves producing the standing wave pattern.

28. Calculate the speed of the wave in the rope. [Show all work, including the equation and substitution with units.]

Base your answers to questions 29 and 30 on the information below.

Shattering Glass

An old television commercial for audio recording tape showed a singer breaking a wine glass with her voice. The question was then asked if this was actually her voice or a recording. The inference is that the tape is of such high quality that the excellent reproduction of the sound is able to break glass.

This is a demonstration of resonance. It is certainly possibly to break a wine glass with an amplified singing voice. If the frequency of the voice is the same as the natural frequency of the glass, and the sound is loud enough, the glass can be set into a resonant vibration whose amplitude is large enough to surpass the elastic limit of the glass. But the inference that high-quality reproduction is necessary is not justified. All that is important is that the frequency is recorded and played back correctly. The waveform of the sound can be altered as long as the frequency remains the same. Suppose, for example, that the singer sings a perfect sine wave, but the tape records it as a square wave. If the tape player plays the sound back at the right speed, the glass will still receive energy at the resonance frequency and will be set into vibration leading to breakage, even though the tape reproduction was terrible. Thus, this phenomenon does not require high-quality reproduction and, thus, does not demonstrate the quality of the recording tape. What it does demonstrate is the quality of the tape player, in that it played back the tape at an accurate speed!

29. List two properties that a singer's voice must have in order to shatter a glass.

30. Explain why the glass would not break if the tape player did not play back at an accurate speed.

Name: ______________________ Period: __________

Waves-Wave Behaviors

31. The diagram below represents two waves of equal amplitude and frequency approaching point P as they move through the same medium.

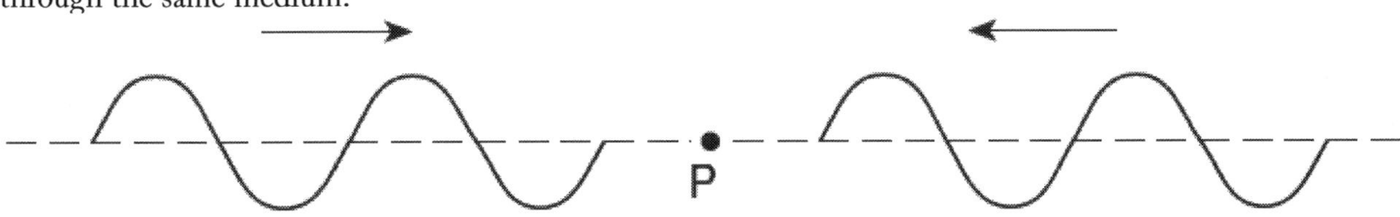

As the two waves pass through each other, the medium at point P will
1. vibrate up and down
2. vibrate left and right
3. vibrate into and out of the page
4. remain stationary

Base your answers to questions 32 and 33 on the information and diagrams below.

The vertical lines in the diagram represent compressions in a sound wave of constant frequency propagating to the right from a speaker toward an observer at point A.

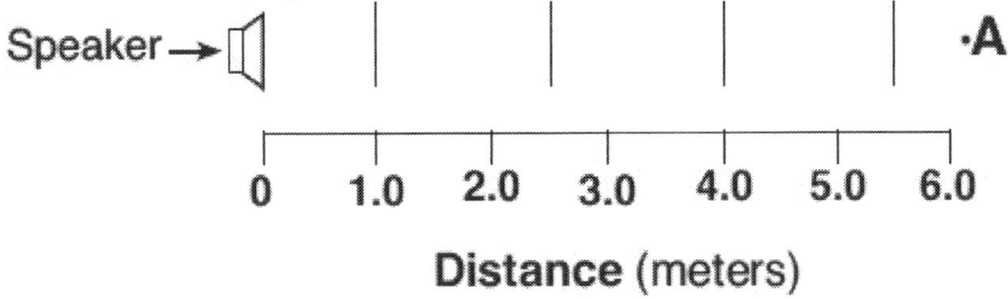

32. Determine the length of this sound wave.

33. The speaker is then moved at constant speed toward the observer at A. Compare the wavelength of the sound wave received by the observer while the speaker is moving to the wavelength observed when the speaker was at rest.

---

34. The diagram below represents two pulses approaching each other.

Which diagram best represents the resultant pulse at the instant the pulses are passing through each other?

(1) (2) (3) (4)

Name: ______________________________ Period: ____________

Waves-Wave Behaviors

35. The diagram below shows two pulses of equal amplitude, A, approaching point P along a uniform string.

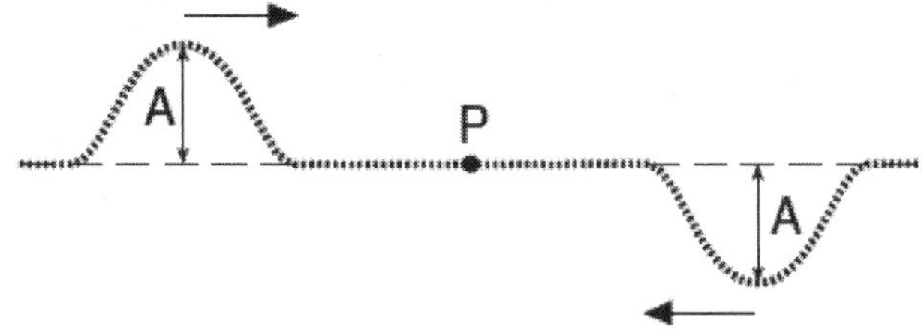

When the two pulses meet at P, the vertical displacement of the string at P will be
1. A
2. 2A
3. 0
4. A/2

36. A student in a band notices that a drum vibrates when another instrument emits a certain frequency note. This phenomenon illustrates
1. reflection
2. resonance
3. refraction
4. diffraction

37. A train sounds a whistle of constant frequency as it leaves the train station. Compared to the sound emitted by the whistle, the sound that the passengers standing on the platform hear has a frequency that is
1. lower, because the sound-wave fronts reach the platform at a frequency lower than the frequency at which they are produced
2. lower, because the sound waves travel more slowly in the still air above the platform than in the rushing air near the train
3. higher, because the sound-wave fronts reach the platform at a frequency higher than the frequency at which they are produced
4. higher, because the sound waves travel faster in the still air above the platform than in the rushing air near the train

38. A girl on a swing may increase the amplitude of the swing's oscillations if she moves her legs at the natural frequency of the swing. This is an example of
1. the Doppler effect
2. destructive interference
3. wave transmission
4. resonance

39. Wave X travels eastward with a frequency f and amplitude A. Wave Y, traveling in the same medium, interacts with wave X and produces a standing wave. Which statement about wave Y is correct?
1. Wave Y must have a frequency of f, an amplitude of A, and be traveling eastward.
2. Wave Y must have a frequency of 2f, an amplitude of 3A, and be traveling eastward.
3. Wave Y must have a frequency of 3f, an amplitude of 2A, and be traveling westward.
4. Wave Y must have a frequency of f, an amplitude of A, and be traveling westward.

40. Two waves traveling in the same medium and having the same wavelength (λ) interfere to create a standing wave. What is the distance between two consecutive nodes on this standing wave?

1. λ
2. $3\lambda/4$
3. $\lambda/2$
4. $\lambda/4$

41. Two waves having the same amplitude and frequency are traveling in the same medium. Maximum destructive interference will occur when the phase difference between the waves is
1. 0°
2. 90°
3. 180°
4. 270°

Name: ______________________ Period: __________

Waves-Wave Behaviors

42. Which phenomenon occurs when an object absorbs wave energy that matches the object's natural frequency?
    1. reflection
    2. diffraction
    3. resonance
    4. interference

43. When observed from Earth, the wavelengths of light emitted by a star are shifted toward the red end of the electromagnetic spectrum. This redshift occurs because the star is
    1. at rest relative to Earth
    2. moving away from Earth
    3. moving toward Earth at decreasing speed
    4. moving toward Earth at increasing speed

44. Ultrasound is a medical technique that transmits sound waves through soft tissue in the human body. Ultrasound waves can break kidney stones into tiny fragments, making it easier for them to be excreted without pain. The shattering of kidney stones with specific frequencies of sound waves is an application of which wave phenomenon?
    1. the Doppler effect
    2. reflection
    3. refraction
    4. resonance

45. A wave passes through an opening in a barrier. The amount of diffraction experienced by the wave depends on the size of the opening and the wave's
    1. amplitude
    2. wavelength
    3. velocity
    4. phase

46. In the diagram below, a stationary source located at point S produces sound having a constant frequency of 512 hertz. Observer A, 50 meters to the left of S, hears a frequency of 512 hertz. Observer B, 100 meters to the right of S, hears a frequency lower than 512 hertz.

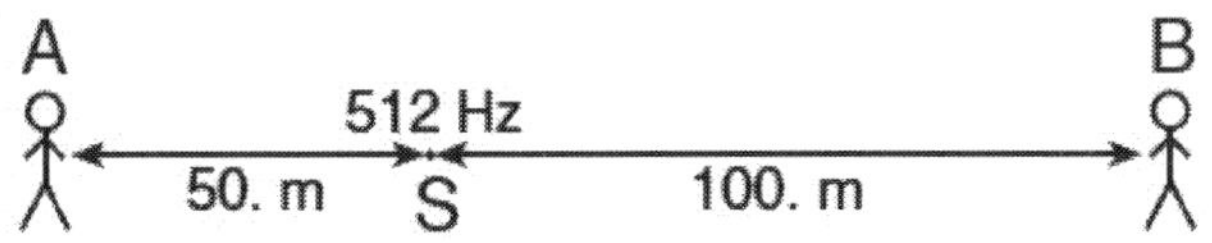

Which statement best describes the motion of the observers?
    1. Observer A is moving toward point S, and observer B is stationary.
    2. Observer A is moving away from point S, and observer B is stationary.
    3. Observer A is stationary and observer B is moving toward point S.
    4. Observer A is stationary, and observer B is moving away from point S.

47. Two speakers, S_1 and S_2, operating in phase in the same medium produce the circular wave patterns shown in the diagram below.

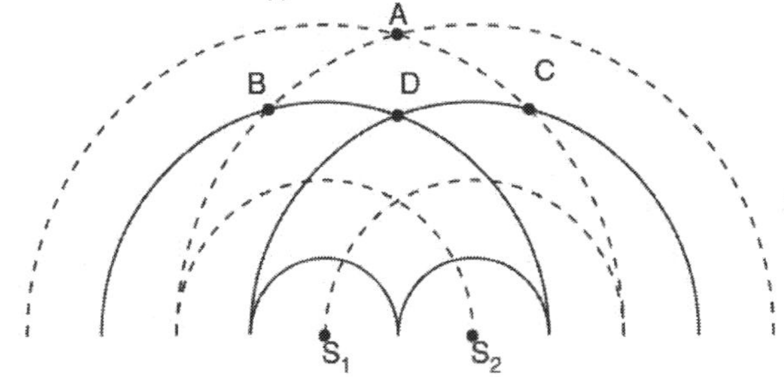

At which two points is constructive interference occurring?
    1. A and B
    2. A and D
    3. B and C
    4. B and D

48. One vibrating 256-hertz tuning fork transfers energy to another 256-hertz tuning fork, causing the second tuning fork to vibrate. This phenomenon is an example of
    1. diffraction
    2. reflection
    3. refraction
    4. resonance

Name: ______________________ Period: __________

Waves-Wave Behaviors

49. The diagram below shows two waves traveling toward each other at equal speed in a uniform medium.

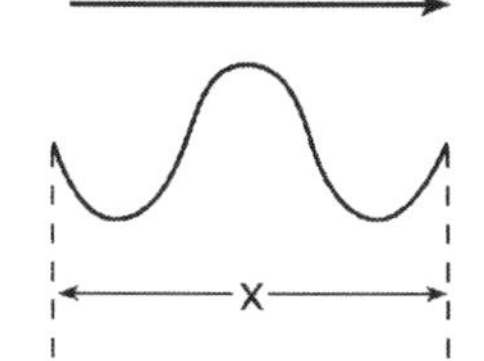

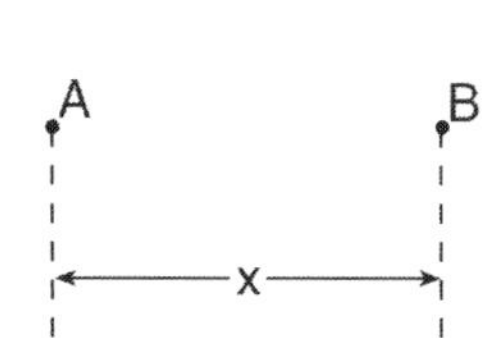

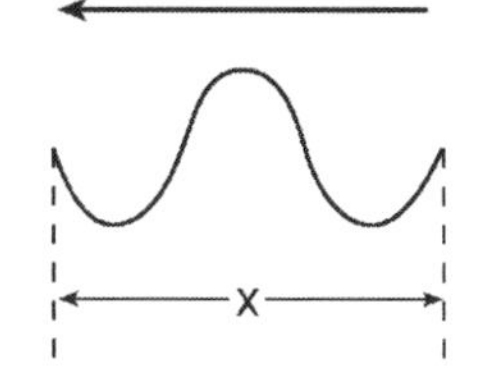

When both waves are in the region between points A and B, they will undergo
1. diffraction
2. the Doppler effect
3. destructive interference
4. constructive interference

50. Sound waves are produced by the horn of a truck that is approaching a stationary observer. Compared to the sound waves detected by the driver of the truck, the sound waves detected by the observer have a greater
1. wavelength
2. frequency
3. period
4. speed

51. The diagram below represents two identical pulses approaching each other in a uniform medium.

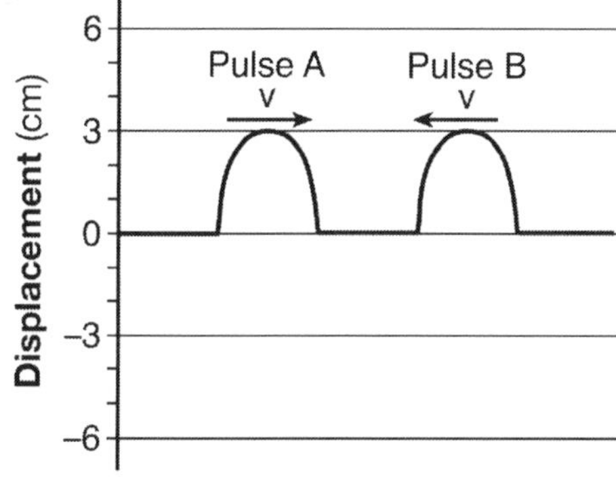

As the pulses meet and are superposed, the maximum displacement of the medium is
1. -6 cm
2. 0 cm
3. 3 cm
4. 6 cm

52. As a car approaches a pedestrian crossing the road, the driver blows the horn. Compared to the sound wave emitted by the horn, the sound wave detected by the pedestrian has a
1. higher frequency and a lower pitch
2. higher frequency and a higher pitch
3. lower frequency and a higher pitch
4. lower frequency and a lower pitch

53. When air is blown across the top of an open water bottle, air molecules in the bottle vibrate at a particular frequency and sound is produced. This phenomenon is called
1. diffraction
2. refraction
3. resonance
4. the Doppler effect

54. Wind blowing across suspended power lines may cause the power lines to vibrate at their natural frequency. This often produces audible sound waves. This phenomenon, often called an Aeolian harp, is an example of
1. diffraction
2. the Doppler effect
3. refraction
4. resonance

55. A student listens to music from a speaker in an adjoining room, as represented in the diagram below.

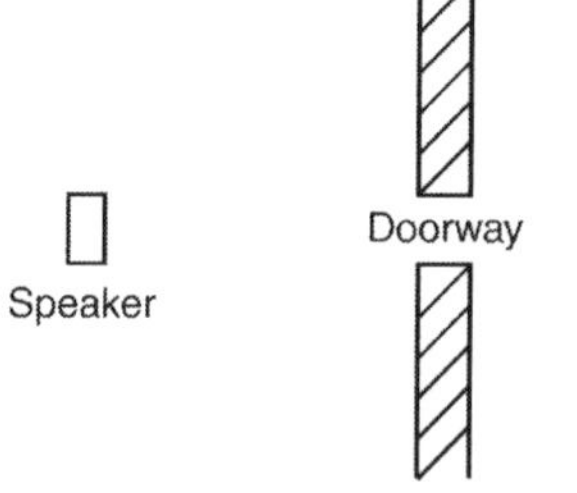

She notices that she does not have to be directly in front of the doorway to hear the music. This spreading of sound waves beyond the doorway is an example of
1. the Doppler effect
2. resonance
3. refraction
4. diffraction

Name: ________________________ Period: __________

Waves-Wave Behaviors

56. The horn of a moving vehicle produces a sound of constant frequency. Two stationary observers, A and C, and the vehicle's driver, B, positioned as represented in the diagram below, hear the sound of the horn.

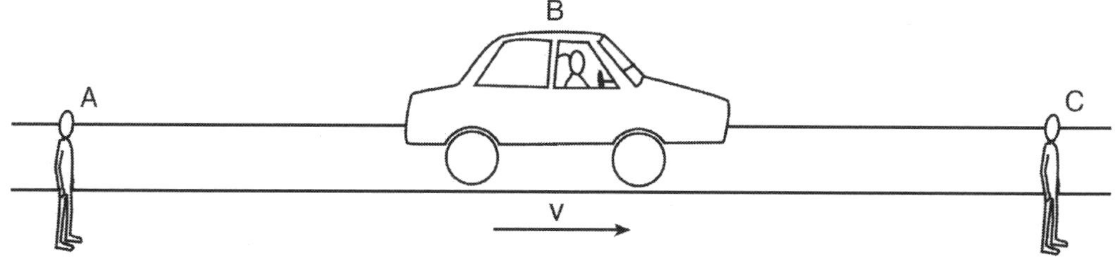

Compared to the frequency of the sound of the horn heard by driver B, the frequency heard by observer A is
1. lower and the frequency heard by observer C is lower
2. lower and the frequency heard by observer C is higher
3. higher and the frequency heard by observer C is lower
4. higher and the frequency heard by observer C is higher

57. Two pulses approach each other in the same medium. The diagram below represents the displacements caused by each pulse.

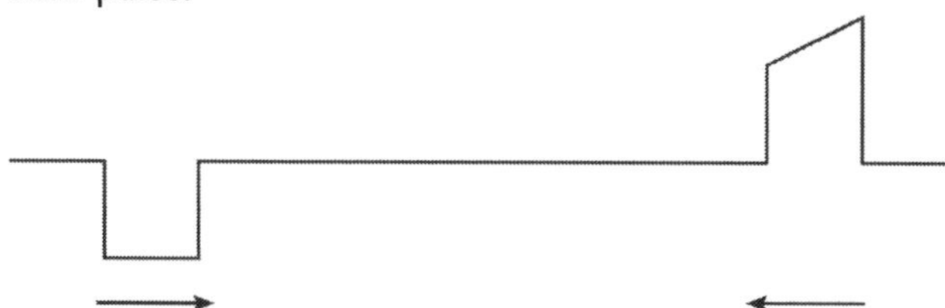

Which diagram best represents the resultant displacement of the medium as the pulses pass through each other?

Name: ______________________________ Period: __________

Waves-Reflection

1. The diagram below represents a light ray striking the boundary between air and glass.

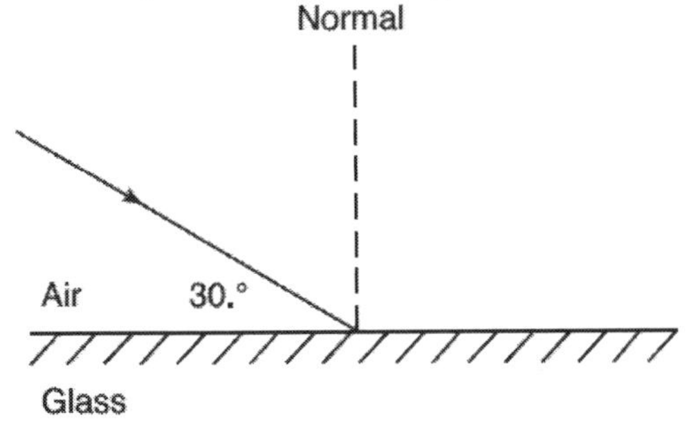

What would be the angle between this light ray and its reflected ray?
1. 30°
2. 60°
3. 120°
4. 150°

2. The diagram below represents a view from above of a tank of water in which parallel wave fronts are traveling toward a barrier.

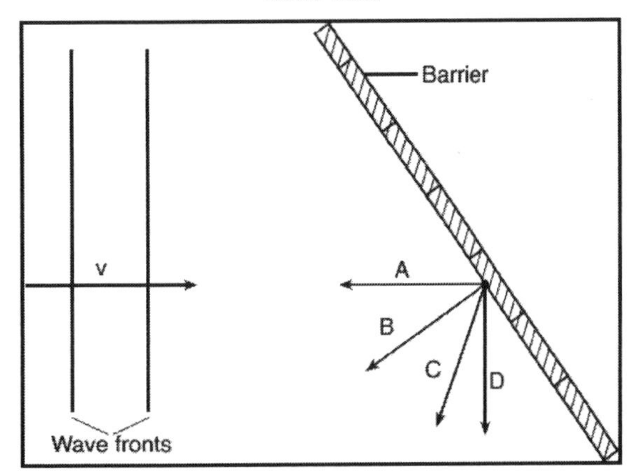

Which arrow represents the direction of travel for the wave fronts after being reflected from the barrier?
1. A
2. B
3. C
4. D

3. A sonar wave is reflected from the ocean floor. For which angles of incidence do the wave's angle of reflection equal its angle of incidence?
1. angles less than 45°, only
2. an angle of 45°, only
3. angles greater than 45°, only
4. all angles of incidence

4. Two plane mirrors are positioned perpendicular to each other as shown. A ray of monochromatic red light is incident on mirror 1 at an angle of 55°. This ray is reflected from mirror 1 and then strikes mirror 2.

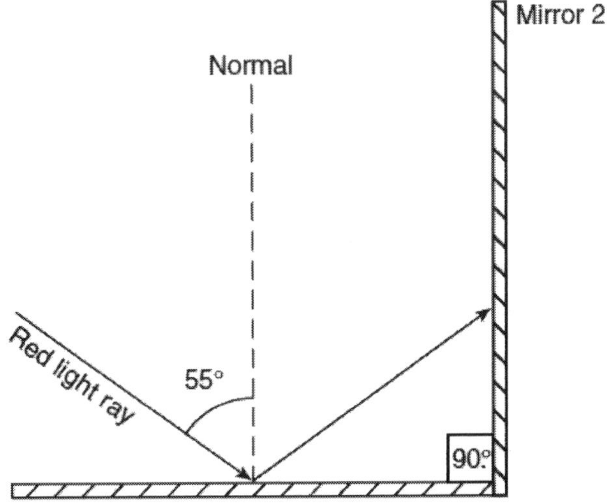

Determine the angle at which the ray is incident on mirror 2 and label the angle on the diagram (in degrees). On the diagram, use a protractor and straightedge to draw the ray of light as it is reflected from mirror 2.

5. The diagram below represents a light ray reflecting from a plane mirror.

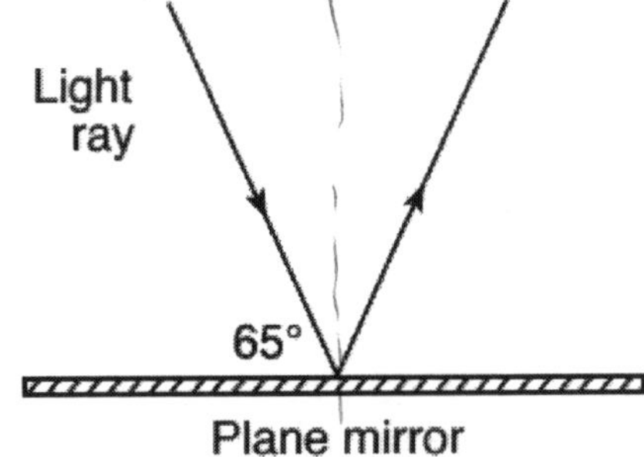

The angle of reflection for the light ray is
1. 25°
2. 35°
3. 50°
4. 65°

Name: ______________________________ Period: ____________

Waves-Reflection

Base your answers to the following questions on the information and diagram below:

In the diagram, a light ray, R, strikes the boundary of air and water.

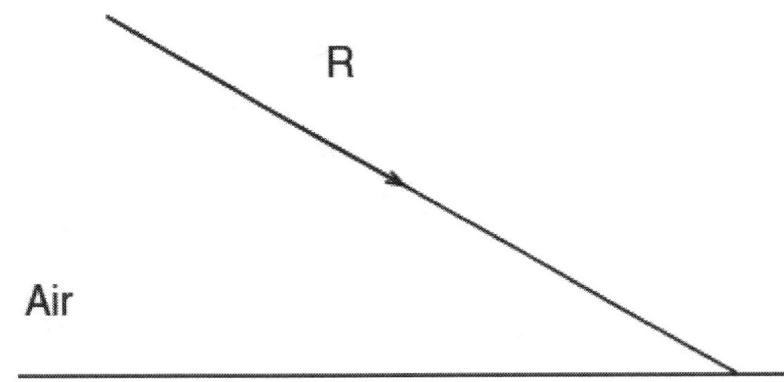

6. Using a protractor, determine the angle of incidence.

7. Using a protractor and straightedge, draw the reflected ray on the diagram above.

8. The diagram below shows a ray of monochromatic light incident on a boundary between air and glass.

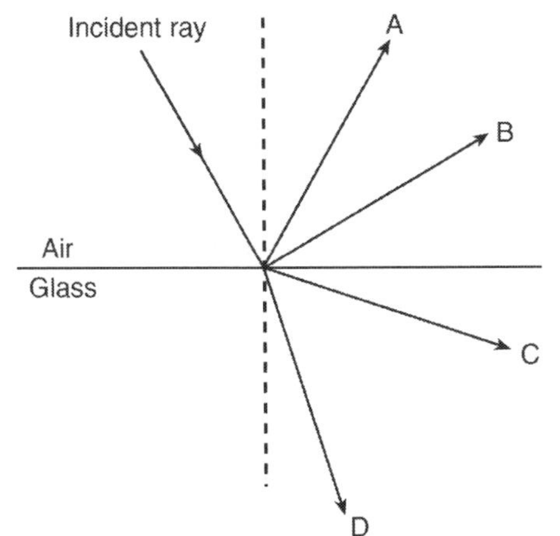

Which ray best represents the path of the reflected light ray?

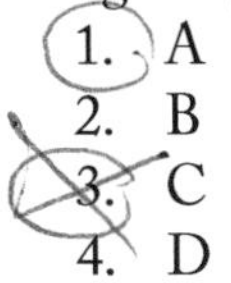

1. A
2. B
3. C
4. D

Name: ______________________________ Period: ____________

Waves-Refraction

1. In which way does blue light change as it travels from diamond into crown glass?
   1. Its frequency decreases.
   2. Its frequency increases.
   3. Its speed decreases.
   4. Its speed increases.

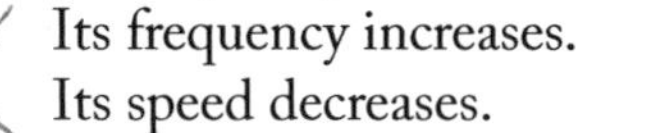

Base your answers to questions 2 through 4 on the information and diagram below.

A monochromatic light ray ($f=5.09 \times 10^{14}$ Hz) traveling in air is incident on the surface of a rectangular block of Lucite ($n=1.50$).

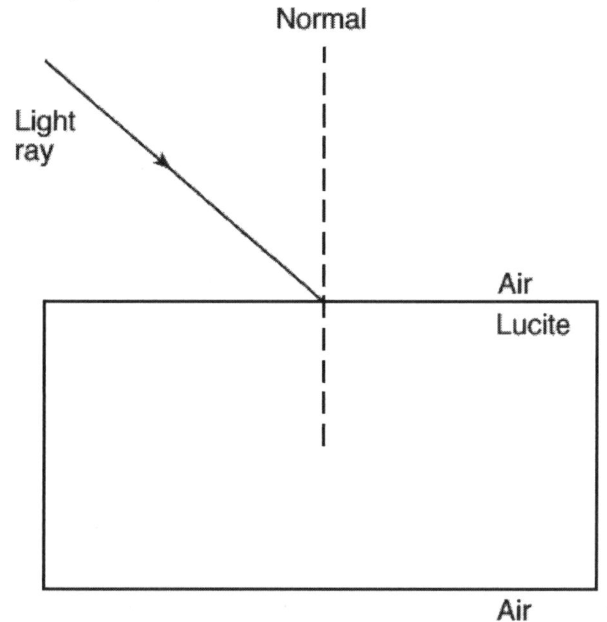

2. Measure the angle of incidence for the light ray to the *nearest degree.*

3. Calculate the angle of refraction of the light ray when it enters the Lucite block. [Show all work, including the equation and substitution with units.]

4. What is the angle of refraction of the light ray as it emerges from the Lucite block back into the air?

5. A change in the speed of a wave as it enters a new medium produces a change in
   1. frequency
   2. period
   3. wavelength
   4. phase

6. The diagram below represents a ray of monochromatic light ($f=5.09 \times 10^{14}$ Hz) passing from medium X ($n=1.46$) into fused quartz ($n=1.46$).

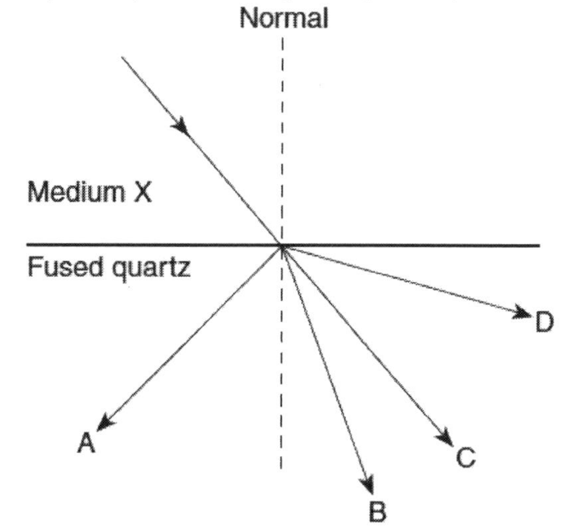

Which path will the ray follow in the quartz?
   1. A
   2. B
   3. C
   4. D

7. A straight glass rod appears to bend when placed in a beaker of water, as shown in the diagram below.

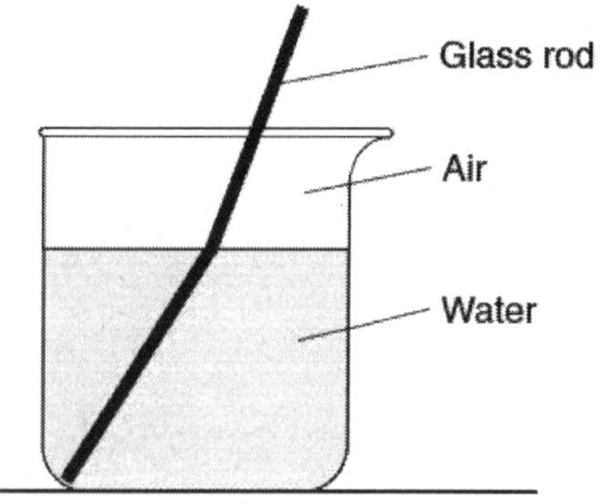

What is the best explanation for this phenomenon?
   1. The water is warmer than the air.
   2. Light travels faster in water than in air.
   3. Light is reflected at the air-water interface.
   4. Light is refracted as it crosses the air-water interface.

Name: ______________________ Period: __________

Waves-Refraction

Base your answers to questions 8 through 10 on the information and diagram below.

A ray of monochromatic light having a frequency of 5.09×10^{14} hertz is incident on an interface of air and corn oil (n=1.47) at an angle of 35° as shown. The ray is transmitted through parallel layers of corn oil and glycerol (n=1.47) and is then reflected from the surface of a plane mirror, located below and parallel to the glycerol layer. The ray then emerges from the corn oil back into the air at point P.

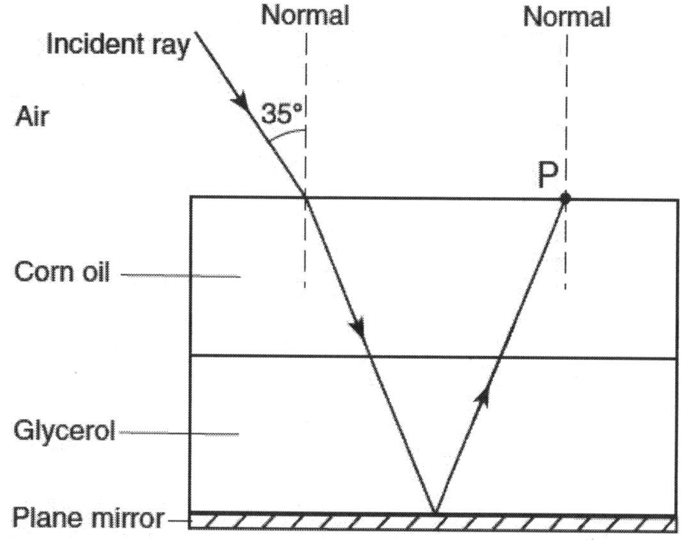

8. Calculate the angle of refraction of the light ray as it enters the corn oil from air. [Show all work, including the equation and the substitution with units.

9. Explain why the ray does not bend at the corn oil-glycerol interface.

10. On the diagram, use a protractor and straightedge to construct the refracted ray representing the light emerging at point P into air.

---

11. Which diagram best represents the behavior of a ray of monochromatic light in air incident on a block of crown glass (n=1.52)?

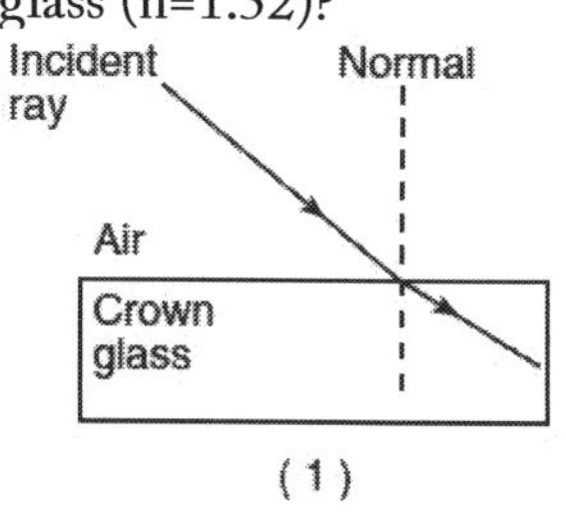

(1)

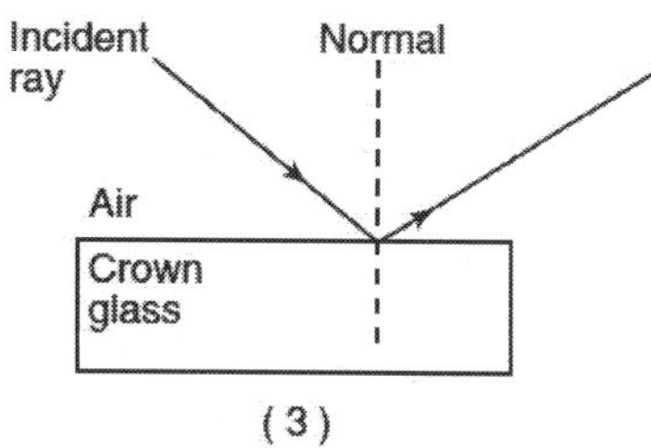

(3)

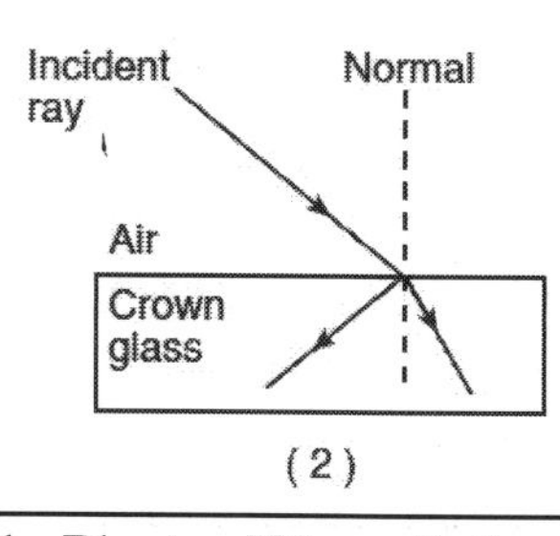

(2)

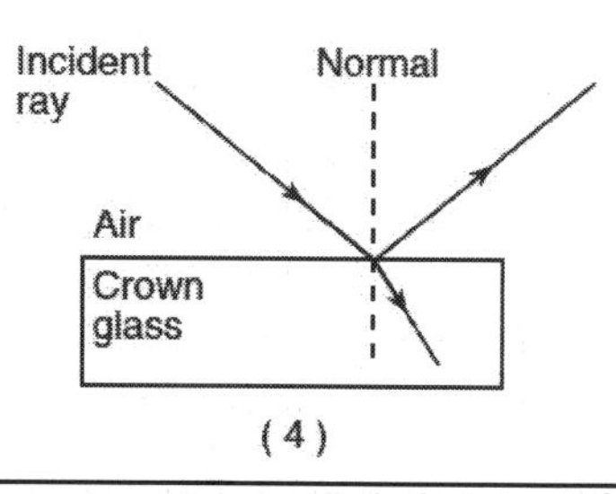

(4)

Name:_______________________________ Period: ___________

Waves-Refraction

Base your answers to questions 12 through 14 on the information below.

A ray of monochromatic light ($f = 5.09 \times 10^{14}$ Hz) passes through air and a rectangular transparent block, as shown in the diagram below.

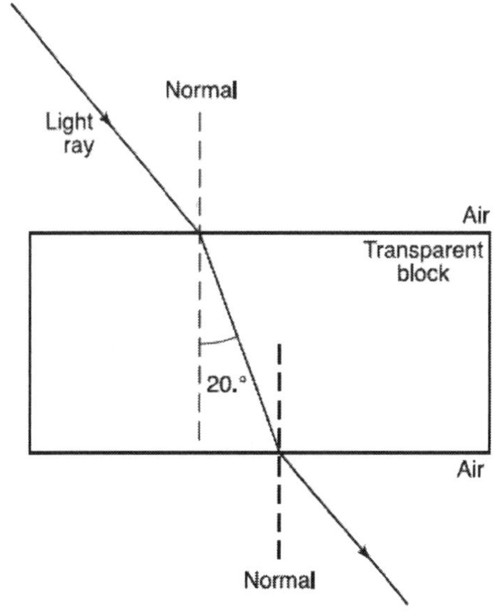

12. Using a protractor, determine the angle of incidence of the light ray as it enters the transparent block from air.

13. Calculate the absolute index of refraction for the medium of the transparent block. [Show all work, including the equation and substitution with units.]

14. Calculate the speed of the light ray in the transparent block. [Show all work, including the equation and substitution with units.]

---

15. A wave generator having a constant frequency produces parallel wave fronts in a tank of water of two different depths. The diagram below represents the wave fronts in the deep water.

    As the wave travels from the deep water into the shallow water, the speed of the waves decreases. On the diagram at right, use a straightedge to draw *at least three* lines to represent the wave fronts, with appropriate spacing, in the shallow water.

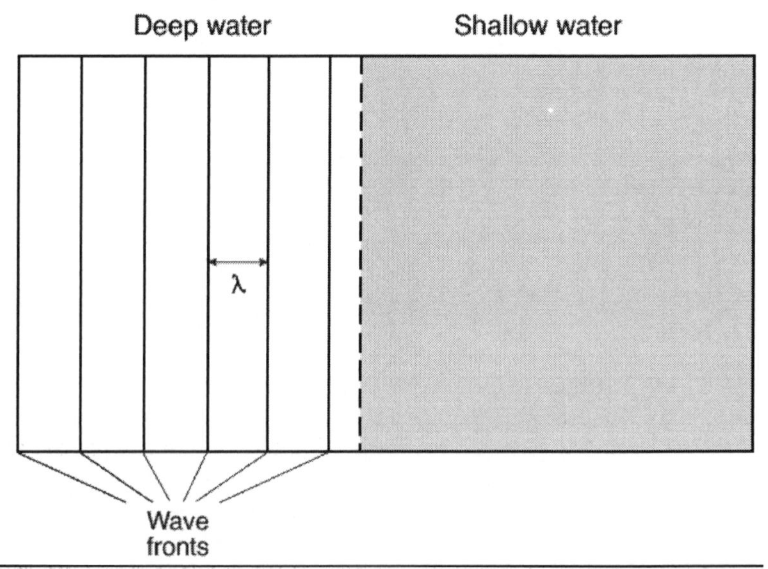

Name: ______________________________ Period: ____________

Waves-Refraction

16. A laser beam is directed at the surface of a smooth, calm pond as represented in the diagram below.

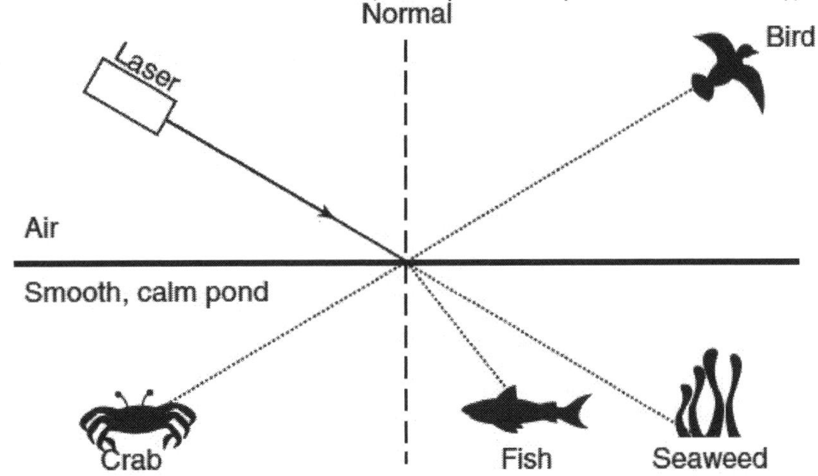

Which organisms could be illuminated by the laser light?
1. the bird and the fish
2. the bird and the seaweed
3. the crab and the seaweed
4. the crab and the fish

Base your answers to questions 17 through 19 on the information and diagram below.

A ray of light (f= 5.09×10^{14} Hz) is incident on the boundary between air and an unknown material X at an angle of incidence of 55°, as shown. The absolute index of refraction of material X is 1.66.

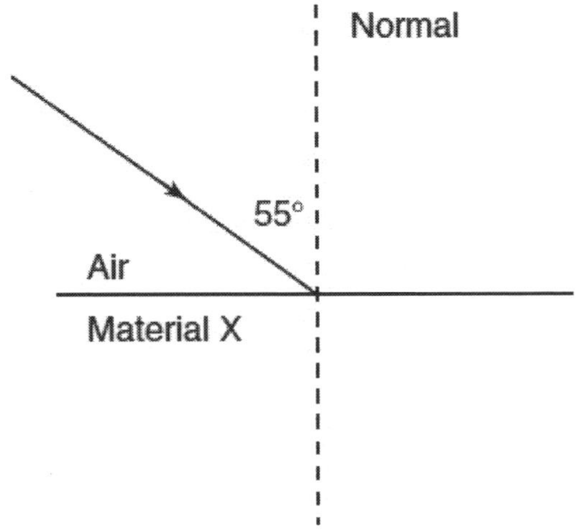

17. Determine the speed of this ray of light in material X.

18. Calculate the angle of refraction of the ray of light in material X.

19. On the diagram above, use a straightedge and protractor to draw the refracted ray of light in material X.

Name: ______________________________ Period: ______________

Waves-Refraction

20. A ray of monochromatic light (f= 5.09×10^{14} Hz) passes from water through flint glass (n=1.66) and into medium X, as shown below.

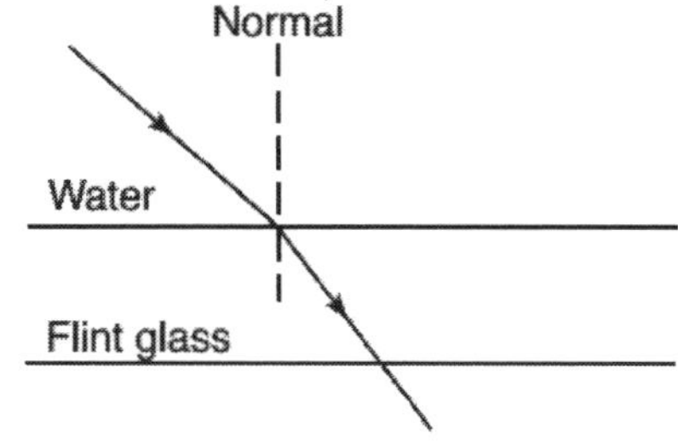

The absolute index of refraction of medium X is
1. less than 1.33
2. greater than 1.33 and less than 1.52
3. greater than 1.52 and less than 1.66
4. equal to 1.66

21. A beam of light travels through medium X with a speed of 1.80×10^{8} meters per second. Calculate the absolute index of refraction of medium X. [Show all work, including the equation and substitution with units.]

22. What happens to the speed and frequency of a light ray when it passes from air into water?
1. The speed decreases and the frequency increases.
2. The speed decreases and the frequency remains the same.
3. The speed increases and the frequency increases.
4. The speed increases and the frequency remains the same.

23. A ray of monochromatic light (f= 5.09×10^{14} Hz) in air is incident at an angle of 30° on a boundary with corn oil (n=1.47). What is the angle of refraction, to the nearest degree, for this light ray in the corn oil?
1. 6°
2. 20°
3. 30°
4. 47°

Base your answers to questions 24 through 26 on the information and diagram below.

A ray of light passes from air into a block of transparent material X as shown in the diagram below.

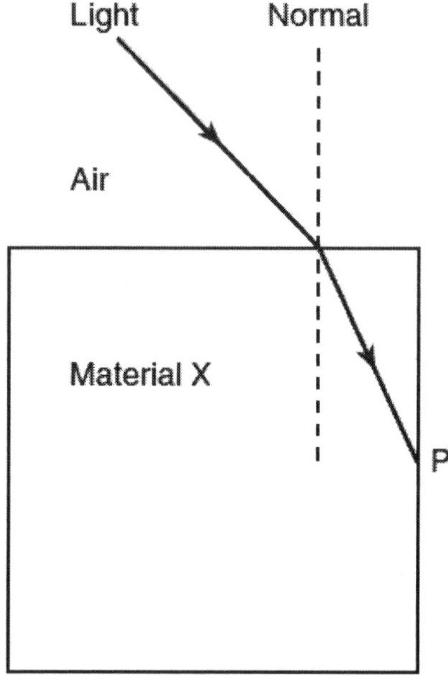

24. Measure the angles of incidence and refraction to the nearest degree for this light ray at the air into material X boundary.

θ_i= θ_r=

25. Calculate the absolute index of refraction of material X. [Show all work, including the equation and substitution with units.]

26. The refracted light ray is reflected from the material X–air boundary at point P. Using a protractor and straightedge, on the diagram in your answer booklet, draw the reflected ray from point P.

27. If the speed of a wave doubles as it passes from shallow water into deeper water, its wavelength will be
1. unchanged
2. doubled
3. halved
4. quadrupled

Name: ______________________________ Period: ____________

Waves-Refraction

Base your answers to questions 28 and 29 on the information and diagram below.

A ray of monochromatic light (f= 5.09×10^{14} Hz) passes from air into Lucite at an angle of incidence of 30°.

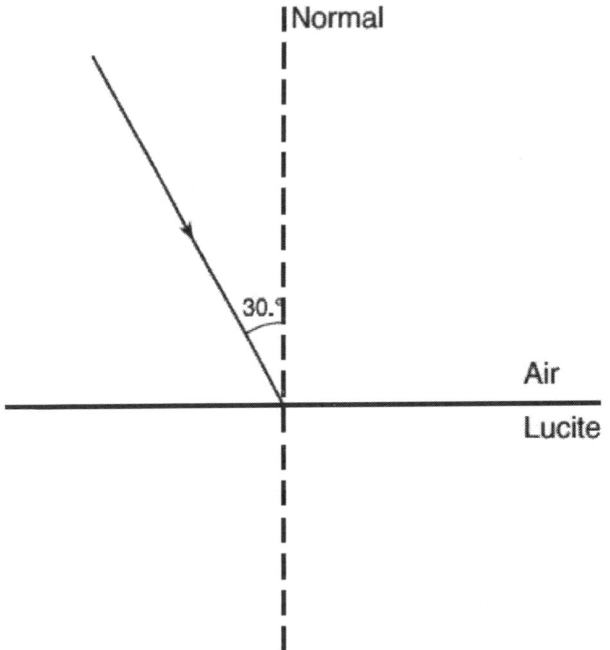

28. Calculate the angle of refraction in the Lucite. [Show all work, including the equation and substitution with units.]

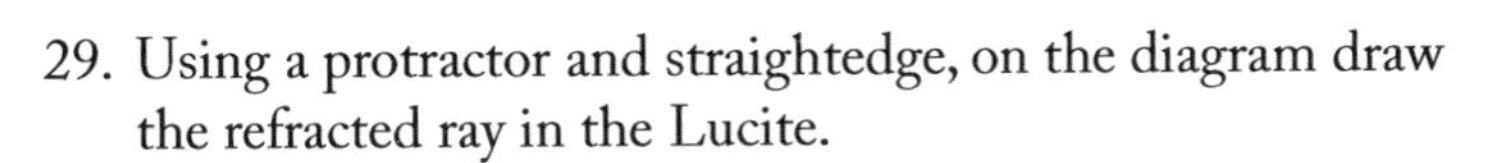

29. Using a protractor and straightedge, on the diagram draw the refracted ray in the Lucite.

30. Which ray diagram best represents the phenomenon of refraction?

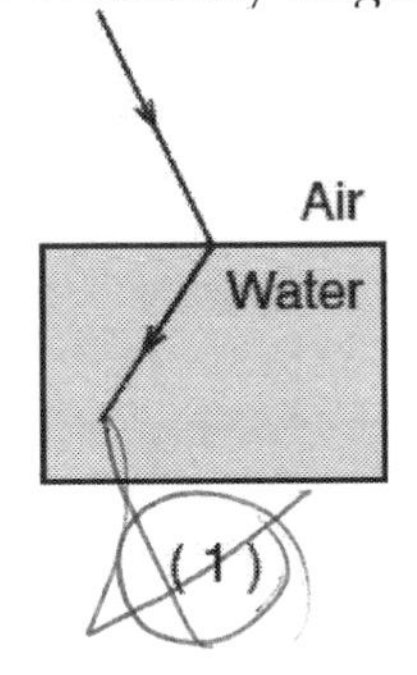

(1)

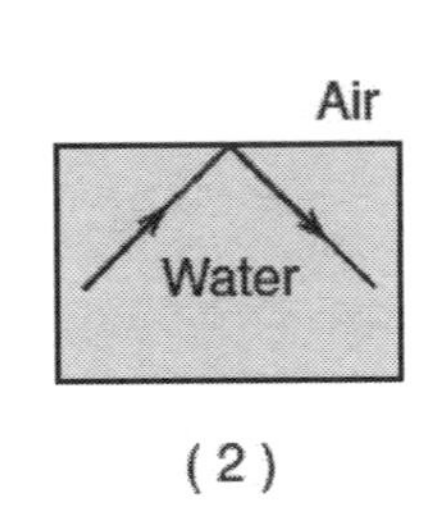

(2)

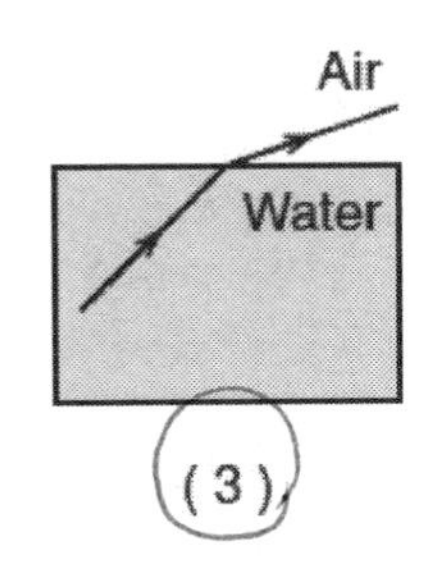

(3)

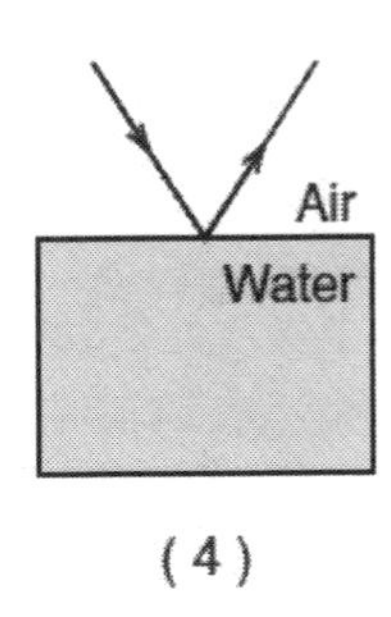

(4)

31. The diagram at right represents straight wave fronts passing from deep water into shallow water, with a change in speed and direction.

    Which phenomenon is illustrated in the diagram?
    1. reflection
    2. refraction
    3. diffraction
    4. interference

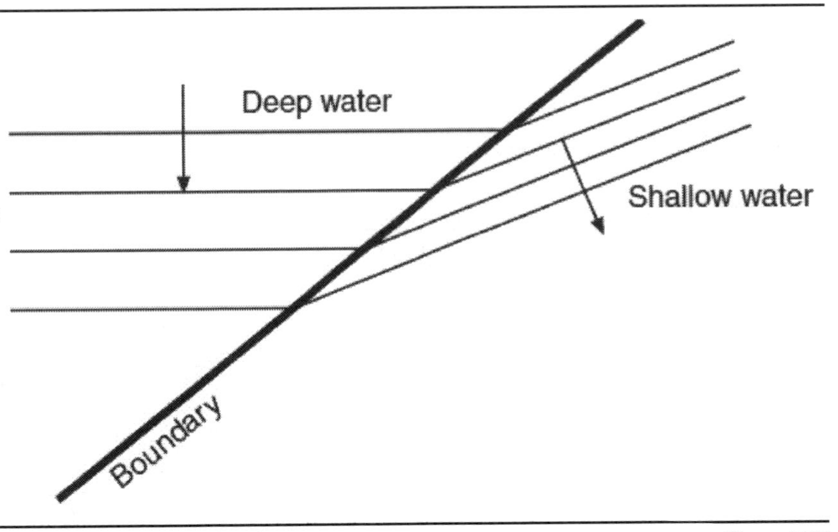

Name: ______________________________ Period: ____________

Waves-Refraction

Base your answers to questions 32 through 34 on the information and diagram below.

A light ray with a frequency of 5.09×10^{14} hertz traveling in air is incident at an angle of 40° on an air-water interface as shown. At the interface, part of the ray is refracted as it enters the water and part of the ray is reflected from the interface.

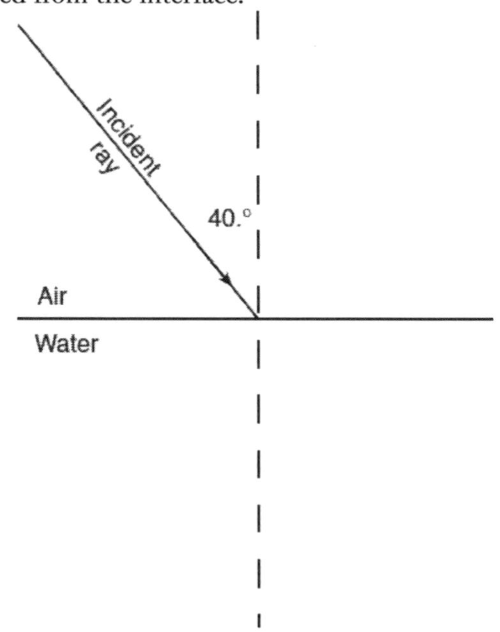

32. Calculate the angle of refraction of the light ray as it enters the water. [Show all work, including the equation and substitution with units.]

33. On the diagram above, using a protractor and straightedge, draw the refracted ray. Label this ray "Refracted ray."

34. On the diagram above, using a protractor and straightedge, draw the reflected ray. Label this ray "Reflected ray."

35. An electromagnetic wave of wavelength 5.89×10^{-7} meter traveling through air is incident on an interface with corn oil (n=1.47). Calculate the wavelength of the EM wave in corn oil.

36. The speed of light in a piece of plastic is 2.00×10^8 meters per second. What is the absolute index of refraction of this plastic?
    1. 1.00
    2. 0.67
    3. 1.33
    4. 1.50

37. A ray of monochromatic light is incident on an air-sodium chloride (n=1.54) boundary as shown in the diagram below. At the boundary, part of the ray is reflected back into the air and part is refracted as it enters the sodium chloride.

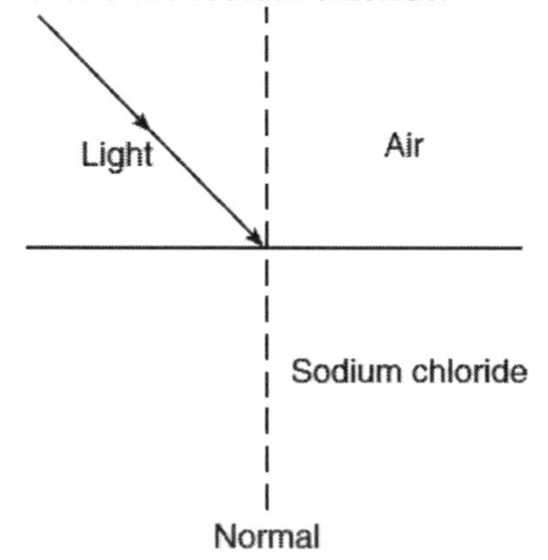

Compared to the ray's angle of refraction in the sodium chloride, the ray's angle of reflection in the air is
    1. smaller
    2. larger
    3. the same

38. The diagram below shows a ray of light passing from air into glass at an angle of incidence of 0°.

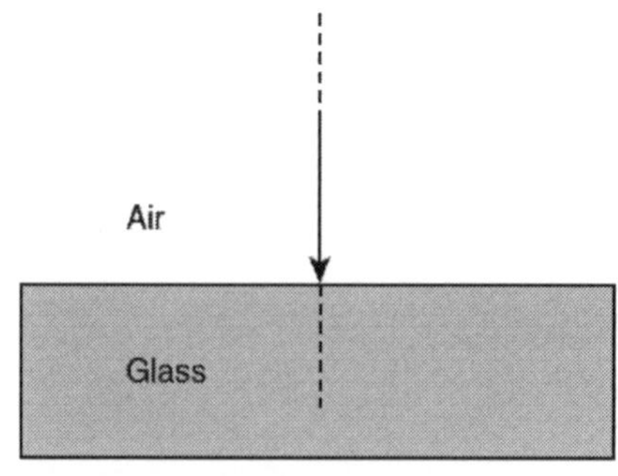

Which statement best describes the speed and direction of the light ray as it passes into the glass?
    1. Only speed changes.
    2. Only direction changes.
    3. Both speed and direction change.
    4. Neither speed nor direction changes.

Name:______________________________ Period: ____________

Waves-Refraction

Base your answers to questions 39 through 42 on the diagram below, which represents a ray of monochromatic light (5.09×10^{14} Hz) in air incident on flint glass (n=1.66).

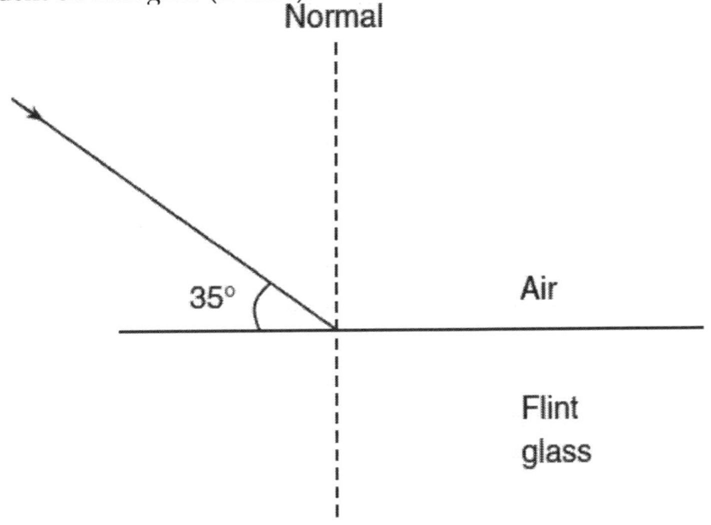

39. Determine the angle of incidence of the light ray in air.

40. Calculate the angle of refraction of the light ray in the flint glass. [Show all work, including the equation and substitution with units.]

41. Using a protractor and straightedge, draw the refracted ray on the diagram.

42. What happens to the light from the incident ray that is *not* refracted or absorbed?

---

43. The diagram below represents a wave.

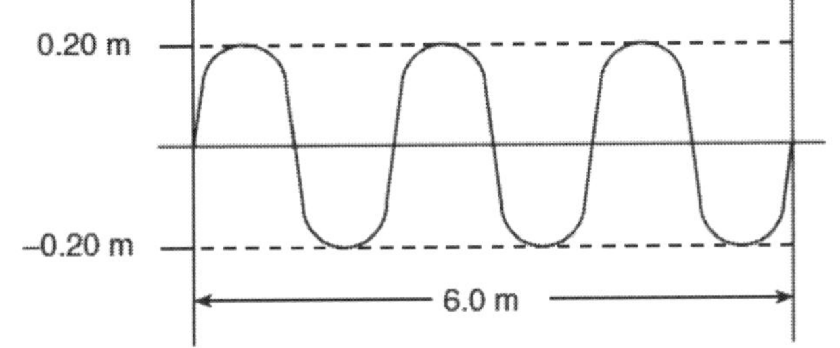

What is the speed of the wave if its frequency is 8.0 hertz?
1. 48 m/s
2. 16 m/s
3. 3.2 m/s
4. 1.6 m/s

44. What is the wavelength of a light ray with frequency 5.09×10^{14} hertz as it travels through Lucite (n=1.50)?
1. 3.93×10^{-7} m
2. 5.89×10^{-7} m
3. 3.39×10^{14} m
4. 7.64×10^{14} m

45. The speed of light ($f=5.09 \times 10^{14}$ Hz) in a transparent material is 0.75 times its speed in air. The absolute index of refraction of the material is approximately
1. 0.75
2. 1.3
3. 2.3
4. 4.0

Name:______________________________ Period: __________

Waves-Refraction

46. A light ray traveling in air enters a second medium and its speed slows to 1.71×10^8 meters per second. What is the absolute index of refraction of the second medium?
    1. 1.00
    2. 0.570
    3. 1.75
    4. 1.94

Base your answers to questions 47 and 48 on the diagram below, which represents a light ray traveling from air to Lucite (n=1.50) to medium Y and back into air.

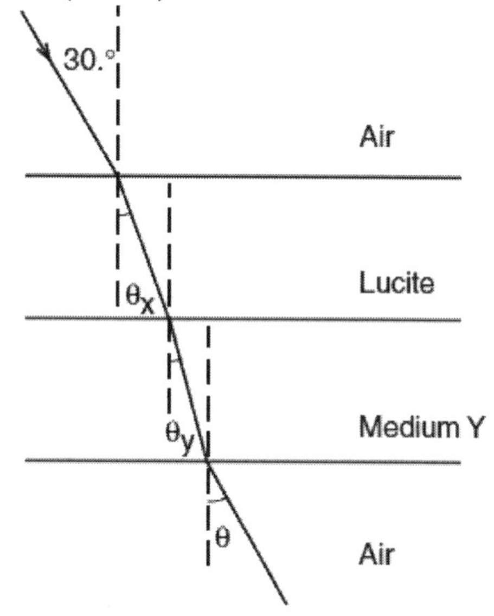

47. The sine of angle θ_x is
    1. 0.333
    2. 0.500
    3. 0.707
    4. 0.886

48. Light travels *slowest* in
    1. air, only
    2. Lucite, only
    3. medium Y, only
    4. air, Lucite, and medium Y

---

49. Which quantity is equivalent to the product of the absolute index of refraction of water and the speed of light in water?
    1. wavelength of light in a vacuum
    2. frequency of light in water
    3. sine of the angle of incidence
    4. speed of light in a vacuum

50. What is the speed of light ($f=5.09 \times 10^{14}$ Hz) in flint glass?
    1. 1.81×10^8 m/s
    2. 1.97×10^8 m/s
    3. 3.00×10^8 m/s
    4. 4.98×10^8 m/s

51. What happens to the frequency and the speed of an electromagnetic wave as it passes from air into glass?
    1. The frequency decreases and the speed increases.
    2. The frequency increases and the speed decreases.
    3. The frequency remains the same and the speed increases.
    4. The frequency remains the same and the speed decreases.

52. When a light wave enters a new medium and is refracted, there must be a change in the light wave's
    1. color
    2. frequency
    3. period
    4. speed

53. As a sound wave passes from water, where the speed is 1.49×10^3 meters per second, into air, the wave's speed
    1. decreases and its frequency remains the same
    2. increases and its frequency remains the same
    3. remains the same and its frequency decreases
    4. remains the same and its frequency increases

54. In a certain material, a beam of monochromatic light ($f=5.09 \times 10^{14}$ Hz) has a speed of 2.25×10^8 meters per second. The material could be
    1. crown glass (n=1.52)
    2. flint glass (n=1.66)
    3. glycerol (n=1.47)
    4. water (n=1.33)

55. A ray of monochromatic light with frequency 5.09×10^{14} Hz is transmitted through four different media: corn oil, ethyl alcohol, flint glass, and water. Rank the four media from the one through which the light travels at the slowest speed to the one through which light travels at the fastest speed.

Name: ______________________________ Period: ____________

Waves-Refraction

Base your answers to questions 56 through 59 on the information below.

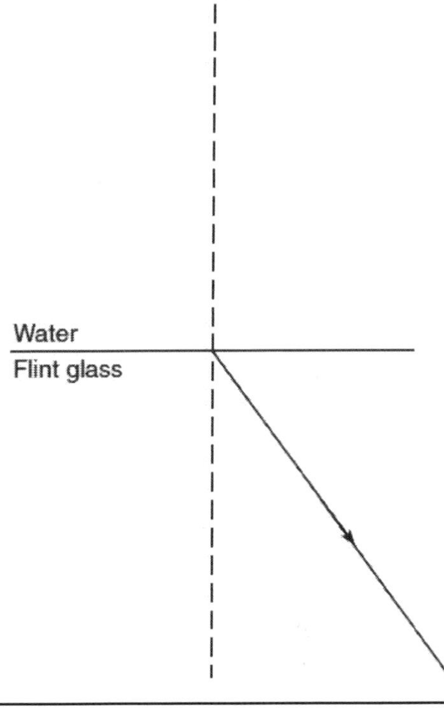

A light ray (f=5.09×10^{14} Hz) is refracted as it travels from water into flint glass. The path of the light ray in the flint glass is shown in the diagram.

56. Using a protractor, measure the angle of refraction of the light ray in the flint glass.

57. Calculate the angle of incidence for the light ray in water. [Show all work, including the equation and substitution with units.]

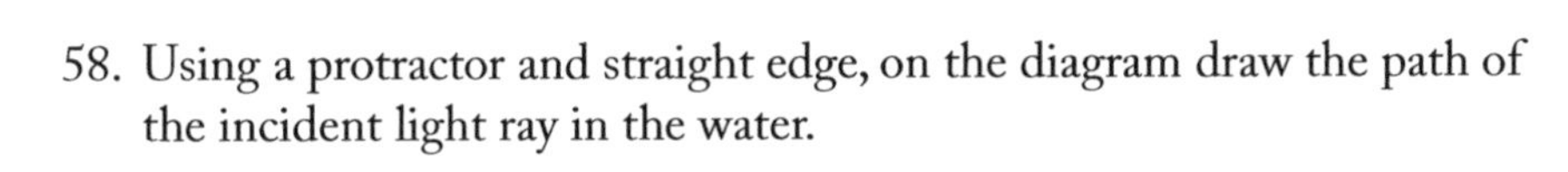

58. Using a protractor and straight edge, on the diagram draw the path of the incident light ray in the water.

59. Identify one physical event, other than transmission or refraction, that occurs as the light interacts with the water-flint glass boundary.

60. The wavelength of a wave doubles as it travels from medium A into medium B. Compared to the wave in medium A, the wave in medium B has
    1. half the speed
    2. twice the speed
    3. half the frequency
    4. twice the frequency

61. A ray of light (f=5.09×10^{14} Hz) travels through various substances. Which graph best represents the relationship between the absolute index of refraction of these substances and the corresponding speed of light in these substances?

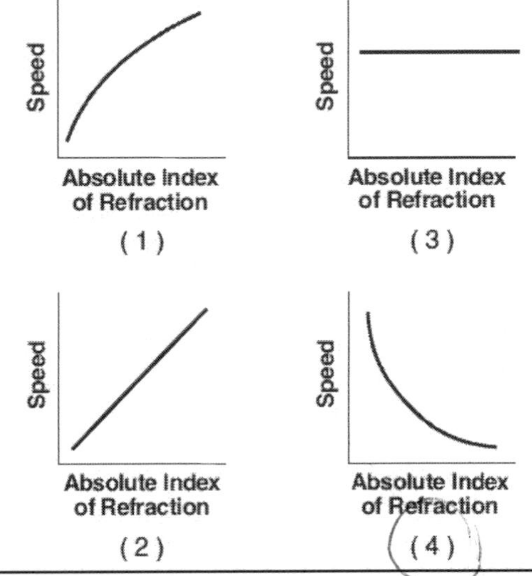

62. What is the speed of light (f=5.09×10^{14} Hz) in ethyl alcohol?
    1. 4.53×10^{-9} m/s
    2. 2.43×10^{2} m/s
    3. 1.24×10^{8} m/s
    4. 2.21×10^{8} m/s

Name:______________________________ Period: __________

Waves-Refraction

Base your answers to questions 63 through 66 on the information below.

A light ray ($f=5.09 \times 10^{14}$ Hz) traveling in water has an angle of incidence of 35° on a water-air interface. At the interface, part of the ray is reflected from the interface and part of the ray is refracted as it enters the air.

63. What is the angle of reflection of the light ray at the interface?

64. On the diagram below, using a protractor and a straightedge, draw the reflected ray.

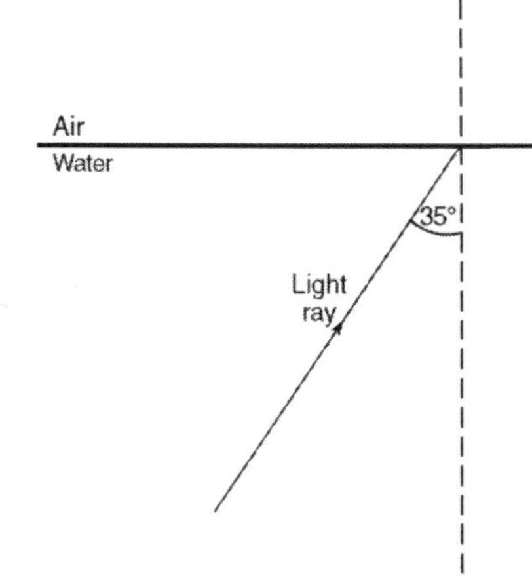

65. Calculate the angle of refraction of the light ray as it enters the air. [Show all work, including the equation and substitution with units.]

66. Identify one characteristic of this light ray that is the same in both the water and the air.

67. As a monochromatic light ray passes from air into water, two characteristics of the ray that will not change are
    1. wavelength and period
    2. frequency and period
    3. wavelength and speed
    4. frequency and speed

68. Which graph best represents the relationship between the absolute index of refraction and the speed of light ($f=5.09 \times 10^{14}$ Hz) in various media?

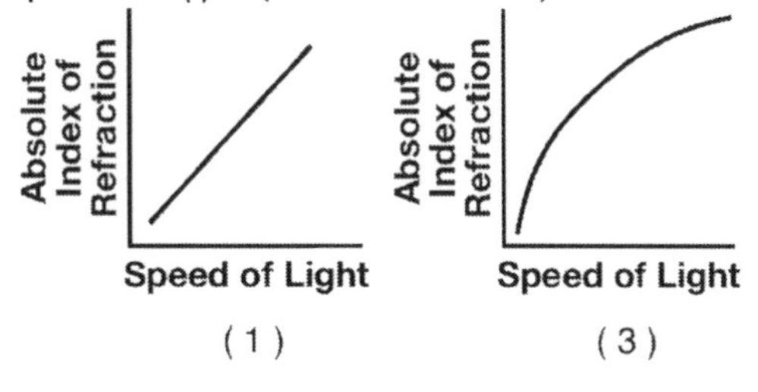

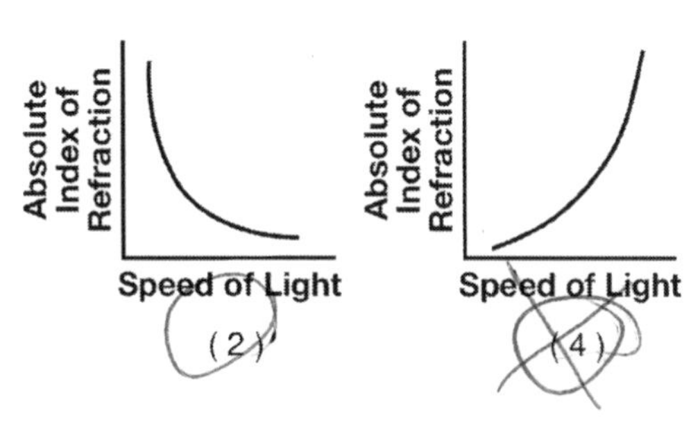

69. When a ray of light traveling in water reaches a boundary with air, part of the light ray is reflected and part is refracted. Which ray diagram best represents the paths of the reflected and refracted rays?

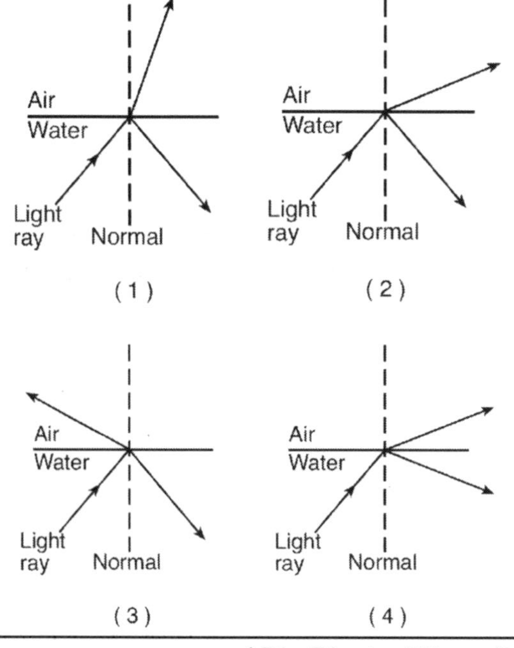

Name: ______________________ Period: __________

Waves-Refraction

70. Which characteristic of a light wave must increase as the light wave passes from glass into air?
    1. amplitude
    2. frequency
    3. period
    4. wavelength

71. A ray of yellow light ($f = 5.09 \times 10^{14}$ Hz) travels at a speed of 2.04×10^8 meters per second in
    1. ethyl alcohol
    2. water
    3. Lucite
    4. glycerol

Base your answers to questions 72 and 73 on the information and diagram below.

A ray of light ($f=5.09 \times 10^{14}$ Hz) traveling through a block of an unknown material, passes at an angle of incidence of 30° into air, as shown in the diagram below.

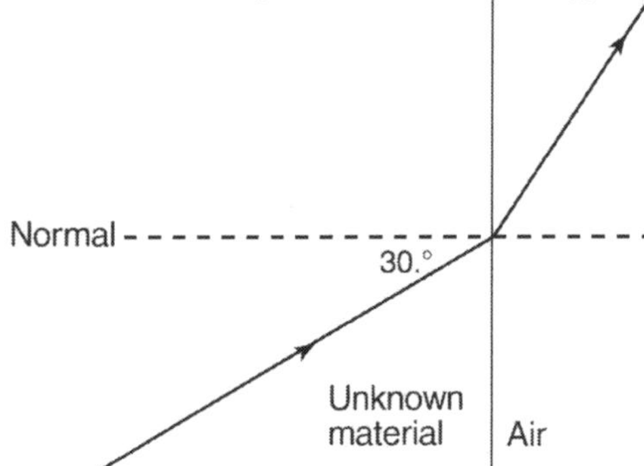

72. Use a protractor to determine the angle of refraction of the light ray as it passes from the unknown material into air.

73. Calculate the index of refraction of the unknown material. [Show all work, including the equation and substitution with units.]

Name:______________________________ Period: ____________

Waves-Diffraction

1. A wave of constant wavelength diffracts as it passes through an opening in a barrier. As the size of the opening is increased, the diffraction effects
   1. decrease
   2. increase
   3. remain the same

2. The diagram below shows a series of wave fronts approaching an opening in a barrier. Point P is located on the opposite side of the barrier.

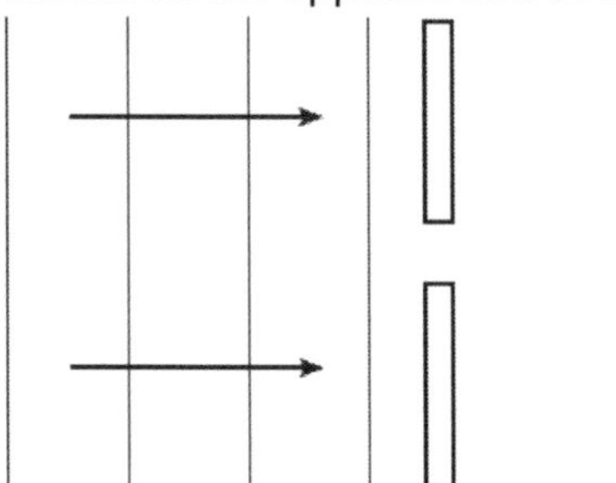

   The wave fronts reach point P as a result of
   1. resonance
   2. refraction
   3. reflection
   4. diffraction

3. Which wave phenomenon makes it possible for a player to hear the sound from a referee's whistle in an open field even when standing behind the referee?
   1. diffraction
   2. Doppler effect
   3. reflection
   4. refraction

4. A wave is diffracted as it passes through an opening in a barrier. The amount of diffraction that the wave undergoes depends on both the
   1. amplitude and frequency of the incident wave
   2. wavelength and speed of the incident wave
   3. wavelength of the incident wave and the size of the opening
   4. amplitude of the incident wave and the size of the opening

5. The diagram below shows a plane wave passing through a small opening in a barrier.

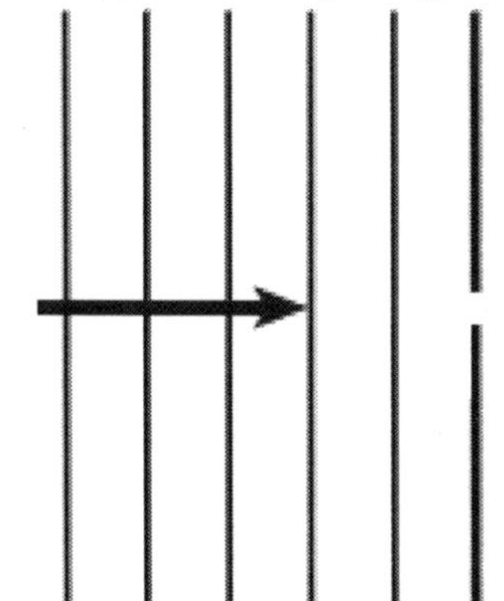

   On the diagram above, sketch four wave fronts after they have passed through the barrier.

6. Which diagram best represents the shape and direction of a series of wave fronts after they have passed through a small opening in a barrier?

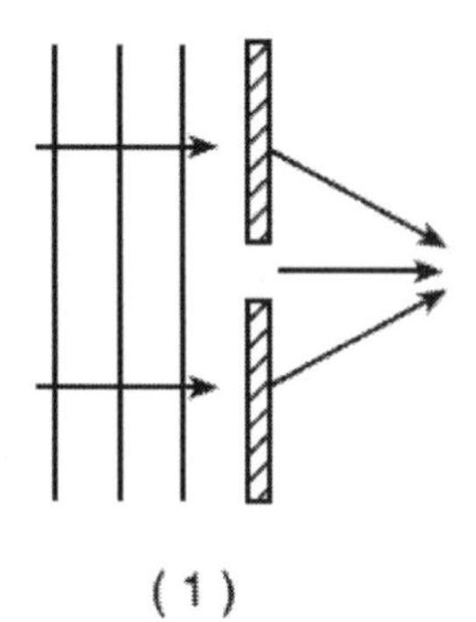

(1)

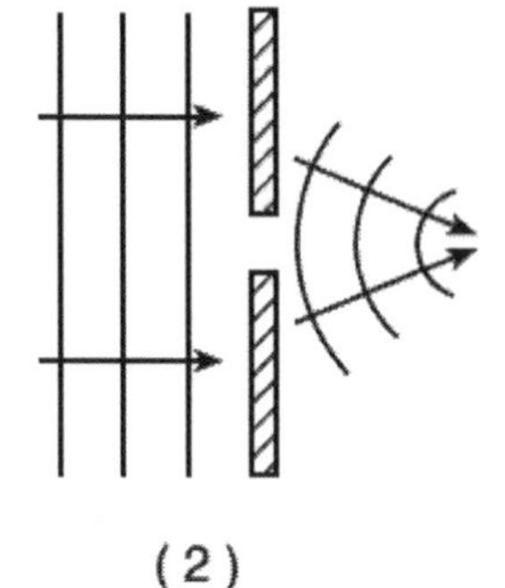

(2)

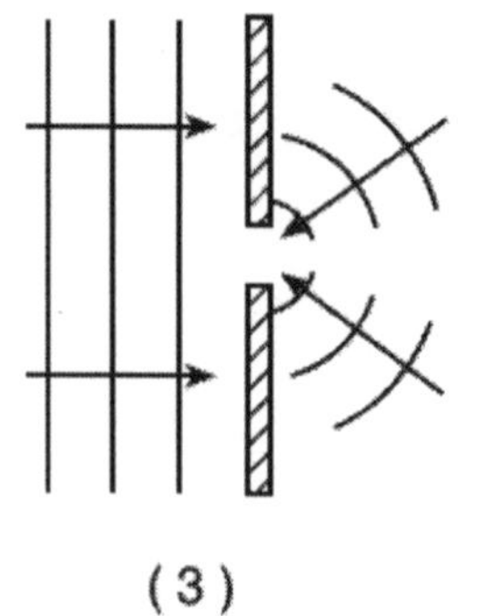

(3)

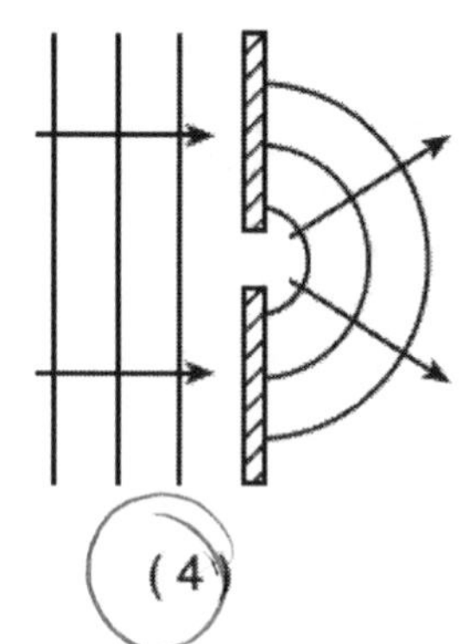

(4)

Name: ______________________ Period: __________

Waves-Diffraction

7. Parallel wave fronts incident on an opening in a barrier are diffracted. For which combination of wavelength and size of opening will diffraction effects be greatest?
   1. short wavelength and narrow opening
   2. short wavelength and wide opening
   3. long wavelength and narrow opening
   4. long wavelength and wide opening

8. A beam of monochromatic light approaches a barrier having four openings, A, B, C, and D, of different sizes as shown below.

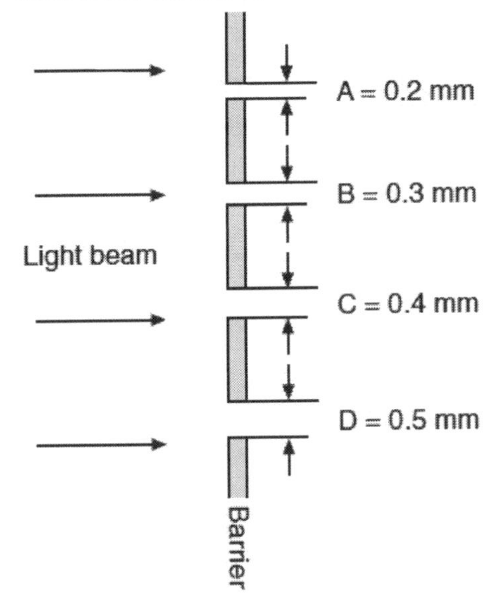

   Which opening will cause the greatest diffraction?
   1. A
   2. B
   3. C
   4. D

9. Radio waves diffract around buildings more than light waves do because, compared to light waves, radio waves
   1. move faster
   2. move slower
   3. have a higher frequency
   4. have a longer wavelength

10. Waves pass through a 10-centimeter opening in a barrier without being diffracted. This observation provides evidence that the wavelength of the waves is
    1. much shorter than 10 cm
    2. equal to 10 cm
    3. longer than 10 cm, but shorter than 20 cm
    4. longer than 20 cm.

11. The spreading of a wave into the region behind an obstruction is called
    1. diffraction
    2. absorption
    3. reflection
    4. refraction

12. The diagram below shows a series of straight wave fronts produced in a shallow tank of water approaching a small opening in a barrier.

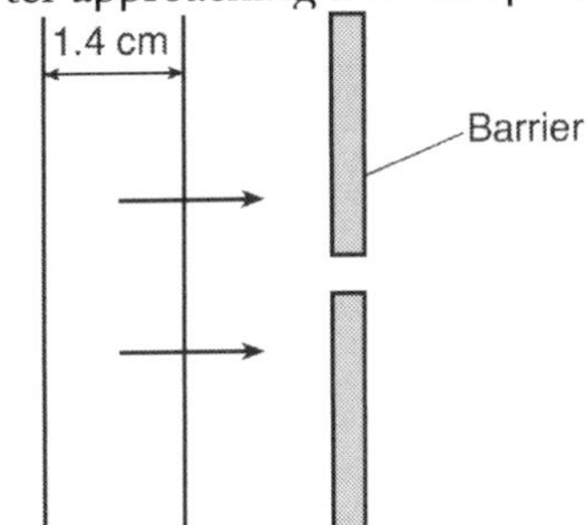

    Which diagram represents the appearance of the wave fronts after passing through the opening in the barrier?

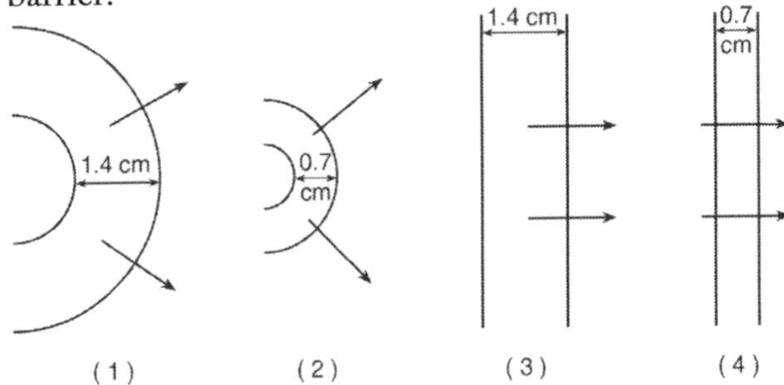

Name:________________________________ Period: ____________

Waves-Diffraction

13. The diagram below shows wave fronts approaching an opening in a barrier. The size of the opening is approximately equal to one-half the wavelength of the waves. On the diagram, draw the shape of at least three of the wave fronts after they have passed through this opening.

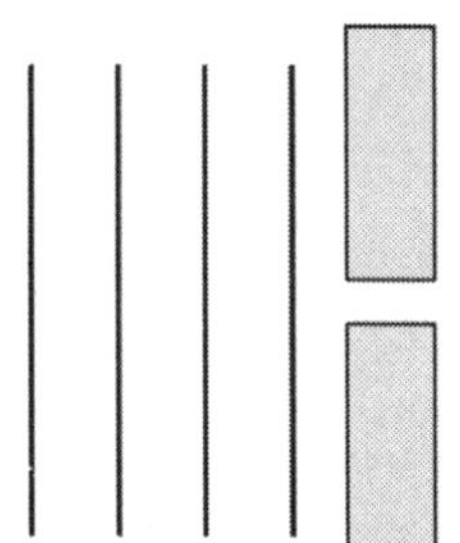

Name: ______________________________ Period: ____________

Waves-EM Spectrum

1. Compared to the speed of a sound wave in air, the speed of a radio wave in air is
   1. less
   2. greater
   3. the same

2. An electromagnetic AM-band radio wave could have a wavelength of
   1. 0.005 m
   2. 5 m
   3. 500 m
   4. 5,000,000 m

3. Which color of light has a wavelength of 5.0×10^{-7} meter in air?
   1. blue
   2. green
   3. orange
   4. violet

4. In a vacuum, all electromagnetic waves have the same
   1. speed
   2. phase
   3. frequency
   4. wavelength

5. An electromagnetic wave traveling through a vacuum has a wavelength of 1.5×10^{-1} meter. What is the period of this electromagnetic wave?
   1. 5.0×10^{-10} s
   2. 1.5×10^{-1} s
   3. 4.5×10^{7} s
   4. 2.0×10^{9} s

6. Which characteristic is the same for every color of light in a vacuum?
   1. energy
   2. frequency
   3. speed
   4. period

7. Explosure to ultraviolet radiation can damage skin. Exposure to visible light does not damage skin. State one possible reason for this difference.

8. An FM radio station broadcasts its signal at a frequency of 9.15×10^{7} hertz. Determine the wavelength of the signal in air.

Base your answers to questions 9 and 10 on the information and graph below.

Sunlight is composed of various intensities of all frequencies of visible light. The graph represents the relationship between light intensity and frequency.

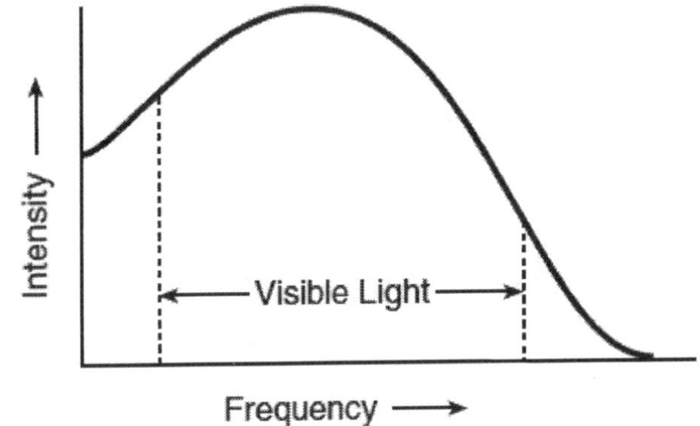

9. Based on the graph, which color of visible light has the lowest intensity?

10. It has been suggested that fire trucks be painted yellow green instead of red. Using information from the graph, explain the advantage of using yellow-green paint.

---

11. Which wave characteristics is the same for all types of electromagnetic radiation traveling in a vacuum?
    1. speed
    2. wavelength
    3. period
    4. frequency

12. Calculate the wavelength in a vacuum of a radio wave having a frequency of 2.2×10^{6} hertz. [Show all work, including the equation and substitution with units.]

Name:_______________ Period: ________

Waves-EM Spectrum

Base your answers to questions 13 and 14 on the information and diagram below.

A 1.50×10^{-6}-meter-long segment of an electromagnetic wave having a frequency of 6.00×10^{14} hertz is represented below.

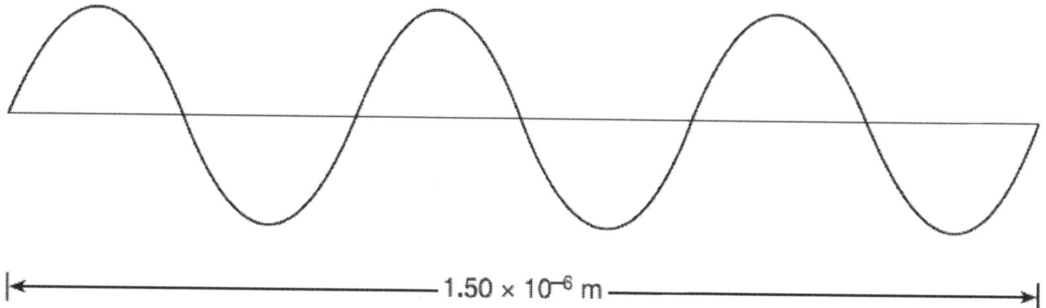

13. On the diagram above, mark two points on the wave that are in phase with each other. Label each point with the letter P.

14. Which type of electromagnetic wave does the segment in the diagram represent?

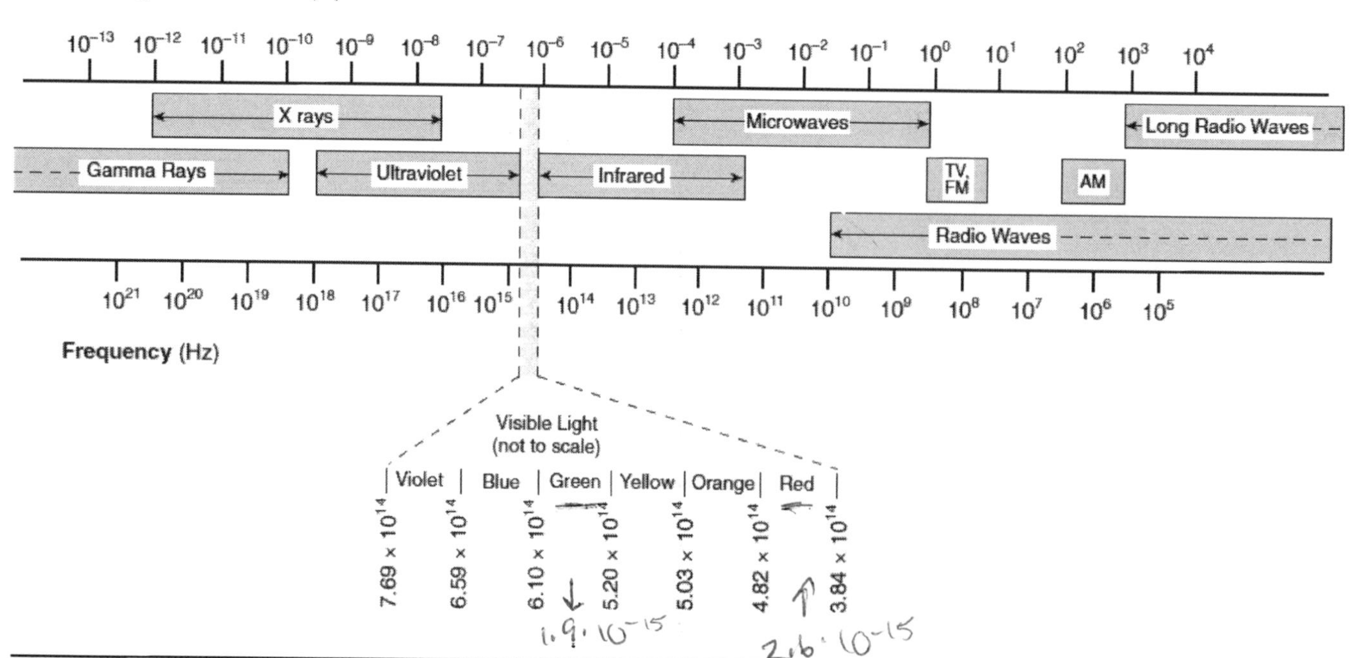

Name: ______________________ Period: ____________

Waves-EM Spectrum

15. Radio waves and gamma rays traveling in space have the same
    1. frequency
    2. wavelength
    3. period
    4. speed

16. Radio waves are propagated through the interaction of
    1. nuclear and electric fields
    2. electric and magnetic fields
    3. gravitational and magnetic fields
    4. gravitational and electric fields

17. Which pair of terms best describes light waves traveling from the Sun to Earth?
    1. electromagnetic and transverse
    2. electromagnetic and longitudinal
    3. mechanical and transverse
    4. mechanical and longitudinal

18. Which wavelength is in the infrared range of the electromagnetic spectrum?
    1. 100 nm
    2. 100 mm
    3. 100 m
    4. 100 μm

19. Compared to the period of a wave of red light the period of a wave of green light is
    1. less
    2. greater
    3. the same

20. Orange light has a frequency of 5.0×10^{14} hertz in a vacuum. What is the wavelength of this light?
    1. 1.5×10^{23} m
    2. 1.7×10^{6} m
    3. 6.0×10^{-7} m
    4. 2.0×10^{-15} m

21. What is the speed of a radio wave in a vacuum?
    1. 0 m/s
    2. 3.31×10^{2} m/s
    3. 1.13×10^{3} m/s
    4. 3.00×10^{8} m/s

22. How much time does it take light from a flash camera to reach a subject 6.0 meters across a room?
    1. 5.0×10^{-9} s
    2. 2.0×10^{-8} s
    3. 5.0×10^{-8} s
    4. 2.0×10^{-7} s

23. Which statement best describes a proton that is being accelerated?
    1. It produces electromagnetic radiation.
    2. The magnitude of its charge increases.
    3. It absorbs a neutron to become an electron.
    4. It is attracted to other protons.

24. An electromagnetic wave is produced by charged particles vibrating at a rate of 3.9×10^{8} vibrations per second. The electromagnetic wave is classified as
    1. a radio wave
    2. an infrared wave
    3. an x ray
    4. visible light

25. When x-ray radiation and infrared radiation are traveling in a vacuum, they have the same
    1. speed
    2. frequency
    3. wavelength
    4. energy per photon

26. A gamma ray and a microwave traveling in a vacuum have the same
    1. frequency
    2. period
    3. speed
    4. wavelength

Name: ______________________________ Period: __________

Modern-Wave Particle Duality

1. Compared to a photon of red light, a photon of blue light has a
   1. greater energy
   2. longer wavelength
   3. smaller momentum
   4. lower frequency

2. Exposure to ultraviolet radiation can damage skin. Exposure to visible light does not damage skin. State *one* possible reason for this difference.

Base your answers to questions 3 and 4 on the information below.

Louis de Broglie extended the idea of wave-particle duality to all of nature with his matter-wave equation:

$$\lambda = \frac{h}{mv}$$

where λ is the particle's wavelength, m is its mass, v is its velocity, and h is Planck's constant.

3. Using this equation, calculate the de Broglie wavelength of a helium nucleus (mass=6.7×10^{-27} kg) moving with a speed of 2.0×10^{6} meters per second.

4. The wavelength of this particle is of the same order of magnitude as which type of electromagnetic radiation?

5. A photon of light carries
   1. energy, but not momentum
   2. momentum, but not energy
   3. both energy and momentum
   4. neither energy nor momentum

6. Wave-particle duality is most apparent in analyzing the motion of
   1. a baseball
   2. a space shuttle
   3. a galaxy
   4. an electron

7. A photon of which electromagnetic radiation has the most energy?
   1. ultraviolet
   2. x ray
   3. infrared
   4. microwave

8. Light of wavelength 5.0×10^{-7} meter consists of photons having an energy of
   1. 1.1×10^{-48} J
   2. 1.3×10^{-27} J
   3. 4.0×10^{-19} J
   4. 1.7×10^{-5} J

9. Electrons oscillating with a frequency of 2.0×10^{10} hertz produce electromagnetic waves. These waves would be classified as
   1. infrared
   2. visible
   3. microwave
   4. x ray

10. The energy of a photon is inversely proportional to its
    1. wavelength
    2. speed
    3. frequency
    4. phase

11. A photon has a wavelength of 9.00×10^{-10} meter. Calculate the energy of this photon in joules. [Show all work, including the equation and substitution with units.]

Name: ______________________________ Period: ____________

Modern-Wave Particle Duality

Base your answers to questions 12 and 13 on the data table at right. The data table lists the energy and corresponding frequency of five photons.

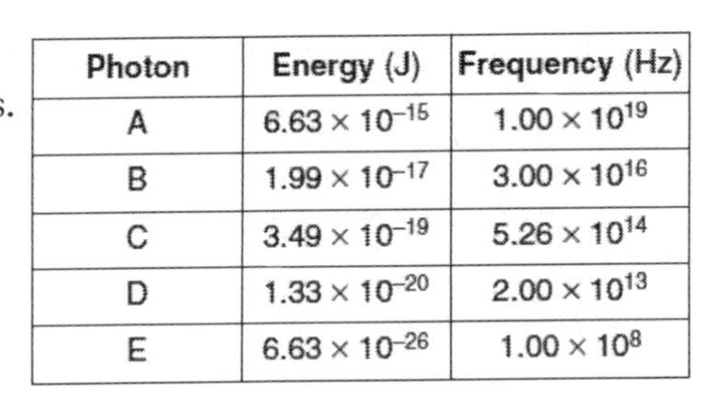

Photon	Energy (J)	Frequency (Hz)
A	6.63×10^{-15}	1.00×10^{19}
B	1.99×10^{-17}	3.00×10^{16}
C	3.49×10^{-19}	5.26×10^{14}
D	1.33×10^{-20}	2.00×10^{13}
E	6.63×10^{-26}	1.00×10^{8}

12. In which part of the electromagnetic spectrum would photon D be found?
   1. infrared
   2. visible
   3. ultraviolet
   4. x ray

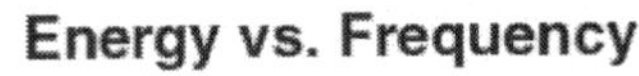

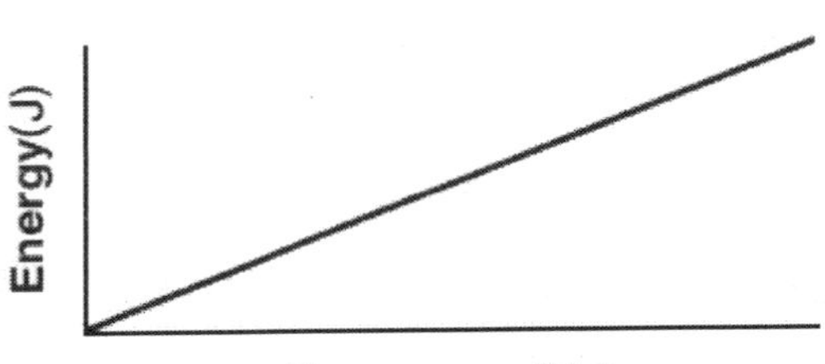

13. The graph at right represents the relationship between the energy and the frequency of photons. The slope of the graph would be
   1. 6.63×10^{-34} J·s
   2. 6.67×10^{-11} N·m²/kg²
   3. 1.60×10^{-19} J
   4. 1.60×10^{-19} C

Base your answers to questions 14 through 16 on the information below.

The alpha line in the Balmer series of the hydrogen spectrum consists of light having a wavelength of 6.56×10^{-7} meter.

14. Calculate the frequency of this light. [Show all work, including the equation and substitution with units.]

15. Determine the energy in joules of a photon of this light.

16. Determine the energy in electronvolts of a photon of this light.

17. Which phenomenon provides evidence that light has a wave nature?
   1. emission of light from an energy-level transition in a hydrogen atom
   2. diffraction of light passing through a narrow opening
   3. absorption of light by a black sheet of paper
   4. reflection of light from a mirror

18. The momentum of a photon, p, is given by the equation $p = \frac{h}{\lambda}$ where h is Planck's constant and λ is the photon's wavelength. Which equation correctly represents the energy of a photon in terms of its momentum?
   1. $E_{photon} = phc$
   2. $E_{photon} = \frac{hp}{c}$
   3. $E_{photon} = \frac{p}{c}$
   4. $E_{photon} = pc$

Name:______________________________ Period: __________

Modern-Wave Particle Duality

19. Which graph best represents the relationship between photon energy and photon frequency?

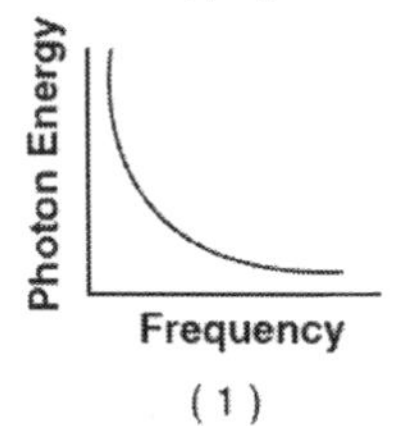

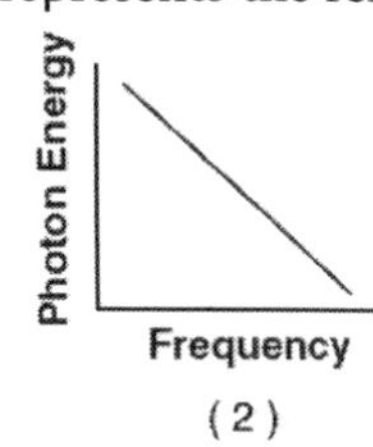

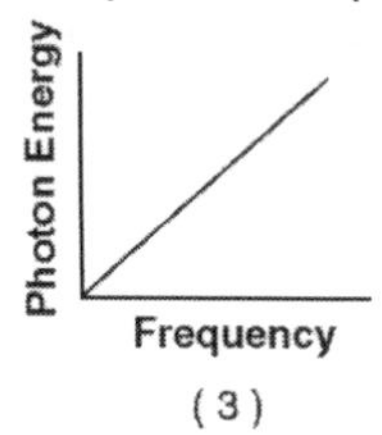

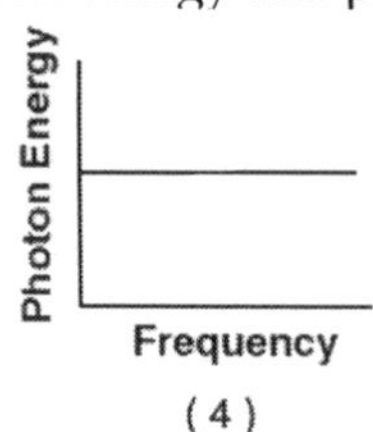

20. Light demonstrates the characteristics of
    1. particles, only
    2. waves, only
    3. both particles and waves
    4. neither particles nor waves

21. The slope of a graph of photon energy versus photon frequency represents
    1. Planck's constant
    2. the mass of a photon
    3. the speed of light
    4. the speed of light squared

22. A photon of light traveling through space with a wavelength of 6.0×10^{-7} meter has an energy of
    1. 4.0×10^{-40} J
    2. 3.3×10^{-19} J
    3. 5.4×10^{10} J
    4. 5.0×10^{14} J

23. On the atomic level, energy and matter exhibit the characteristics of
    1. particles, only
    2. waves, only
    3. neither particles nor waves
    4. both particles and waves

24. A variable-frequency light source emits a series of photons. As the frequency of the photon increases, what happens to the energy and wavelength of the photon?
    1. The energy decreases and the wavelength decreases.
    2. The energy decreases and the wavelength increases.
    3. The energy increases and the wavelength decreases.
    4. The enery increases and the wavelength increases.

25. Calculate the wavelength of a photon having 3.26×10^{-19} joule of energy. [Show all work, including the equation and substitution with units.]

$6.1 \cdot 10^{-7}$ m

26. All photons in a vacuum have the same
    1. speed
    2. wavelength
    3. energy
    4. frequency

27. Which phenomenon best supports the theory that matter has a wave nature?
    1. electron momentum
    2. electron diffraction
    3. photon momentum
    4. photon diffraction

28. Moving electrons are found to exhibit properties of
    1. particles, only
    2. waves, only
    3. both particles and waves
    4. neither particles nor waves

29. Determine the frequency of a photon whose energy is 3.00×10^{-19} joule.

Name:________________________ Period: ____________

Modern-Wave Particle Duality

Base your answers to questions 30 through 33 on the information below and your knowledge of physics.

An electron traveling with a speed of 2.50×10^6 meters per second collides with a photon having a frequency of 1.00×10^{16} hertz. After the collision, the photon has 3.18×10^{-18} joule of energy.

30. Calculate the original kinetic energy of the electron. [Show all work, including the equation and substitution with units.]

31. Determine the energy in joules of the photon before the collision.

32. Determine the energy lost by the photon during the collision.

33. Name *two* physical quantities conserved in the collision..

34. A monochromatic beam of light has a frequency of 7.69×10^{14} hertz. What is the energy of a photon of this light?
    1. 2.59×10^{-40} J
    2. 6.92×10^{-31} J
    3. 5.10×10^{-19} J
    4. 3.90×10^{-7} J

35. Which graph best represents the relationship between photon energy and photon wavelength?

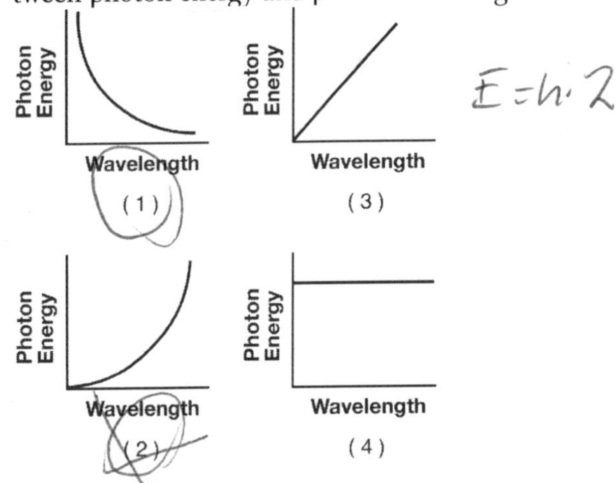

E=h·λ

36. A blue-light photon has a wavelength of 4.80×10^{-7} meter. What is the energy of the photon?
    1. 1.86×10^{22} J
    2. 1.44×10^{2} J
    3. 4.14×10^{-19} J
    4. 3.18×10^{-26} J

Name:____________________ Period: ____________

Modern-Energy Levels

1. An electron in a mercury atom drops from energy level i to the ground state by emitting a single photon. This photon has an energy of
   1. 1.56 eV
   2. 8.82 eV
   3. 10.38 eV
   4. 11.94 eV

down
↓
emits

2. White light passes through a cloud of cool hydrogen gas and is examined with a spectroscope. The dark lines observed on a bright background are caused by
   1. the hydrogen emitting all frequencies in white light
   2. the hydrogen absorbing certain frequencies of the white light
   3. diffraction of the white light
   4. constructive interference

3. The bright-line emission spectrum of an element can best be explained by
   1. electrons transitioning between discrete energy levels in the atoms of that element
   2. protons acting as both particles and waves
   3. electrons being located in the nucleus
   4. protons being dispersed uniformly throughout the atoms of that element

4. Explain why a hydrogen atom in the ground state can absorb a 10.2-electronvolt photon, but can *not* absorb an 11.0-electronvolt photon.

5. Excited hydrogen atoms are all in the *n=3* state. How many different photon energies could possibly be emitted as these atoms return to the ground state?
   1. 1
   2. 2
   3. 3
   4. 4

6. How much energy is required to move an electron in a mercury atom from the ground state to energy level *h*?
   1. 1.57 eV
   2. 8.81 eV
   3. 10.38 eV
   4. 11.95 eV

Base your answers to questions 7 through 10 on the information below.

An electron in a hydrogen atom drops from the n=3 energy level to the n=2 energy level.

7. What is the energy, in electronvolts, of the emitted photon?

8. What is the energy, in joules, of the emitted photon?

9. Calculate the frequency of the emitted radiation. [Show all work, including the equation and substitution with units.]

10. Calculate the wavelength of the emitted radiation. [Show all work, including the equation and substitution with units.]

11. A hydrogen atom with an electron initially in the n=2 level is excited further until the electron is in the n=4 level. This energy level change occurs because the atom has
    1. absorbed a 0.85-eV photon
    2. emitted a 0.85-eV photon
    3. absorbed a 2.55-eV photon
    4. emitted a 2.55-eV photon

Name: ______________________________ Period: ____________

Modern-Energy Levels

Base your answers to questions 12 through 14 on the information below.

The light of the "alpha line" in the Balmer series of the hydrogen spectrum has a wavelength of 6.58×10^{-7} m.

12. Calculate the energy of an "alpha line" photon in joules. [Show all work, including the equation and substitution with units.]

13. What is the energy of an "alpha line" photon in electronvolts?

14. Using your answer to question 13, explain whether or not this result verifies that the "alpha line" corresponds to a transition from the energy level n=3 to energy level n=2 in a hydrogen atom.

15. An electron in the c level of a mercury atom returns to the ground state. Which photon energy could *not* be emitted by the atom during this process?
    1. 0.22 eV
    2. 4.64 eV
    3. 4.86 eV
    4. 5.43 eV

Base your answers to questions 16 through 18 on the information below.

A photon with a frequency of 5.02×10^{14} hertz is absorbed by an excited hydrogen atom. This causes the electron to be ejected from the atom, forming an ion.

16. Calculate the energy of this photon in joules. [Show all work, including the equation and substitution with units.]

17. Determine the energy of this photon in electron-volts.

18. What is the number of the lowest energy level (closest to the ground state) of a hydrogen atom that contains an electron that would be ejected by the absorption of this photon?

19. A photon having an energy of 9.40 electronvolts strikes a hydrogen atom in the ground state. Why is the photon not absorbed by the hydrogen atom?
    1. The atom's orbital electron is moving too fast.
    2. The photon striking the atom is moving too fast.
    3. The photon's energy is too small.
    4. The photon is being repelled by electrostatic force.

Name: ____________________ Period: __________

Modern-Energy Levels

Base your answers on questions 20 through 22 on the information below.

A photon with a frequency of 5.48×10^{14} hertz is emitted when an electron in a mercury atom falls to a lower energy level.

20. Identify the color of light associated with this photon.

21. Calculate the energy of this photon in joules. [Show all work, including the equation and substitution with units.]

22. Determine the energy of this photon in electronvolts.

Base your answers to questions 23 through 25 on the information below.

A photon with a wavelength of 2.29×10^{-7} meter strikes a mercury atom in the ground state.

23. Calculate the energy, in joules, of this photon. [Show all work, including the equation and substitution with units.]

24. Determine the energy, in electronvolts, of this photon.

25. Based on your answer to question 24, state if this photon can be absorbed by the mercury atom. Explain your answer.

Base your answers to questions 26 through 29 on the information below.

As a mercury atom absorbs a photon of energy, an electron in the atom changes from energy level *d* to energy level *e*.

26. Determine the energy of the absorbed photon in electronvolts.

27. Express the energy of the absorbed photon in joules.

28. Calculate the frequency of the absorbed photon. [Show all work, including the equation and substitution with units.]

29. Based on your calculated value of the frequency of the absorbed photon, determine its classification in the electromagnetic spectrum.

---

30. Which type of photon is emitted when an electron in a hydrogen atom drops from the n=2 to the n=1 energy level?
    1. ultraviolet
    2. visible light
    3. infrared
    4. radio wave

31. An electron in a mercury atom drops from energy level *f* to energy level *c* by emitting a photon having an energy of
    1. 8.20 eV
    2. 5.52 eV
    3. 2.84 eV
    4. 2.68 eV

Name: ______________________ Period: __________

Modern-Energy Levels

32. The diagram below represents the bright-line spectra of four elements, A, B, C, and D, and the spectrum of an unknown gaseous sample.

Unknown sample

Element A

Element B

Element C

Element D

Based on comparisons of these spectra, which two elements are found in the unknown sample?
1. A and B
2. A and D
3. B and C
4. C and D

33. A mercury atom in the ground state absorbs 20.00 electronvolts of energy and is ionized by losing an electron. How much kinetic energy does this electron have after ionization?
1. 6.40 eV
2. 9.62 eV
3. 10.38 eV
4. 13.60 eV

Base your answers to questions 34 and 35 on the information below.

In a mercury atom, as an electron moves from energy level *i* to energy level *a*, a single photon is emitted.

34. Determine the energy, in electronvolts, of this emitted photon.

35. Determine this photon's energy, in joules.

Base your answers to questions 36 through 39 on the Energy Level Diagram for Hydrogen in the *Reference Tables for Physical Settings/Physics.*

36. Determine the energy, in electronvolts, of a photon emitted by an electron as it moves from the n=6 to n=2 energy level in a hydrogen atom.

37. Convert the energy of the photon to joules.

38. Calculate the frequency of the emitted photon. [Show all work, including the equation and substitution with units.]

39. Is this the only energy and/or frequency that an electron in the n=6 energy level of a hydrogen atom could emit? Explain your answer.

40. Electrons in excited hydrogen atoms are in the n=3 energy level. How many different photon frequencies could be emitted as the atoms return to the ground state?
1. 1
2. 2
3. 3
4. 4

Name:______________________________ Period: ____________

Modern-Energy Levels

Base your answers to questions 41 through 43 on the information below.

Auroras over the polar regions of Earth are caused by collisions between charged particles from the Sun and atoms in Earth's atmosphere. The charged particles give energy to the atoms, exciting them from their lowest available energy level, the ground state, to higher energy levels, excited states. Most atoms return to their ground state within 10 nanoseconds.

In the higher regions of the Earth's atmosphere, where there are fewer interatom collisions, a few of the atoms remain in excited states for longer times. For example, oxygen atoms remain in an excited state for up to 1.0 second. These atoms account for the greenish and red glows of the auroras. As these oxygen atoms return to their ground state, they emit green photons ($f=5.38 \times 10^{14}$ Hz) and red photons ($f=4.76 \times 10^{14}$ Hz). These emissions last long enough to produce the changing aurora phenomenon.

41. What is the order of magnitude of the time, in seconds, that most atoms spend in an excited state?

42. Calculate the energy of a photon, in joules, that accounts for the red glow of the aurora. [Show all work, including the equation and substitution with units.]

43. Explain what is meant by an atom being in its ground state.

44. A photon is emitted as the electron in a hydrogen atom drops from the n=5 energy level directly to the n=3 energy level. What is the energy of the emitted photon?
 1. 0.85 eV
 2. 0.97 eV
 3. 1.51 eV
 4. 2.05 eV

45. Which electron transition between the energy levels of hydrogen causes the emission of a photon of visible light?
 1. n=6 to n=5
 2. ~~n=5 to n=6~~
 3. n=5 to n=2
 4. ~~n=2 to n=5~~

down

46. What is the minimum energy required to ionize a hydrogen atom in the n=3 state?
 1. 0.00 eV
 2. 0.66 eV
 3. 1.51 eV
 4. 12.09 eV

Name: ______________________________ Period: __________

Modern-Energy Levels

Base your answers to questions 47 through 50 on the information below and on your knowledge of physics.

An electron in a mercury atom changes from energy level b to a higher energy level when the atom absorbs a single photon with an energy of 3.06 electronvolts.

47. Determine the letter that identifies the energy level to which the electron jumped when the mercury atom absorbed the photon.

48. Determine the energy of the photon, in joules.

49. Calculate the frequency of the photon. [Show all work, including the equation and substitution with units.]

50. Classify the photon as one of the types of electromagnetic radiation listed in the electromagnetic spectrum.

Name: ______________________________ Period: ____________

Modern-Mass Energy Equivalence

1. If a deuterium nucleus has a mass of 1.53×10^{-3} universal mass units less than its components, this mass represents an energy of
   1. 1.38 MeV
   2. 1.42 MeV
   3. 1.53 MeV
   4. 3.16 MeV

2. The energy equivalent of 5.0×10^{-3} kilogram is
   1. 8.0×10^{5} J
   2. 1.5×10^{6} J
   3. 4.5×10^{14} J
   4. 3.0×10^{19} J

3. How much energy, in megaelectronvolts, is produced when 0.250 universal mass unit of matter is completely converted into energy?

   233 Mev

4. The energy equivalent of the rest mass of an electron is approximately
   1. 5.1×10^{5} J
   2. 8.2×10^{-14} J
   3. 2.7×10^{-22} J
   4. 8.5×10^{-28} J

5. The energy produced by the complete conversion of 2.0×10^{-5} kilogram of mass into energy is
   1. 1.8 TJ
   2. 6.0 GJ
   3. 1.8 MJ
   4. 6.0 kJ

6. What is the minimum total energy released when an electron and its antiparticle (positron) annihilate each other?
   1. 1.64×10^{-13} J
   2. 8.20×10^{-14} J
   3. 5.47×10^{-22} J
   4. 2.73×10^{-22} J

7. The energy required to separate the 3 protons and 4 neutrons in the nucleus of a lithium atom is 39.3 megaelectronvolts. Determine the mass equivalent of this energy, in universal mass units.

8. Which graph best represents the relationship between enery and mass when matter is converted into energy?

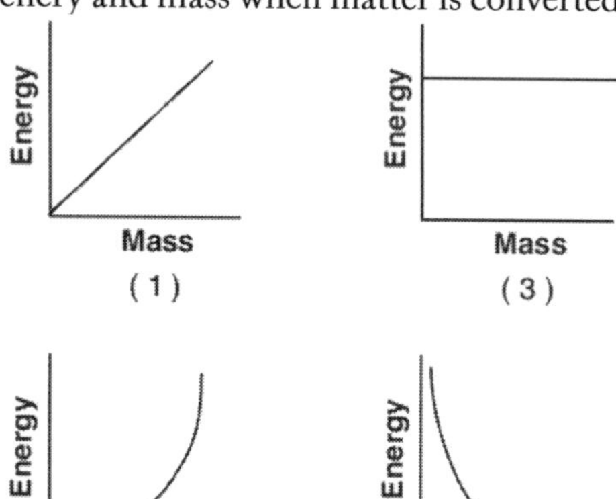

Energy
Mass
(2)
Energy
Mass
(4)

9. The total conversion of 1.00 kilograms of the Sun's mass into energy yields
   1. 9.31×10^{2} MeV
   2. 8.38×10^{19} MeV
   3. 3.00×10^{8} J
   4. 9.00×10^{16} J

10. What total mass must be converted into energy to produce a gamma photon with an energy of 1.03×10^{-13} joule?
    1. 1.14×10^{-30} kg
    2. 3.43×10^{-22} kg
    3. 3.09×10^{-5} kg
    4. 8.75×10^{29} kg

11. A tritium nucleus is formed by combining two neutrons and a proton. The mass of this nucleus is 9.106×10^{-3} universal mass unit less than the combined mass of the particles from which it is formed. Approximately how much energy is released when this nucleus is formed?
    1. 8.48×10^{-2} MeV
    2. 2.73 MeV
    3. 8.48 MeV
    4. 273 MeV

12. After a uranium nucleus emits an alpha particle, the total mass of the new nucleus and the alpha particle is less than the mass of the original uranium nucleus. Explain what happens to the missing mass.

Name: ______________________________ Period: ____________

Modern-Mass Energy Equivalence

Base your answers to questions 13 and 14 on the information and data table below.

In the first nuclear reaction using a particle accelerator, accelerated protons bombarded lithium atoms, producing alpha particles and energy. The energy resulted from the conversion of mass into energy. The reaction can be written as shown below.

$$^{1}_{1}H + ^{7}_{3}Li \rightarrow ^{4}_{2}He + ^{4}_{2}He + energy$$

Data Table

Particle	Symbol	Mass (u)
proton	$^{1}_{1}H$	1.007 83
lithium atom	$^{7}_{3}Li$	7.016 00
alpha particle	$^{4}_{2}He$	4.002 60

13. Determine the difference between the total mass of a proton plus a lithium atom, $^{1}_{1}H + ^{7}_{3}Li$, and the total mass of two alpha particles, $^{4}_{2}He + ^{4}_{2}He$, in universal mass units.

14. Determine the energy in megaelectronvolts produced in the reaction of a proton with a lithium atom.

15. If a proton were to combine with an antiproton, they would annihilate each other and become energy. Calculate the amount of energy that would be released by this annihilation. [Show all work, including the equation and substitution with units.]

16. The graph below represents the relationship between energy and the equivalent mass from which it can be converted.

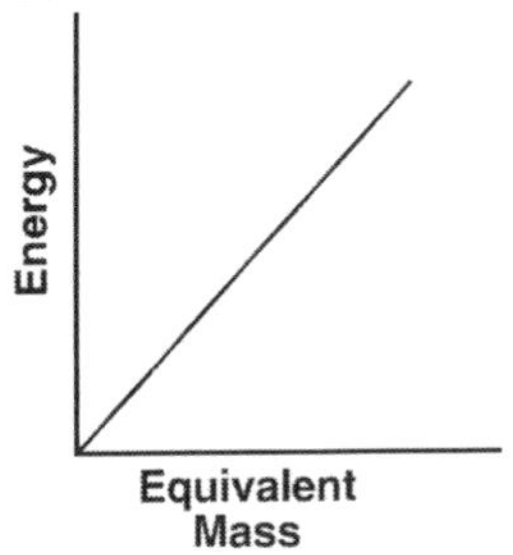

The slope of this graph represents
1. c
2. c^2
3. g
4. g^2

17. What is the total energy released when 9.11×10^{-31} kilogram of mass is converted into energy?
1. 2.73×10^{-22} J
2. 8.20×10^{-14} J
3. 9.11×10^{-31} J
4. 1.01×10^{-47} J

18. Calculate the energy equivalent in joules of the mass of a proton. [Show all work, including the equation and substitution with units.]

Name:____________________________ Period: __________

Modern-Standard Model

1. The strong force is the force of
   1. repulsion between protons
   2. attraction between protons and electrons
   3. repulsion between nucleons
   4. attraction between nucleons

2. The tau neutrino, the muon neutrino, and the electron neutrino are all
   1. leptons
   2. hadrons
   3. baryons
   4. mesons

Base your answers to questions 3 and 4 on the information below.

A lambda particle consists of an up, a down, and a strange quark.

3. A lambda particle can be classified as a
   1. baryon
   2. lepton
   3. meson
   4. photon

4. What is the charge of a lambda particle in elementary charges?

5. According to the Standard Model, a proton is constructed of two up quarks and one down quark (*uud*) and a neutron is constructed of one up quark and two down quarks (*udd*). During beta decay, a neutron decays into a proton, an electron, and an electron antineutrino. During this process there is a conversion of a
   1. *u* quark to a *d* quark
   2. *d* quark to a meson
   3. baryon to another baryon
   4. lepton to another lepton

6. Which statement is true of the strong nuclear force?
   1. It acts over very great distances
   2. It holds protons and neutrons together
   3. It is much weaker than gravitational forces
   4. It repels neutral charges

7. Which combination of quarks could produce a neutral baryon?
   1. *cdt*
   2. *cts*
   3. *cdb*
   4. *cdu*

8. A meson may *not* have a charge of
   1. +1e
   2. +2e
   3. 0e
   4. -1e

Base your answers to questions 9 and 10 on the information and equation below.

During the process of beta (β^-) emission, a neutron in the nucleus of an atom is converted into a proton, an electron, an electron antineutrino, and energy.

$$neutron \rightarrow proton + electron + electron\ antineutrino + energy$$

9. Based on conservation laws, how does the mass of the neutron compare to the mass of the proton?

10. Since charge must be conserved in the reaction shown, what charge must an electron antineutrino carry?

11. Protons and neutrons are examples of
    1. positrons
    2. baryons
    3. mesons
    4. quarks

12. The force that holds protons and neutrons together is known as the
    1. gravitational force
    2. strong force
    3. magnetic force
    4. electrostatic force

Name:________________________________ Period: ____________

Modern-Standard Model

Base your answers to questions 13 through 16 on the passage below and on your knowledge of physics.

More Sci- Than Fi, Physicists Create Antimatter

Physicists working in Europe announced yesterday that they had passed through nature's looking glass and had created atoms made of antimatter, or antiatoms, opening up the possibility of experiments in a realm once reserved for science fiction writers. Such experiments, theorists say, could test some of the basic tenets of modern physics and light the way to a deeper understanding of nature.

By corralling [holding together in groups] clouds of antimatter particles in a cylindrical chamber laced with detectors and electric and magnetic fields, the physicists assembled antihydrogen atoms, the looking glass equivalent of hydrogen, the most simple atom in nature. Whereas hydrogen consists of a positively charged proton circled by a negatively charged electron, in antihydrogen the proton's counterpart, a positively charged antiproton, is circled by an antielectron, otherwise known as a positron.

According to the standard theories of physics, the antimatter universe should look identical to our own. Antihydrogen and hydrogen atoms should have the same properties, emitting the exact same frequencies of light, for example. . . .

Antimatter has been part of physics since 1927 when its existence was predicted by the British physicist Paul Dirac. The antielectron, or positron, was discovered in 1932. According to the theory, matter can only be created in particle-antiparticle pairs. It is still a mystery, cosmologists say, why the universe seems to be overwhelmingly composed of normal matter.

Dennis Overbye, "More Sci- Than Fi, Physicists Create Antimatter," New York Times, Sept. 19, 2002

13. The author of the passage concerning antimatter incorrectly reported the findings of the experiment on antimatter. Which particle mentioned in the article has the charge incorrectly identified?

14. How should the emission spectrum of antihydrogen compare to the emission spectrum of hydrogen?

15. Identify one characteristic that antimatter particles must possess if clouds of them can be corralled by electric and magnetic fields.

16. According to the article, why is it a mystery that "the universe seems to be overwhelmingly composed of normal matter?"

17. The particles in the nucleus are held together primarily by the
    1. strong force
    2. gravitational force
    3. electrostatic force
    4. magnetic force

18. Baryons may have charges of
    1. +1 e and +4/3 e
    2. +2 e and +3 e
    3. -1 e and +1 e
    4. -2 e and -2/3 e

Name:______________________________ Period: ____________

Modern-Standard Model

19. The charge of an antistrange quark is approximately
    1. $+5.33 \times 10^{-20}$ C
    2. -5.33×10^{-20} C
    3. $+5.33 \times 10^{20}$ C
    4. -5.33×10^{20} C

20. Which fundamental force holds quarks together to form particles such as protons and neutrons?
    1. electromagnetic force
    2. gravitational force
    3. strong force
    4. weak force

21. What is the total number of quarks in a helium nucleus consisting of 2 protons and 2 neutrons?
    1. 16
    2. 12
    3. 8
    4. 4

Base your answers to questions 22 and 23 on the statement below.

The spectrum of visible light emitted during transitions in excited hydrogen atoms is composed of blue, green, red, and violet lines.

22. What characteristics of light determines the amount of energy carried by a photon of that light?
    1. amplitude
    2. frequency
    3. phase
    4. velocity

23. Which color of light in the visible hydrogen spectrum has photons of the shortest wavelength?
    1. blue
    2. green
    3. red
    4. violet

24. What are the sign and charge, in coulombs, of an antiproton?

25. A deuterium nucleus consists of one proton and one neutron. The quark composition of a deuterium nucleus is
    1. 2 up quarks and 2 down quarks
    2. 2 up quarks and 4 down quarks
    3. 3 up quarks and 3 down quarks
    4. 4 up quarks and 2 down quarks

26. Which particles are not affected by the strong force?
    1. hadrons
    2. protons
    3. neutrons
    4. electrons

27. A tau lepton decays into an electron, an electron antineutrino, and a tau neutrino, as represented in the reaction below.

$$\tau \rightarrow e + \bar{v}_e + v_\tau$$

On the equation above, show how this reaction obeys the Law of Conservation of Charge by indicating the amount of charge on each particle.

28. Compared to the mass and charge of a proton, an antiproton has
    1. the same mass and the same charge
    2. greater mass and the same charge
    3. the same mass and the opposite charge
    4. greater mass and the opposite charge

29. Which fundamental force is primarily responsible for the attraction between protons and electrons?
    1. strong
    2. weak
    3. gravitational
    4. electromagnetic

30. A subatomic particle could have a charge of
    1. 5.0×10^{-20} C
    2. 8.0×10^{-20} C
    3. 3.2×10^{-19} C
    4. 5.0×10^{-19} C

31. A particle that is composed of two up quarks and one down quark is a
    1. meson
    2. neutron
    3. proton
    4. positron

Name: ______________________ Period: __________

Modern-Standard Model

Base your answers to questions 32 through 34 on the passage below.

For years, theoretical physicists have been refining a mathematical method called lattice quantum chromodynamics to enable them to predict the masses of particles consisting of various combinations of quarks and antiquarks. They recently used the theory to calculate the mass of the rare B_c particle, consisting of a charm quark and a bottom antiquark. The predicted mass of the B_c particle was about six times the mass of a proton.

Shortly after the prediction was made, physicists working at the Fermi National Accelerator Laboratory, Fermilab, were able to measure the mass of the B_c particle experimentally and found it to agree with the theoretical prediction to within a few tenths of a percent. In the experiment, the physicists sent beams of protons and antiprotons moving at 99.999% the speed of light in opposite directions around a ring 1.0 kilometer in radius. The protons and antiprotons were kept in their circular paths by powerful electromagnets. When the protons and antiprotons collided, their energy produced numerous new particles, including the elusive B_c.

These results indicate that lattice quantum chromodynamics is a powerful tool not only for confirming the masses of existing particles, but also for predicting the masses of particles that have yet to be discovered in the laboratory.

32. Identify the class of matter to which the B_c particle belongs.

33. Determine both the sign and the magnitude of the charge of the B_c particle in elementary charges.

34. Explain how it is possible for a colliding proton and antiproton to produce a particle with six times the mass of either.

---

35. The diagram below represents the sequence of events (steps 1 through 10) resulting in the production of a D^- meson and a D^+ meson. An electron and a positron (antielectron) collide (step 1), annihilate each other (step 2), and become energy (step 3). This energy produces an anticharm quark and a charm quark (step 4), which then split apart (steps 5 through 7). As they split, a down quark and an antidown quark are formed, leading to the final production of a D^- meson and a D^+ meson (steps 8 through 10).

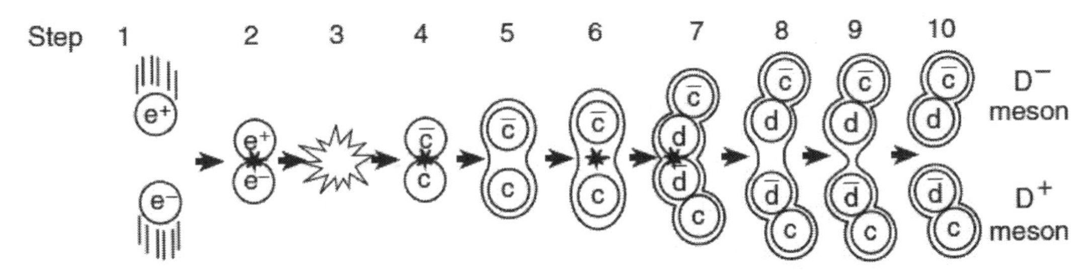

Adapted from: Electon/Positron Annihilation http:/www.particleadventure.org/frameless/eedd.html 7/23/2007

Which statement best describes the changes that occur in this sequence of events?

1. Energy is converted into matter and then matter is converted into energy.
2. Matter is converted into energy and then energy is converted into matter.
3. Isolated quarks are being formed from baryons.
4. Hadrons are being converted into leptons.

Name: ______________________________ Period: ____________

Modern-Standard Model

Base your answers to questions 36 and 37 on the table below, which shows data about various subatomic particles.

Subatomic Particle Table

Symbol	Name	Quark Content	Electric Charge	Mass (GeV/c^2)
p	proton	uud	+1	0.938
$\bar{p}$	antiproton	$\bar{u}\bar{u}\bar{d}$	−1	0.938
n	neutron	udd	0	0.940
λ	lambda	uds	0	1.116
Ω^-	omega	sss	−1	1.672

36. Which particle listed on the table has the opposite charge of, and is more massive than, a proton?
    1. antiproton
    2. neutron
    3. lambda
    4. omega

37. All the particles listed on the table are classified as
    1. mesons
    2. hadrons
    3. antimatter
    4. leptons

38. According to the Standard Model of Particle Physics, a meson is composed of
    1. a quark and a muon neutrino
    2. a quark and an antiquark
    3. three quarks
    4. a lepton and an antilepton

39. A particle unaffected by an electric field could have a quark composition of
    1. css
    2. bbb
    3. udc
    4. uud

40. A helium atom consists of two protons, two electrons, and two neutrons. In the helium atom, the strong force is a fundamental interaction between the
    1. electrons, only
    2. electrons and protons
    3. neutrons and electrons
    4. neutrons and protons

41. A lithium atom consists of 3 protons, 4 neutrons, and 3 electrons. This atom contains a total of
    1. 9 quarks and 7 leptons
    2. 12 quarks and 6 leptons
    3. 14 quarks and 3 leptons
    4. 21 quarks and 3 leptons

42. A top quark has an approximate charge of
    1. -1.07×10^{-19} C
    2. -2.40×10^{-19} C
    3. $+1.07 \times 10^{-19}$ C
    4. $+2.40 \times 10^{-19}$ C

43. The composition of a meson with a charge of -1 elementary charge could be
    1. $s\bar{c}$
    2. dss
    3. $u\bar{b}$
    4. $\overline{ucd}$

44. In a process called pair production, an energetic gamma ray is converted into an electron and a positron. It is not possible for a gamma ray to be converted into two electrons because
    1. charge must be conserved
    2. momentum must be conserved
    3. mass-energy must be conserved
    4. baryon number must be conserved

45. An antibaryon composed of two antiup quarks and one antidown quark would have a charge of
    1. +1e
    2. -1e
    3. 0e
    4. -3e

46. Which force is responsible for producing a stable nucleus by opposing the electrostatic force of repulsion between protons?
    1. strong
    2. weak
    3. frictional
    4. gravitational

Name: ______________________ Period: ____________

Modern-Standard Model

Base your answers to questions 47 through 50 on the information below.

Two experiments running simultaneously at the Fermi National Accelerator Laboratory in Batavia, Ill., have observed a new particle called the cascade baryon. It is one of the most massive examples yet of a baryon -- a class of particles made of three quarks held together by the strong nuclear force -- and the first to contain one quark from each of the three known families, or generations, of these elementary particles.

Protons and neutrons are made of up and down quarks, the two first-generation quarks. Strange and charm quarks constitute the second generation, while the top and bottom varieties make up the third. Physicists had long conjectured that a down quark could combine with a strange and a bottom quark to form the three-generation cascade baryon.

On June 13, the scientists running Dzero, one of two detectors at Fermilab's Tevatron accelerator, announced that they had detected characteristic showers of particles from the decay of cascade baryons. The baryons formed in proton-antiproton collisions and lived no more than a trillionth of a second. A week later, physicists at CDF, the Tevatron's other detector, reported their own sighting of the baryon...

Source: D.C., "Pas de deux for a three-scoop particle," Science News, Vol. 172, July 7, 2007

47. Which combination of three quarks will produce a neutron?

48. What is the magnitude and sign of the charge, in elementary charges, of a cascade baryon?

49. The Tevatron derives its name from teraelectronvolt, the maximum energy it can impart to a particle. Determine the energy, in joules, equivalent to 1.00 teraelectronvolt.

50. Calculate the maximum total mass, in kilograms, of particles that could be created in the head-on collision of a proton and an antiproton, each having an energy of 1.60×10^{-7} joule. [Show all work, including the equation and substitution with units.]

51. What is the quark composition of a proton?
    1. uud
    2. udd
    3. csb
    4. uds

Solutions

Metric Estimation

1. 3) 0.01 m
2. 3) 10^3 kg
3. 3) 10^{-2} m
4. 2) 10^1 m
5. 2) 1.3×10^{-1} m
6. 3) 1×10^{-3} kg
7. 2) 1.5×10^{-1} m
8. 2) 10^0 m
9. 2) 2×10^{-1} m
10. 2) 1000 dm
11. 2) 2×10^{-1} m
12. 3) 10^2 m
13. 2) 10^0 m

Kinematics-Defining Motion

1. 1) less
2. 3) 6.0 m/s
3. 1) vector quantity that has a direction associated with it
4. 2) 2.21 s
5. 2) 2.00×10^4 kg
6. 2) 650 km/h
7. 4) average speed
8. magnitude and direction
9. 1 cm = 0.19 m/s
10. vector up and to the right
11. v=1.66 m/s
12. 65°
13. 2) 6m shorter
14. 1) 0.16 km/min
15. 1) 20 m south
16. 4) displacement
17. 5.66 m
18. 1 cm = 2.1 m
19. vector up and to the right
20. 10 m
21. 3) 85 km/h
22. scalars have magnitude only; vectors have magnitude and direction
23. 1
24. 2) 2.5 m/s
25. 3) 15 s
26. 120 m east
27. 4) 83.3 m/s
28. 2) 5.0 m
29. 1) speed is to velocity
30. 1) 2.5 m/s
31. 4) distance
32. 8.6 km
33. 12 km
34. 4) 180 m
35. 3) 3.0 m/s
36. 25 s
37. vector starting at P pointing to the right with length 4 cm
38. 3.6 ms
39. 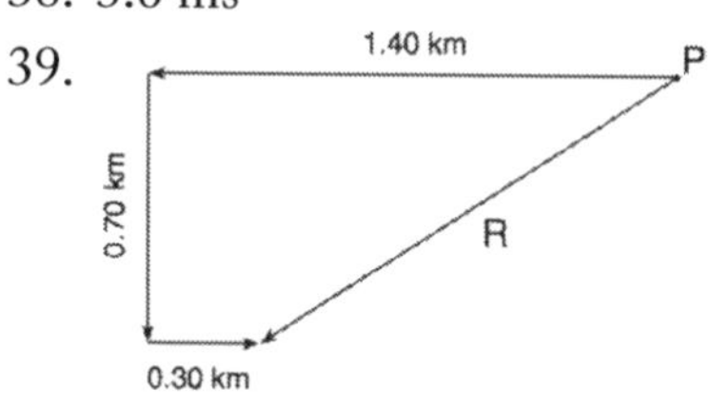

40. 0.2 km/min or 3.3 m/s
41. 1.3 km
42. 32°
43. 3) 226 m/s
44. 1) 14 m/s

Kinematics-Motion Graphs

1. 1
2. 10 m/s
3. 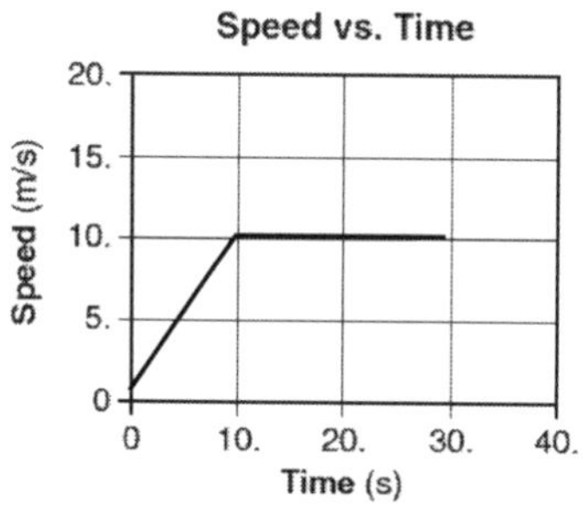 Speed vs. Time
4. 50 m
5. 4
6. 4) 24 m
7. 4
8. 1
9. 50 m
10. 4
11. 1) 0.0 m/s^2
12. 3) 40 m
13. 4
14. 1) 12 m
15. 3) 40 m
16. 1.25 m/s^2
17. 30 m
18. 15 m/s
19. 2.5 m/s^2
20. displacement, distance traveled, etc.

Solutions

Kinematics-Horizontal Kinematics

1. 1225 m
2. vector pointing toward bottom of page with a length of 4 cm
3. 3) 1.5 m/s^2
4. 3) 216 m
5. 3) 44.3 m/s
6. 5.0 cm
7. 1.11 m/s^2
8. 0.167 m/s
9. four evenly spaced dots
10. 2) 2.0 m/s^2
11. 1) 2.5 m/s^2
12. Start at rest and time how long it takes for skater to reach a set distance. Measure distance.
13. $a=2d/t^2$
14. 1) 0.40 m/s^2
15. 1) directed northward
16. 3) 1.2×10^2 m
17. 2.4 m/s^2
18. 19 m/s
19. 2) 1.1 m/s^2
20. 2) 2.2 m/s^2
21. 1) 1.9 m/s^2
22. 2 m/s^2 west
23. 4) 1.5×10^2 m
24. 3) 3.0 m/s
25. 1) 3.5 m/s^2 east
26. 2) 1.5 m/s^2
27. 1) 8.0 m/s
28. 4) 83 m
29. 4) 2400 m
30. 1) 2.5 m/s^2

Kinematics-Free Fall

1. make graph
2. make graph
3. make graph
4. 15.5 m/s
5. 4) Acceleration remains the same and speed increases
6. 3) 3.6 m/s^2
7. 2) 7.4 m/s
8. 3) 1.6 s
9. 2) at the end of its first second of fall
10. 3) 44.1 m
11. 3) 3.06 s
12. 2) 15 m/s
13. 3) 47 m
14. 2) 2 s
15. 3) 4.9 m
16. 2) 0.78 m
17. 3) 44 m
18. 1) 0.0 m/s
19. 1) 3.0×10^1 s
20. air resistance
21. 1) 2.9 s
22. 2) 46 m
23. 3) 78.3 m
24. 1) 1.6 m/s^2
25. 3) 10.1 s
26. 2) 4.90 m
27. 2)

Kinematics-Projectiles

1. 1
2. vector up and to the right at an angle of 60 degrees above the horizontal with a length of 5 cm
3. 125 m/s
4. there are no horizontal forces to cause acceleration, as gravity only pulls down
5. 1) the same
6. 3) the same
7. the same
8. $v_{yA} < v_{yB}$
9. parabolic path but lands closer to the base of the cliff
10. 4) 48 m/s
11. 1) 8.6 m/s
12. 1) 3.19 s
13. 4) increase the launch angle and increase the ball's initial speed
14. 16.1 m/s
15. 13.2 m
16. parabolic path with launch and landing angles of 40 degrees above the horizontal
17. 2) Both spheres hit the ground at the same time, but sphere A lands twice as far as sphere B from the base of the tower.
18. 1) v_x=17.0 m/s and v_y=9.80 m/s
19. 2s
20. 4) D
21. 2
22. 6.96 m/s
23. 1.23 m
24. 16 m/s
25. 4.9 m/s
26. 3.7 m

Solutions

27. symmetric parabolic path with launch and landing angles of 30 degrees above the horizontal
28. increases
29. increases
30. 25 m/s
31. 38.7°
32. 31.2 m/s
33. 2) 100 m/s
34. 3) It remains the same.
35. 3) the same
36. 4) It remains the same.
37. parabolic path
38. 75 m
39. $t = \sqrt{\frac{2h}{g}}$
40. 3) 78 m
41. 2) 45°
42. 2) 9.8 m/s² downward
43. 2) 8.7 m/s
44. 2) 45°
45. 2) Ball A hits the tabletop at the same time as ball B.
46. 1) lower and shorter
47. 3) 63°
48. 3) 15.7 m/s
49. As the launch angle increases from 45° to 60°, time in the air increases and total horizontal distance decreases.
50. 1) v_x=40 m/s and v_y=10 m/s
51. 4) 90°
52. 2) 13 m/s vertical and 7.5 m/s horizontal
53. 4)
54. 4) less than 38 m horizontally and less than 6.7 m vertically
55.

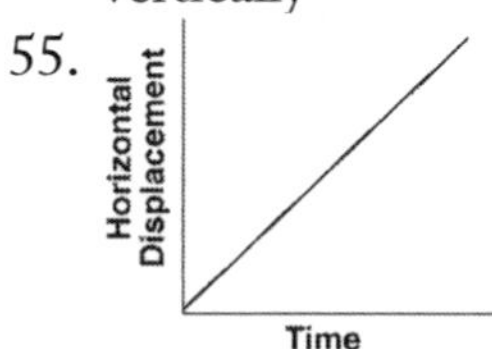

56. 1.5s

Dynamics-Newton's 1st Law

1. 2) mass of the contents of the box
2. 1) m/s^2
3. 4) A 10-kilogram sled at rest
4. 4) continue moving with constant velocity
5. 3) 10^2 kg
6. 4) A 20-kg mass moving at 1 m/s
7. 4) D
8. 4) unchanged
9. 1) A 110-kg wrestler resting on a mat
10. 3) a seated high school student
11. 4
12. 4) A 20-kg object at rest
13. 1) $kg{\cdot}m/s^2$
14. 4) a 15-kg object at rest
15. 4) a 4-kg cart traveling at 1 m/s
16. 1) a 15-kg mass traveling at 5 m/s
17. 4) a 1500-kg car at rest in a parking lot
18. 1) more mass and more inertia
19. 4)

Dynamics-Newton's 2nd Law

1. 4
2. 4) 4.0 m/s^2
3. 4) a block sliding at constant velocity across a table-top
4. 2) 600 N
5. 3) 0 N
6. 1) inertia
7. 1) less than the weight of the student when at rest
8. 3
9. 4) 28 N, southwest
10. 1) accelerating upward
11. 2) 5 kg
12. draw arrow down at position X and label
13. draw arrow down at position Y and label
14. 3) 3 N and 4 N
15. 1
16. 2
17. 1
18. 8000 m/s^2 east
19. 1120 N
20. 2
21. 4) 4 N toward the left
22. 3) 9 N
23. 3) both magnitude and direction
24. 1 cm = 2 N
25.

B
9.4 N
P
A
5.6 N

Solutions

26. 7.2 N
27. 3
28. 9 N
29. 3 N
30. 0.75 m/s^2
31. 1) decreases
32. 2
33. 10 m/s^2
34. 5 N
35. 4) F_H=20 N and F_V=14 N
36. 1) 0°
37. 1.5 N
38. 2
39. 4) 13 N
40. 4
41. 3) 12 N to 2 N
42. 4) 0 m/s^2
43. 3) 600 N
44. 2) increases
45. 4) 600 N
46. 3) 10^0 N
47. 4) 6.0 N
48. 3) a car moving with a constant speed along a straight, level road
49. 1
50. 1) 0.67 m/s^2
51. 4) upward at increasing speed
52. 2) 2.0 N
53. 1) decreases
54. 3) a hockey puck moving at constant velocity across ice
55. 1) 0°
56. 3) a man standing still on a bathroom scale
57. 1)
58. 3) the same as the magnitude of the rock's weight
59. 4) accelerates to the left
60. 1 cm = 10 N
61. draw resultant vector 10cm long at an angle of 37° east of north
62. 100 N
63. 37°
64. 2) 6.0 m/s^2
65. 2) less than 750 N
66. 1) 0 N
67. 3) 5.0 N, down
68. 4)

Dynamics-Newton's 3rd Law

1. 1) 20 N
2. 4) 50 N
3. 2) 100 N
4. 3) the same
5. 4) the same
6. west
7. 1
8. 1) the same
9. 1) F
10. 4) 1,000 N

Dynamics-Friction

1. 2
2. 850 N
3. 42.5 N
4. 156 N
5. 28 N
6. vector to right of length 2.8 cm
7. 196 N
8. 55 N
9. 7.15 m/s^2
10. 8930 N
11. 12,300 N
12. 9840 N
13. The claim is reasonable since the frictional force can provide 9840 N, and the car needs 8930 N to meet the manufacturer's claim.
14. 52 N
15. 52 N
16. 3) 18 N
17. vector pointing upward of length 2 cm, labeled
18. 6 N
19. 2 N
20. 2 kg
21. 1 m/s^2
22. 2) 10 N
23. 1) dry concrete
24. 3) 12 N
25. 1) left
26. 780 N
27. 7.5 N
28. force of gravity (mg) down, normal force up, equal in magnitude (25 N)
29. applied force of 10 N to the right, frictional force of 7.5 N to the left
30. 2.5 N
31. yes because the net force is not equal to zero
32. 20 N
33. vector beginning at P with length of 2 cm pointing to the left

Solutions

34. 0.204
35. 14.7 N
36. 1) 40 N
37. 2) is less than the force of static friction
38. 3) remain the same
39. forces have the same magnitude
40. 3) 40 N
41. 1) less
42. 7500 N
43. 1) 0.24
44. 1) 2.4 N
45. 20N
46. 49N
47. 0.41
48. 1.96 N
49. 0.71 N
50. 3.3 N
51. the block speeds up

Dynamics-Ramps and Inclines

1. 4
2. 2 assuming B remains at the same height; 4 assuming B is at a greater height
3. 3) remain the same
4. 2) 8 N
5. 3) 1.4×10^4 N
6. 3
7. 4) zero
8. 2) 2.1 N

UCM-Circular Motion

1. 1
2. 2) 1.9×10^3 N
3. 1) radius of the path is increased
4. 4) $4F_C$
5. draw arrow toward center of circular path (down)
6. 4) is quadrupled
7. 2) B
8. 2) Hz·m
9. 2
10. 2
11. construct graph
12. T=0.89s
13. g is equal to $4\pi^2$/slope
14. 1
15. 3 m/s
16. 3) 29 N
17. 2
18. 4) 64 m/s^2
19. 2) 12 m/s
20. 3) Increase the radius of the track.
21. 4) east
22. 6.28 m/s
23. 1.1 N
24. 3) C
25. 2) 4.8 m/s^2
26. 1) toward the center of the circular curve
27. 3
28. 3) 8750 N
29. 27.9 m/s
30. draw arrow from A toward center of circle
31. 4.87 m/s^2
32. 2) doubling V and halving R
33. label axes (length on X, period on Y)
34. plot points
35. draw curve
36. 1 s
37. 1) A
38. 4) weight
39. draw arrow straight up
40. radius would increase
41. radius, period, and mass of weights
42. F_g down, F_C toward center of circle
43. 4) 4.0 m/s
44. 3) 2.5 m/s
45. 4
46. 7.8 m/s^2
47. toward the center
48. 3) 8.0 m/s^2
49. 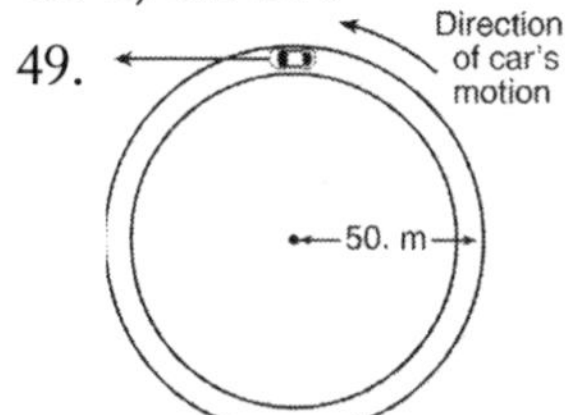

Track, as Viewed from Above

50. 2.9 m/s^2
51. 3)
52. 1) 0.30 Hz

UCM-Gravity

1. 2
2. 3) 8.17×10^{-10} N
3. 4) 26 m/s^2
4. 1) F
5. 1) 25 N
6. 1) 638 N
7. 1) acceleration due to gravity

Solutions

8. 2) 3.72 N/kg
9. gravity
10. 2.26×10^{17} N
11. The sun has a much larger mass than Neptune.
12. 4) 8.0×10^{20} N
13. 2) 2.00×10^{2} N
14. 3) 100 kg
15. 2) 3.8×10^{7} m
16. 1) attractive, only
17. 2×10^{30} kg
18. 3) one-half as great
19. 4
20. 3) one-fouth as great
21. 6.35×10^{22} N
22. 0.71 m/s^2
23. 3
24. There are multiple correct answers, which may include a kinematics approach in which an object is dropped from rest. By measuring the distance fallen (tape measure) and the time required to fall that distance (stopwatch), the acceleration can be calculated as $a=2d/t^2$. Alternately, a pendulum could be created of length L (measured with tape measure) and its period measured (stopwatch). The acceleration could then be determined from $4l\pi^2/T^2$.
25. 3) 64 times as great
26. 3) 50 N
27. 1) 50 kg
28. 3) 3.00 m/s^2
29. 2) mass remains the same
30. 1) 50 N
31. 2) 2.00 kg
32. 3) 4400 N
33. 4) 19.6 N
34. 1) 20 N
35. 3.6×10^{22} N
36. 2) 2.00 N/kg
37. 2×10^{20} N
38. 4)
39. 1)
40. 2) 5 N/kg
41. 4)
42. 3) 9.8 N/kg
43. 3.53×10^{18} N
44. 2.28×10^{-3} m/s^2
45. Pluto has a greater mass than Charon.
46. 3) 8.3 N/kg

Momentum-Impulse

1. 4) 1.2×10^{5} N
2. 1) 8.0 N·s
3. 2) 2.0 N·s
4. 1) 1 N
5. 2) 2.0×10^{1} m/s
6. 4) 2.4×10^{3} N
7. 50 N
8. 3) applying a net force of 5.0 N for 2.0 s
9. 4) 56 N
10. 1.1 N·s
11. 3) speed
12. 2) 9.9 N·s
13. 4) 0.50 s
14. 6000 N·s
15. 4) greater inertia and the same magnitude of momentum
16. 2) 3.3 m/s
17. 3) 9×10^{2} N
18. 2) 2.0 m/s west
19. 3) 25 N
20. 4) 3.0×10^{2} N
21. 3) 65 m/s
22. 3) 690 N
23. 1) 1.2×10^{2} N
24. 2) 50 kg·m/s
25. 2 s
26. 4) time
27. 0.0060 s
28. 1) increasing the length of time the force acts on the driver
29. 2) 15 kg•m/s
30. 3) 10 N
31. 3) 360 m/s

Momentum-Conservation

1. 3) 3.0 m/s
2. 2.4 m/s
3. west
4. 3) 10 kg
5. 2
6. 3) 3.0 m/s
7. 3) smaller magnitude and the same direction
8. 18.2 m/s
9. 1) 1.5 m/s
10. 3) the same before and after the collision
11. 4) impulse and momentum
12. 2) 2.4 m/s
13. 4.11 m/s

Solutions

14. 23,000 N
15. 4) $m_Av/(m_A+m_B)$
16. 4) $(m/(m+M))v$
17. 3) 0.25 m/s
18. 1) 0.50 m/s left
19. 7.4 kg•m/s

WEP-Work and Power

1. 4) kinetic energy
2. 4) power
3. 3) 12 N
4. 4) 9.0×10^3 W
5. 3) 40 m
6. 1) 100 J
7. 4) J/s
8. 2) greater
9. 2) 280 W
10. 3) 48 W
11. 3) watts
12. 2) distance the box is moved
13. 3) 4,900 W
14. 1) 1 J
15. 2) 30 N
16. 3) 1200 W
17. 1) $kg{\cdot}m^2/s^2$
18. 3) 73 J
19. 1
20. 1) the same
21. 1) impulse
22. 2) power
23. 2) 9.65×10^3 W
24. 3) 570 J
25. 1) 5.0×10^4 W
26. 1) N·m
27. 4) 49 W
28. 7.7 m
29. 128 W
30. 3) 3.6×10^2 J
31. 1) the same work but develops more power
32. power
33. 1) 2.4×10^3 J
34. 3) 27 J
35. 2) 1.5 J
36. 2) 5.0 m
37. 2) 1.5 W
38. 4) 6.0×10^4 J
39. 3) 0.3 J
40. 2) energy
41. 2) impulse
42. 4) four times as great
43. 3) 4.16 m
44. 3) 4.9×10^3 W
45. 1) the same
46. 2)
47. 98 W
48. 3) $kg{\cdot}m^2/s^2$
49. 3) 1.5×10^5 J
50. 2) 9.8 s
51. 1.95×10^4 W

WEP-Springs

1. plot points
2. draw line
3. slope=k=4 N/m
4. 1) 1 J
5. 2) 0.5 J
6. 2) 7.5 J
7. k, the spring constant
8. $PE_A < PE_B$
9. 2) 20 N
10. B has the most kinetic energy because all the spring potential energy has been converted into kinetic energy.
11. A has the maximum gravitational potential energy because A is located at the highest height.
12. C because all the kinetic energy and gravitational potential energy has been converted into spring potential energy.
13. 1) 3.6 J
14. 4) 400 N/m
15. 2) larger
16. 3) 120 N/m
17. plot points
18. draw line
19. 0.30 m
20. 0.1875 J
21. 0.96 m
22. 40 N/m
23. 2) 67 N/m
24. 1) 0.18 J
25. 2) 4.0×10^3 N/m
26. 1) 3.75 J
27. plot points
28. draw curve
29. 0.363 J
30. 5.6 N
31. mark scale, plot points, draw line
32. k=slope=55 N/m

Solutions

33. 1) 0.47 J
34. 1) speed
35. 1) 32 N/m
36. $k=mv^2/x^2$
37. 1) A
38. 0.131 m
39. 1.28 J
40. 2
41. 2) 2.0 m
42. 4) 520 N/m
43. 2) 2.0 N/cm
44. 40 N/m
45. 20 N/m
46. 0.9 J
47. 4) 74.9 N/m
48. 2) 15 N

WEP-Energy

1. 2) 200 J
2. 1) quadrupled
3. 2) internal energy, only
4. 2) increases
5. 3) 3.3 J
6. 2) 5.1 m
7. 4) 1920 J
8. 4) joules
9. 5 m/s
10. 750 J
11. 750 J
12. 4
13. 3
14. 1 m/s
15. 3000 J
16. $KE_{after} < KE_{before}$
17. 4) internal energy
18. 11,760 N
19. 7880 N
20. 126,000 J
21. 14.5 m/s
22. 2) 8.00 m/s
23. 1) a decrease in kinetic energy and an increase in internal energy
24. 4) kinetic energy
25. 2) 2.21×10^3 J
26. 3) same
27. 3) position
28. 3) 120 J
29. 3) Internal energy increases.
30. 4
31. 2) 330 J less
32. kinetic energy decreases and internal energy increases
33. 4
34. 63,700 J
35. 19.8 m/s
36. total mechanical energy remains the same
37. 1
38. 3) 30 J
39. 3) remains the same
40. 110 m
41. 46.5 m/s
42. 8.77 m/s^2
43. 4) B and C
44. 3) 72.0 J
45. 4) 9.0×10^3 J
46. 4) $kg \cdot m^2/s^2$
47. 4
48. 88.2 J
49. 58.8 J
50. G
51. 1
52. 55 J
53. 2.5 kg
54. weight
55. draw a line starting at 0,0 with a steeper slope
56. 4
57. 3) Both elastic potential energy and kinetic energy at t_i are converted to internal energy at t_f.
58. 1
59. 2
60. 1.5 m/s
61. 84.4 J
62. 3) 120 J
63. 4) quadrupled
64. 2) 279 J
65. 2) increases and its kinetic energy remains the same
66. Conservation of energy states that unless work is done on the pendulum, its energy can't increase. The pendulum loses some energy due to air resistance (friction) and friction at the pivot, therefore it cannot return to the previous height on the return swing.
67. 3500 J
68. 10.4 m/s
69. 1) work and kinetic energy
70. 4) $KE=p^2/2m$
71. 1) Lubrication decreases friction and minimizes the increase of internal energy.

Solutions

72. 

73. 2) 41 m
74. 1.39×10^{-16} J
75. 4) PE=540 J and KE=1080 J
76. 3) 450 J
77. 3) It remains the same.
78. 4) a boy jumping down from a tree limb
79. 2) 9 J
80. 3) Kinetic energy remains the same and total mechanical energy increases.
81. 3) 5.4×10^3 J
82. 2) mechanical energy to electrical energy
83. 2) 270 J
84. 3) 2.5 m
85. 3) the same
86. 3) 20 m/s
87. 4) 2.0×10^5 J
88. 1) $\sqrt{(2gh)}$
89. 736 J
90. 4) internal energy
91. 3) 110 J
92. plot points
93. draw the curve
94. 70 kg
95. $KE_{soccer\ player} < KE_{runner}$
96. 3)
97. 182 J
98. 120 J
99. KE of the crate is constant.
100. Internal energy of the crate increases.
101. 3) light -> electrical -> mechanical
102. 4) remains the same
103. 3) electromagnetic energy and internal energy
104. 4)
105. 4) 7.50 J
106. 3) The kinetic energy decreases and the gravitational potential energy remains the same.
107. 1) 0.02 J
108. 5.00 N/m
109. Energy is converted into sound/thermal energy, friction, etc. (any of a variety of acceptable answers).
110. 1) speed and work
111. 1) electrical --> mechanical
112. 4) internal (thermal) energy
113. 3) 0.0625 J

Electrostatics-Charge

1. 1) A, only
2. 3) -3 units
3. 2) 2.5×10^{19} more electrons than protons
4. Bring both positive and negative rods near the sphere sequentially. If the sphere is attracted by each rod, the sphere must be neutral.
5. Bring the positive rod near the sphere. If it is repelled, the sphere must be positively charged.
6. 4) 2.6×10^{-19} C
7. 4) D
8. 3) 6.9×10^{12}
9. 1) may be zero or negative
10. 3) 1.76×10^{11} C/kg
11. 4.8×10^{-19} C
12. 4
13. 4) $+3.2 \times 10^{-19}$ C
14. 2) may be positive or neutral
15. 9.6×10^{-13} C
16. 1) loses electrons
17. 2) 4.8×10^{-19} C
18. 3) 3.2×10^{-19} C
19. 3) electrostatic
20. 1) $+4.80 \times 10^{-19}$ C
21. 4)
22. 2) 3.8×10^{13}

Electrostatics-Coulomb's Law

1. 1
2. 3
3. 3) 2.30×10^{-12} N
4. 1
5. 0.9 N
6. 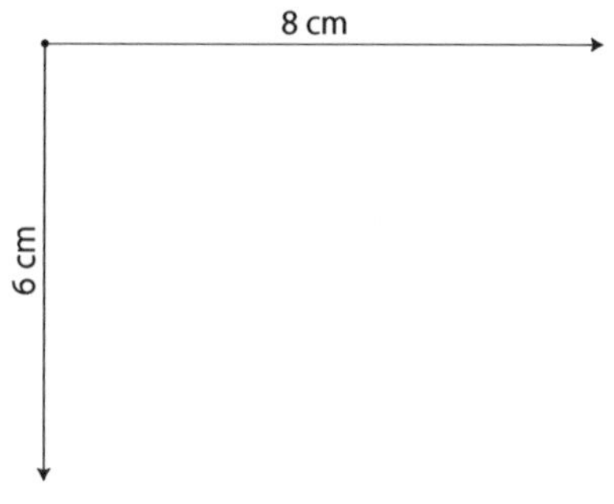

Solutions

7\.

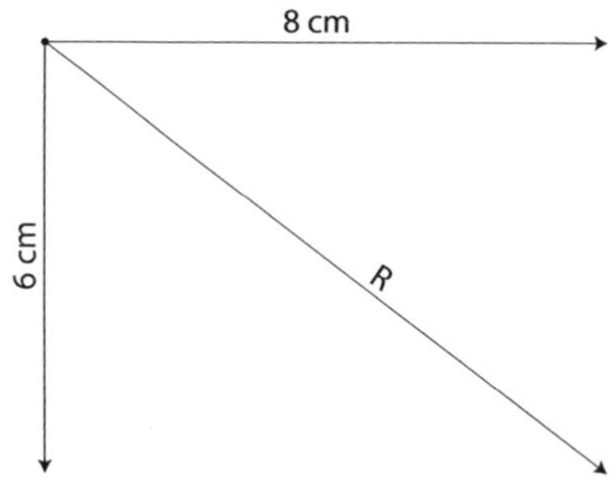

8 cm

6 cm

R

8. 1×10^{-14} N
9. 53.1°
10. 1
11. 4
12. 2
13. 3) 3/4 as great
14. 4) quadrupled
15. 2) 2.4 N
16. 1) stronger and repulsive
17. 4
18. 3) electrostatic forces between the particles of the balloon and the particles of the wall
19. 1) F/4
20. 4) 4F
21. 4) The gravitational force is attractive and the electrostatic force is repulsive
22. 1) $F_g/9$ and $F_e/9$
23. 2) 3.6×10^{-3} N d
24. 1) -3.0×10^{-7} C
25. 1) 2.56×10^{-17} N away from each other
26. 2) 2F
27. 1) F/9

Electrostatics-E Field

1. 2
2. 4
3. 3) 1.25×10^{4} N/C
4. 2.25×10^{4} N/C
5. 1) positively, and the electric field is directed from plate A toward plate B
6. 1) A
7. 4
8. four straight arrows pointing toward negative charge
9. 3
10. 2.4×10^{-19} N

11\.

Magnitude of Electrostatic Force

Distance Between Centers

12\.

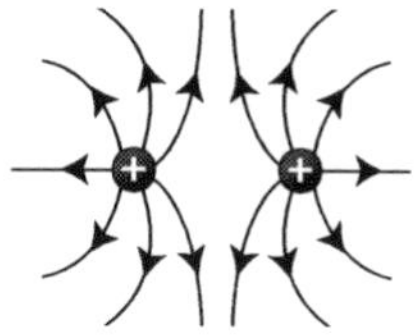

13. 4) toward the top of the page
14. 2) 1.44×10^{-6} N/C
15. 1
16. 2) 1.25×10^{4} N/C directed toward the sphere
17. 2) 5.6×10^{-2} N
18. 3
19. 2) Sphere A is negative and sphere B is positive
20. five arrows between the plates pointing straight down
21. 3.2×10^{-16} N
22. 2) toward plate B
23. 3) C to D
24. 3)
25. 2) 2.00×10^{4} N/C directed toward the sphere
26. 3) 9.60×10^{-17} N
27. 2) 4.80×10^{-16} N
28. 3) one-fourth as great

Electrostatics-Potential

1. 1) 1.0×10^{0} V
2. 4) toward the bottom of the page
3. 2) 1.60×10^{-17} J
4. the forces are the same
5. 4) coulomb·volt
6. 2) 12 V
7. 3) electron, because it has the smallest mass
8. 4.0×10^{-15} J
9. 3) 2.5×10^{2} V
10. 4) It is the same at points A, B, and C
11. 2) 2.0 V
12. 4) 2.0×10^{-3} C
13. 3) 3.0×10^{2} V
14. 2) 1.00×10^{7} m/s
15. 5.22×10^{5} V
16. 2) electric field strength
17. 4) a metal sphere with a charge of 1.0×10^{-9} C moved through a potential difference of 4.0 V

Solutions

18. 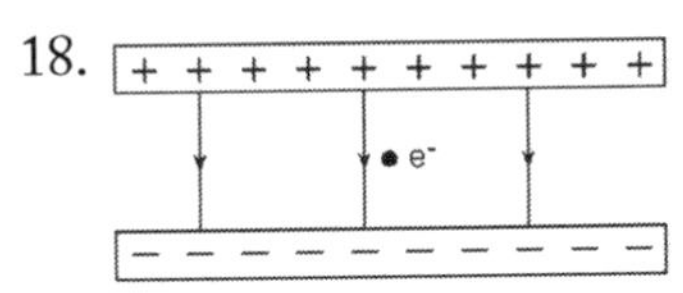

19. toward the top of the page
20. 2.4×10^3 N/C
21. force remains constant
22. 1) energy
23. 2) 27 J
24. 3) volt•coulomb
25. 2) 4.80×10^{-19} J

Circuits-Current

1. 2) 2.0 A
2. 4) four times as great
3. 3) 1.6×10^{15} electrons
4. 3) 4.00×10^{-3} A
5. 3) 120 C
6. 2) 2.0 A
7. 2) 5.0 A
8. 30 C
9. 3) 9.1×10^{-2} A
10. 1) 1.0 s
11. 2) 108 C
12. 4) 1.0×10^2 A

Circuits-Resistance

1. 2) gold
2. 2) 2.0 Ω
3. 2) increase
4. The graph is not linear, and the varying slope indicates a varying resistance.
5. 2) increased
6. temperature likely increased
7. mark scale
8. plot points
9. draw line
10. 0.326 Ω/m
11. 3
12. 1) R
13. 6 Ω
14. 2.5×10^{-8} m^2
15. 3) 3.74×10^{-2} Ω
16. 2) increases
17. 1) increasing the applied potential difference and decreasing the length of wire
18. 0.924 m
19. 4
20. R increases
21. 4) 4R
22. 4) silver
23. 4) electric field strength and N/C
24. 3.14×10^{-6} m^2
25. 5.5×10^{-2} Ω
26. 2) 1.4×10^{-1} Ω
27. 1.91 Ω
28. 0.785 A
29. 3) cross-sectional area
30. 3) nichrome
31. 429 Ω
32. 0.0625 Ω
33. 5.6×10^{-8} Ω·m
34. 4) nichrome
35. 3) aluminum
36. 3) decreasing the wire's temperature
37. 1.06×10^{-7} Ω·m
38. 2) 1.12×10^{-2} Ω
39. 1) 7.9×10^{-8} m^2

Circuits-Ohm's Law

1. 4
2. 3) 10 W
3. 3) 1.44×10^{-4} J
4. 1) current, only
5. 138,000 J
6. 1) decreases
7. 1
8. 2) 0.25 A
9. 4
10. 2) 12 V
11. 2) 2.0 ohms
12. 2) 15.2 mA
13. 3) 9.0×10^3 J
14. 3) 6.0×10^3 J
15. 4) more resistance and draws less current
16. 5250 J
17. mark scale
18. plot points
19. draw line
20. 30 Ω
21. 3) 150 ohms
22. 2) 0.60 A
23. 1
24. 25 Ω
25. 1
26. 4) 1100 W
27. 3) 55 ohms
28. 4) quadruples

Solutions

29. 3) 37.4 ohms
30. 1) 1 V
31. 1) half as great
32. 2) BC
33. 2) 240 ohms
34. 1) 2.16×10^4 J
35. 4) 3.6×10^4 J
36. 2) 0.15 W
37. 1) halved
38. 4) electrical potential difference and joules/coulomb
39. 3) is halved
40. 2) 5.76×10^4 J
41. 4) 18 J
42. 16 ohms
43. 4) 4.0 mA
44. 3) 10 J
45. 1) 1.5 eV
46. 2) increases
47. 120 W
48. 4
49. 3) 12 V
50. 2) 15 A
51. 1) 5.67×10^5 J
52. 1) 1.7×10^4 Ω

Circuits-Circuit Analysis

1. 2.4 ohms
2. 4 A
3. 12 ohms
4. 1) 1.0 A
5. 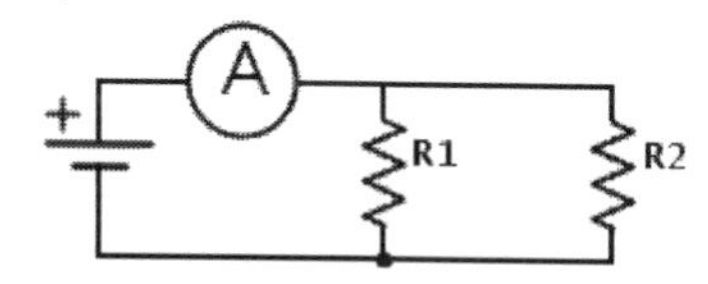

6. 12 ohms
7. 48 W
8. 1
9. 4
10. Same diagram as #5 above
11. 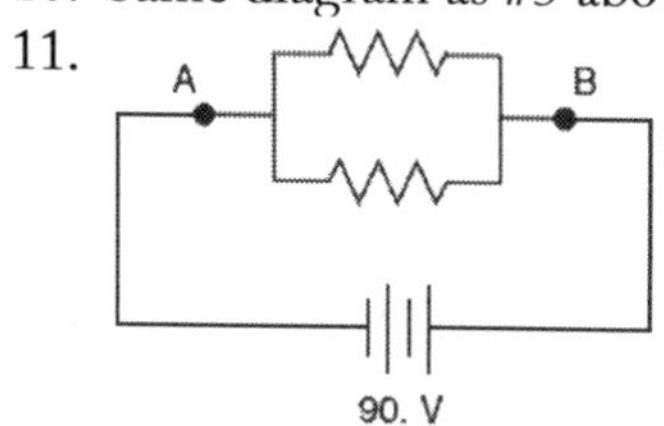

12. 90 V
13. 6 A
14. 240 ohms
15. 190 ohms
16. 12.5 W
17. 4
18. 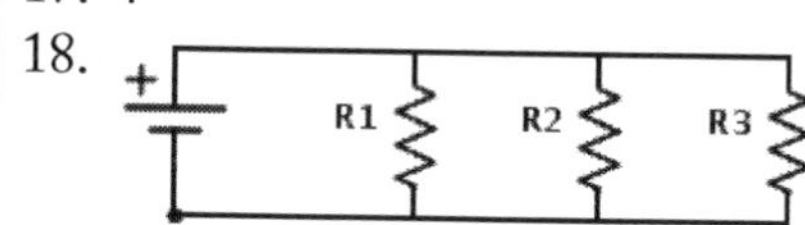

19. 40.1 V
20. 47.7 ohms
21. 40.1 V
22. 0.11 A
23. 4
24. 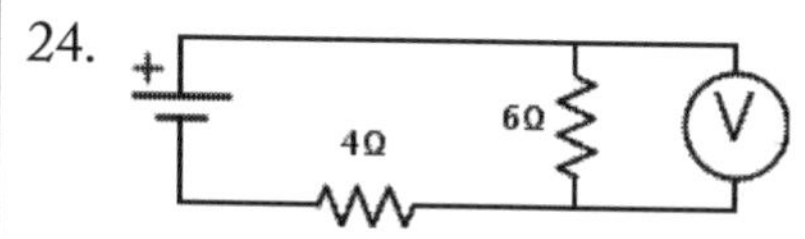

25. 1) 5 A
26. 1) 0.50 A
27. 3) Equivalent resistance decreases and total current increases.
28. 2) 2.0 A
29. 3) 3 ohms
30. 3
31. 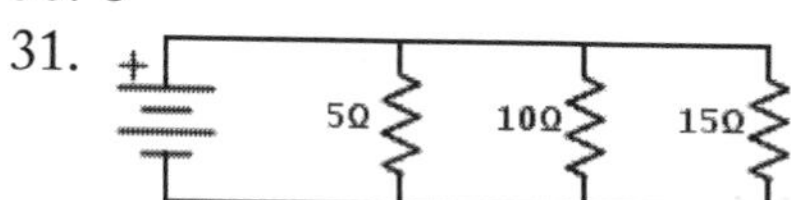

32. 3460 J
33. no effect
34. 3) 32 W
35. 1) 1 ohm
36. 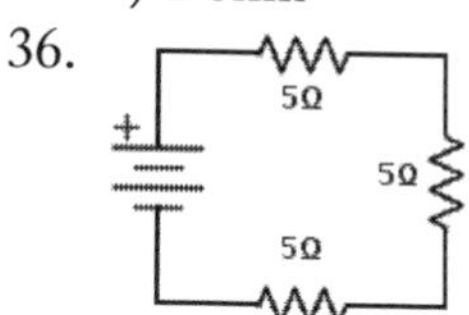

37. 2) 2.0 A
38. 3) 3.0 ohms
39. 4) 4.0 W
40. 1) less than 4 ohms
41. 20 ohms
42. 2) 12 Ω
43. 2) 0.60 A
44. 1) 4.8 W
45. 4) 0.75 A

Solutions

46.

47. 2) 2.4 Ω
48. 2) 9.0 Ω
49. 3) 80 Ω
50. 4
51. 4
52. 3
53. 2) 2 A
54. 3) a voltmeter and an ammeter, only
55. 4) 4.0 A
56. 1) 6.0 A
57. 4) 40 V
58. 1) ammeter at 1 and voltmeter at 4
59. 4) source of potential difference
60. 2) 1.5 Ω
61. 1) The ammeter reading decreases.
62. 2) X/3 Ω
63. 4) 24 V
64. 3) the same
65. 3) 4.62 Ω
66. 2) 2.0 A
67. 1) The potential difference across the 6-oohm resistor is the same as the potential difference across the 3-ohm resistor.
68. 1) 0.018 A
69. 1) less resistance and draws more current
70. same diagram as #5
71. 8.6 Ω
72. 3) 9 A
73. 15 Ω
74. 60 Ω
75.

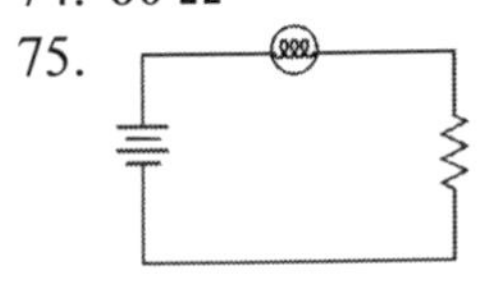

76. 24 Ω
77. 14 Ω
78. 2.5 W
79. 1) varies directly with its resistance
80. 360 Ω
81. no change
82. equivalent resistance would increase
83. equivalent resistance of series circuit is greater

Magnetism

1. 2) moving
2. 3) right
3. 2
4. 3) both an electric field and a magnetic field
5. one is a magnet and the other is a magnetic attractable
6. wire should be moved left and right
7. 2) B
8. 2) increase
9. 1) electric charge
10. 1) The gravitational force is attractive and the magnetic force is repulsive
11. 1) potential difference across it
12. 1) A is a north pole and B is a south pole
13. 1) moving and charged
14. 3) Gravitational force is attractive and magnetic force is repulsive
15.

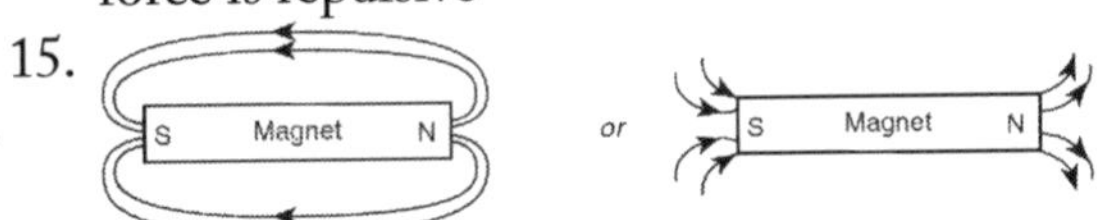

16. 2) bar magnet
17. 4) a moving charged particle
18.

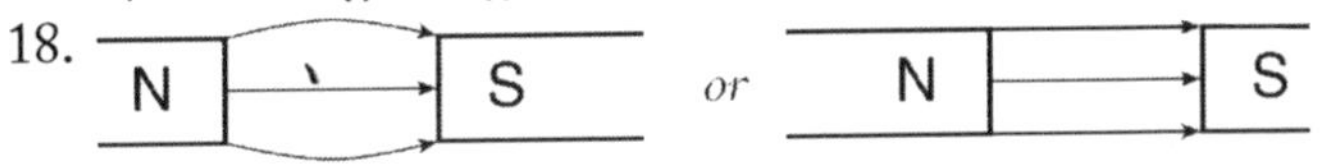

19. 3) both a magnetic and an electric field

Waves-Wave Basics

1. 1) sound
2. 4) a pulse
3. 2) down
4. 1) perpendicular to the direction of wave travel
5. 1) energy, only
6. 4) mechanical wave
7. 1) Light waves can travel through a vacuum, but sound waves cannot.
8. 3) The amount of energy a sound wave transmits is directly related to the wave's amplitude
9. 4) transfers energy without transferring mass
10. 2) down, the, up, then down
11. 1) energy, only
12. 3
13. 3) longitudinal, because the air molecules are vibrating parallel to the direction of wave motion
14. 1) light, only
15. 1) Both have the same frequency as their respective sources.

Solutions

16. 1) a pulse
17. 4
18. 1) transverse
19. 1) perpendicular to the direction of wave travel
20. 4) mechanical waves that require a medium for transmission
21. 4
22. 2) A and C
23. 1) amplitude
24. 3.2 m
25. 0.6 m
26. 2) amplitude
27. 4) Both types of waves transfer energy.
28.

29. 4) sound waves
30. 1) A and C
31.

32. 3) loudness
33. 1) vibrate parallel to the direction of the wave's propagation

Waves-Wave Characteristics

1. 2) 1.29 m
2. 5 m/s
3. 3) 2.5 s
4. 1) halved
5. 0.65 m
6. 3) C
7. 2) 2.0 m
8. 1) Light waves can travel through a vacuum, but sound waves cannot.
9. 2) B and F
10.

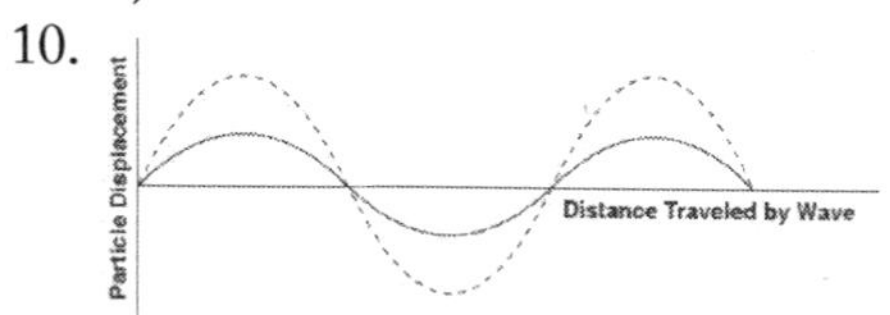

11. 4
12. 3) A and C
13. 1) amplitude
14. 3
15.

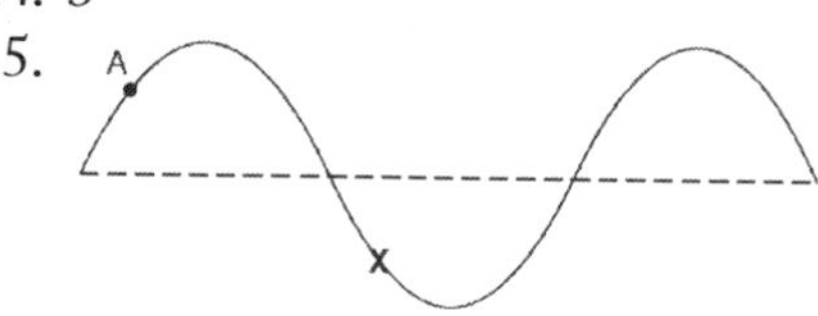

16. 1 m
17. 0.5 s
18. 6 m/s
19. 1525 m/s
20. 1.29 m
21. 8.47×10^{-4} s
22. 4) 4.0 Hz
23. 3) remain the same
24. 2) period
25. 4) 4 cm
26. 2) struck harder
27. 2) B
28.

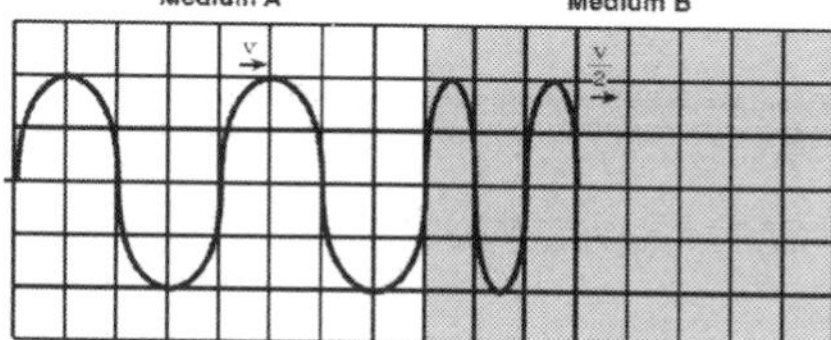

29. 4) amplitude
30. 3) A and E
31. 3) remain the same
32. 4) greater loudness
33. 2) amplitude
34. 2) A and D
35. 2) Hz·m
36. 2) 0.75 m
37. 0.2 Hz
38. 0.4 m/s
39. 3) 8.0 m/s
40. 1) 0°
41. 1) 1.8 m
42.

43. 1) one
44. 2) longitudinal wave of constant frequency
45. 0.509 m

Solutions

46. mark any two points on the wave that are in phase with each other

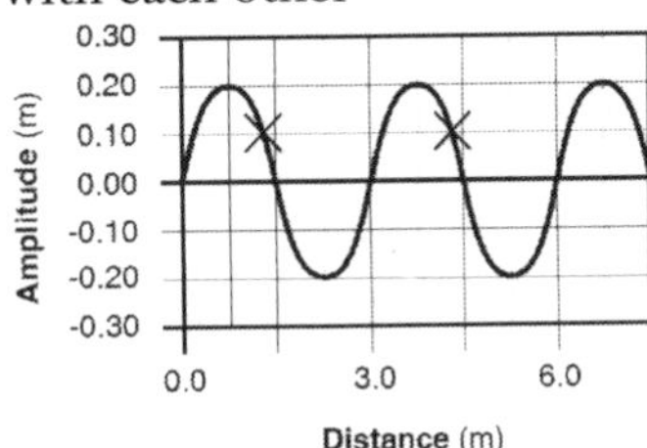

47. 0.75 s
48. 1) longitudinal wave with air molecules vibrating parallel to the direction of travel
49. 2) They transfer energy.
50. 1) 0.130 m
51. 2) 2 s
52. 1) vibrate east and west
53. 4) D and G
54. 3) 2.5×10^{-2} m/s
55. 3) 1 s
56. 4) 2.94×10^{-3} s
57. 4) 6.7×10^{14} Hz
58. 1) amplitude
59. 2) 0.50 Hz
60. 2) halved
61. 2) twice the amplitude and half the wavelength
62. 4) B and D
63. 1.50×10^{10} Hz

Waves-Wave Behaviors

1. 4) 8.60 m
2. 2
3. 1) resonance
4. 2) higher
5. 2) the same frequency, the same amplitude, and travel in opposite directions
6. 3) 6 nodes and 5 antinodes
7. 4) 380 Hz
8. 4) reflecting from a barrier
9. 1) +1 unit
10. 1) decreases in amplitude and decreases in frequency
11. 1) resonance
12. 2) longitudinal
13. 4) amplitude
14. 1) lower
15. 3
16. 2) a node resulting from destructive interference
17. 3) he accelerates toward the source
18. 4
19. 1) resonance
20. 3) 5 nodes and 4 antinodes
21. 2) the Doppler Effect
22. 2) resonance
23. 2
24. 1) A
25. 3
26. 2) resonance
27. 3 m
28. 60 m/s
29. The singer's frequency must match the natural frequency of the glass; and the singer's amplitude (loudness) must be large enough to surpass the elastic limit of the glass.
30. The frequency of the sound from the tape player would not match the natural frequency of the glass.
31. 4) remain stationary
32. 1.5 m
33. The observed frequency is higher while the speaker is moving toward the observer due to the Doppler Effect, so the observed wavelength must be shorter.
34. 2
35. 3) 0
36. 2) resonance
37. 1) lower, because the sound-wave fronts reach the platform at a frequency lower than the frequency at which they are produced
38. 4) resonance
39. 4) Wave Y must have a frequency of f, an amplitude of A, and be traveling westward
40. 3) $\lambda/2$
41. 3) 180°
42. 3) resonance
43. 2) moving away from Earth
44. 4) resonance
45. 2) wavelength
46. 4) Observer A is stationary, and observer B is moving away from point S.
47. 2) A and D
48. 4) resonance
49. 4) constructive interference
50. 2) frequency
51. 4) 6 cm
52. 2) higher frequency and a higher pitch
53. 3) resonance
54. 4) resonance
55. 4) diffraction
56. 2) lower and the frequency heard by observer C is higher
57. 2)

Solutions

Waves-Reflection

1. 3) 120°
2. 3) C
3. 4) all angles of incidence
4. 35°

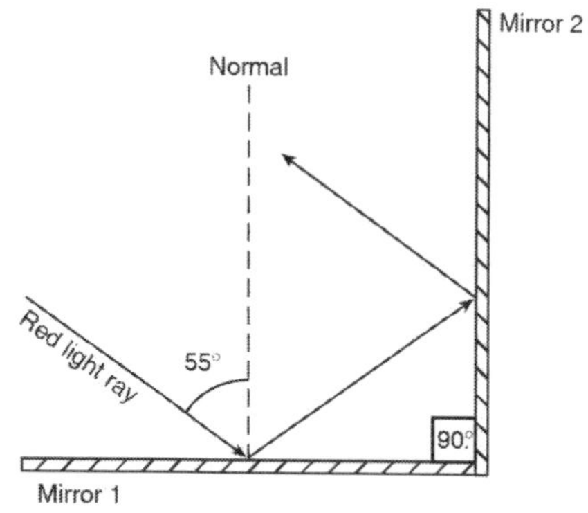

5. 1) 25°
6. 60°
7.

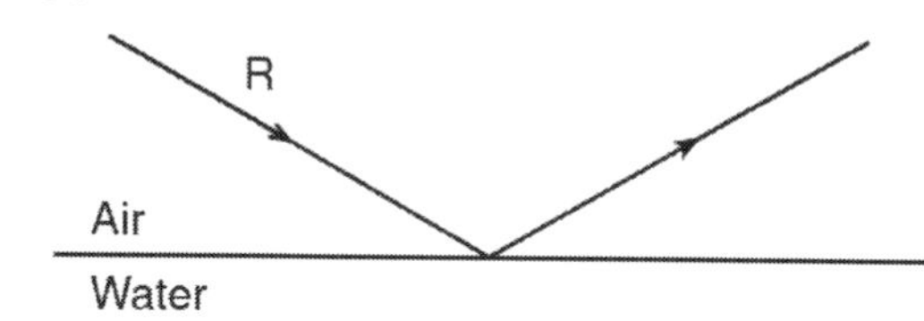

8. 1) A

Waves-Refraction

1. 4) Its speed increases.
2. 50°
3. 30.7°
4. 50°
5. 3) wavelength
6. 3) C
7. 4) Light is refracted as it crosses the air-water interface.
8. 23°
9. There is no change in index of refraction, therefore there is no change in wave speed, and therefore no refraction.
10.

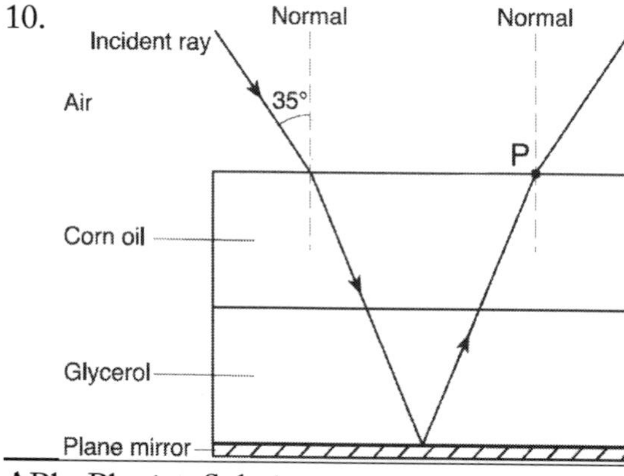

11. 4
12. 40°
13. 1.88
14. 1.6×10^8 m/s
15.

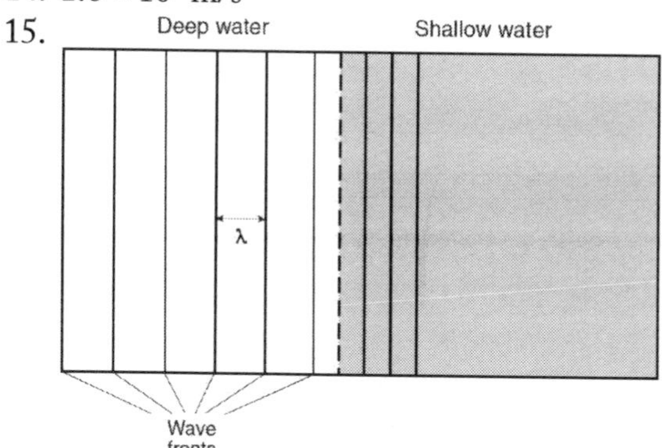

16. 1) the bird and the fish
17. 1.81×10^8 m/s
18. 29.6°
19.

Normal
55°
Air
Material X

20. 4) equal to 1.66
21. 1.67
22. 2) The speed decreases and the frequency remains the same.
23. 20°
24. θ_1=45°, θ_2=26°
25. 1.61
26.

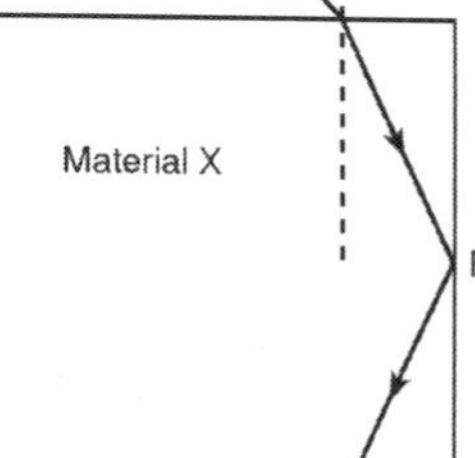

27. 2) doubled

Solutions

28. 19.5°
29.

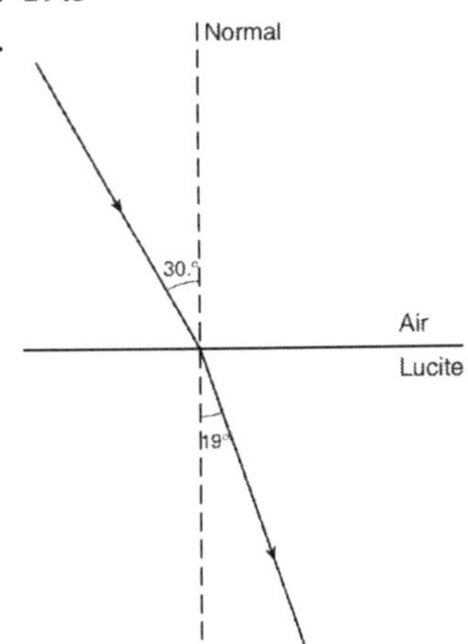

30. 3
31. 2) refraction
32. 28.9°
33.
34.

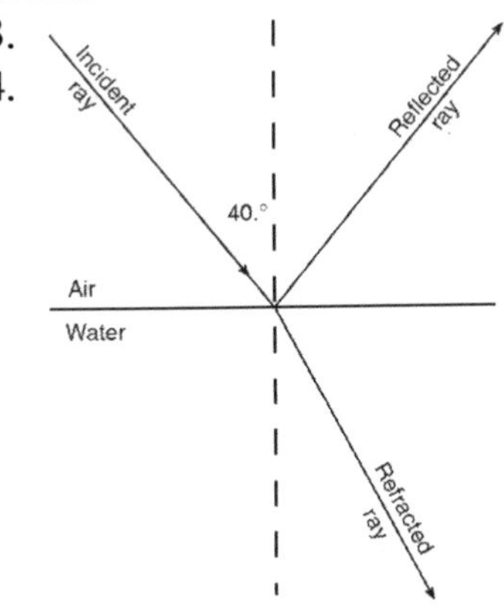

35. 4.01×10^{-7} m
36. 4) 1.50
37. 2) larger
38. 1) Only speed changes
39. 55°
40. 29.6°
41.

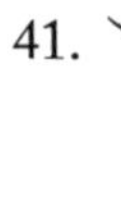

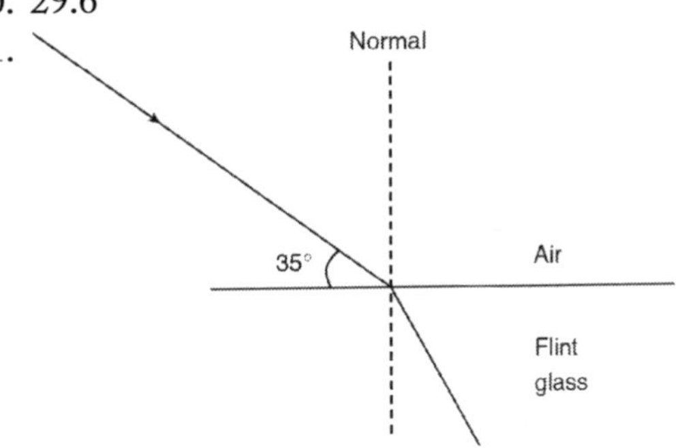

42. It is reflected and/or scattered.
43. 2) 16 m/s
44. 1) 3.93×10^{-7} m
45. 2) 1.3
46. 3) 1.75
47. 1) 0.333
48. 3) Medium Y, only
49. 4) speed of light in a vacuum
50. 1) 1.81×10^{8} m/s
51. 4) The frequency remains the same and the speed decreases.
52. 4) speed
53. 1) decreases and its frequency remains the same
54. 4) water (n=1.33)
55. flint glass, corn oil, ethyl alcohol, water
56. 37°
57. 49°
58.

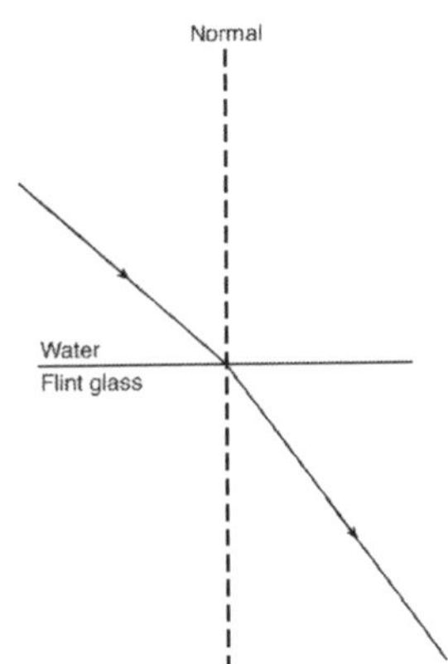

59. reflection, absorption, scattering, decrease in speed, decrease in wavelength, etc.
60. 2) twice the speed
61. 4
62. 4) 2.21×10^{8} m/s
63. 35°
64.

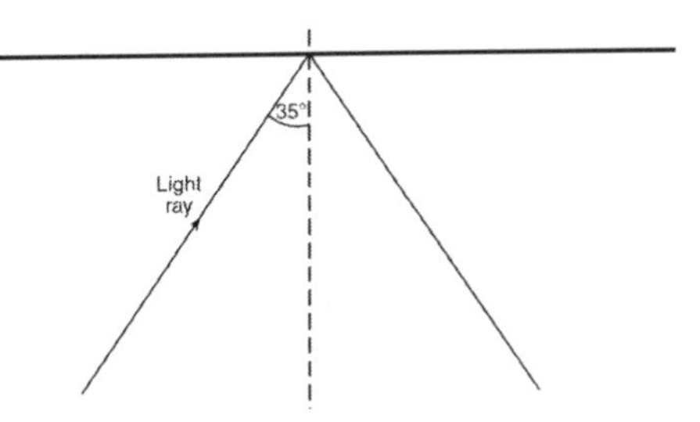

65. 49° or 50°
66. frequency, period, phase, color or transverse
67. 2) frequency and period
68. 2)
69. 2)
70. 4) wavelength
71. 4) glycerol
72. 56°
73. 1.7

Solutions

Waves-Diffraction

1. 1) decrease
2. 4) diffraction
3. 1) diffraction
4. 3) wavelength of the incident wave and the size of the opening
5. 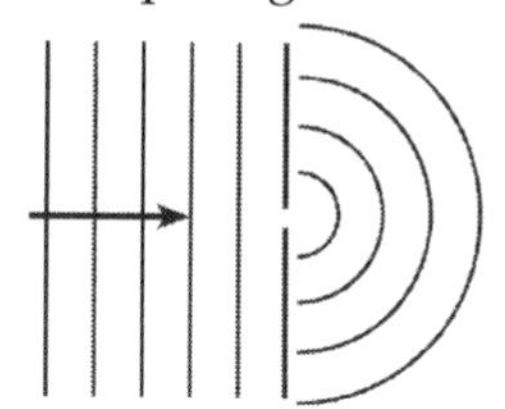
6. 4
7. 3) long wavelength and narrow opening
8. 1) A
9. 4) have a longer wavelength
10. 1) much shorter than 10 cm
11. 1) diffraction
12. 1)
13. 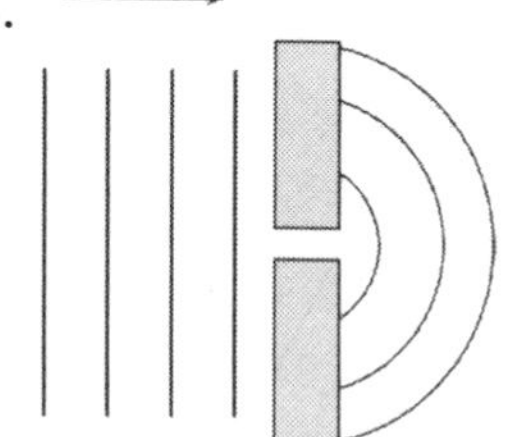

Waves-EM Spectrum

1. 2) greater
2. 3) 500 m
3. 2) green
4. 1) speed
5. 1) 5.0×10^{-10} s
6. 3) speed
7. UV radiation has a higher energy than visible light and therefore imparts more damage.
8. 3.28 m
9. violet
10. Yellow-green paint allows the most light to reflect off the truck during daylight hours.
11. 1) speed
12. 136 m
13. 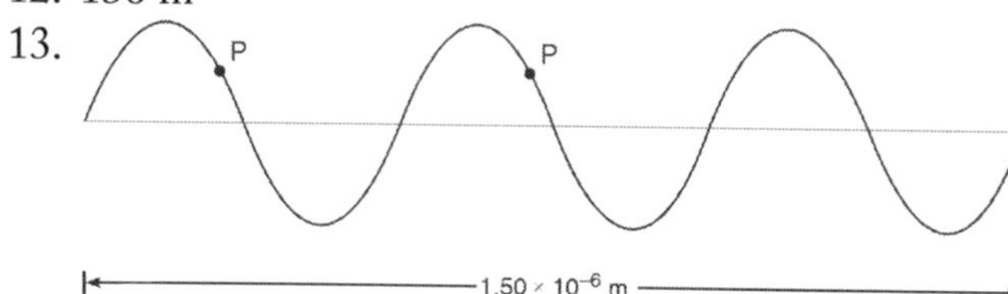

14. green light
15. 4) speed
16. 2) electric and magnetic fields
17. 1) electromagnetic and transverse
18. 4) 100 μm
19. 1) less
20. 3) 6.0×10^{-7} m
21. 4) 3.00×10^{8} m/s
22. 2) 2.0×10^{-8} s
23. 1) It produces electromagnetic radiation.
24. 1) a radio wave
25. 1) speed
26. 3) speed

Modern-Wave Particle Duality

1. 1) greater energy
2. UV radiation has a higher energy than visible light and therefore imparts more damage.
3. 4.94×10^{-14} m
4. gamma rays
5. 3) both energy and momentum
6. 4) an electron
7. 2) x ray
8. 3) 4.0×10^{-19} J
9. 3) microwave
10. 1) wavelength
11. 2.21×10^{-16} J
12. 1) infrared
13. 1) 6.63×10^{-34} J·s
14. 4.57×10^{14} Hz
15. 3.03×10^{-19} J
16. 1.89 eV
17. 2) diffraction of light passing through a narrow opening
18. 4) $E_{photon} = pc$
19. 3
20. 3) both particles and waves
21. 1) Planck's constant
22. 2) 3.3×10^{-19} J
23. 4) both particles and waves
24. 3) The energy increases and the wavelength decreases.
25. 6.09×10^{-7} m
26. 1) speed
27. 2) electron diffraction
28. 3) both particles and waves
29. 4.53×10^{14} Hz
30. 2.85×10^{-18} J
31. 6.63×10^{-18} J
32. 3.45×10^{-18} J

Solutions

33. mass, charge, momentum, energy
34. 3) 5.10×10^{-19} J
35. 1)
36. 3) 4.14×10^{-19} J

Modern-Energy Levels

1. 2) 8.82 eV
2. the hydrogen absorbing certain frequencies of the white light
3. 1) electrons transitioning between discrete energy levels i the atoms of that element
4. Absorbing a 10.2 eV photon allows the electron to jump from the ground state to n=2, while an 11 eV photon cannot be absorbed because there is no available state 11 eV higher than the ground state.
5. 3) 3
6. 2) 8.81 eV
7. 1.89 eV
8. 3.02×10^{-19} J
9. 4.57×10^{14} Hz
10. 6.56×10^{-7} m
11. 3) absorbed a 2.55-eV photon
12. 3.02×10^{-19} J
13. 1.89 eV
14. This result verifies that the alpha line corresponds to a transition from energy level n=3 to energy level n=2 because the difference in energies between those two levels is 1.89 eV, the energy of the emitted photon.
15. 4) 5.43 eV
16. 3.33×10^{-19} J
17. 2.08 eV
18. n=3
19. 3) The photon's energy is too small.
20. green
21. 3.63×10^{-19} J
22. 2.27 eV
23. 8.69×10^{-19} J
24. 5.43 eV
25. This photon can be absorbed by te mercury atom because an electron in the ground state can absorb a 5.43 eV-photon to jump to energy level d.
26. 1.24 eV
27. 1.98×10^{-19} J
28. 2.99×10^{14} Hz
29. infrared
30. 1) ultraviolet
31. 3) 2.84 eV
32. 3) B and C
33. 2) 9.62 eV
34. 8.82 eV
35. 1.41×10^{-18} J
36. 3.02 eV
37. 4.83×10^{-19} J
38. 7.29×10^{14} Hz
39. No, the electron could jump to other energy levels by emitting photons of other energies / frequencies.
40. 3) 3
41. -8 or 10^{-8} Do not allow credit for 10 nanoseconds or a decimal form such as 0.0000000010 s.
42. 3.16×10^{-19} J
43. The ground state is the lowest available energy level that an atom can have *or* the ground state is the most stable energy state.
44. 2) 0.97 eV
45. 3) n=5 to n=2
46. 3) 1.51 eV
47. energy level f
48. 4.90×10^{-19} J
49. 7.39×10^{14} Hz
50. visible light / violet

Modern-Mass Energy Equivalence

1. 2) 1.42 MeV
2. 3) 4.5×10^{14} J
3. 233 MeV
4. 2) 8.2×10^{-14} J
5. 1) 1.8 TJ
6. 1) 1.64×10^{-13} J
7. 0.042 u
8. 1
9. 4) 9.00×10^{16} J
10. 1) 1.14×10^{-30} J
11. 3) 8.48 MeV
12. some mass was converted into energy
13. 0.01863 u
14. 17.3 MeV
15. 3×10^{-10} J
16. 2) c^2
17. 2) 8.20×10^{-14} J
18. 1.50×10^{-10} J

Modern-Standard Model

1. 4) attraction between nucleons
2. 1) leptons
3. 1) baryon
4. 0 (neutral)
5. 3) baryon to another baryon

Solutions

6. 2) It holds protons and neutrons together
7. 3) cdb
8. 2) +2e
9. $m_{neutron} > m_{proton}$
10. 0 (neutral)
11. 2) baryons
12. 2) strong force
13. antiproton
14. the same
15. charge
16. Although matter is only created in matter-antimatter pairs, most known matter is normal.
17. 1) strong force
18. 3) -1 e and +1 e
19. 1) $+5.33 \times 10^{-20}$ C
20. 3) strong force
21. 2) 12
22. 2) frequency
23. 4) violet
24. -1.60×10^{-19} C
25. 3) 3 up quarks and 3 down quarks
26. 4) electrons
27. -1e -> -1e + 0e + 0e
28. 3) the same mass and the opposite charge
29. 4) electromagnetic
30. 3) 3.2×10^{-19} C
31. 3) proton
32. meson OR hadron
33. +1e OR -1e
34. The particles have enough (kinetic) energy to be converted to that much mass.
35. 2) Matter is converted into energy and then energy is converted into matter.
36. 4) omega
37. 2) hadrons
38. 2) a quark and an antiquark
39. 1) css
40. 4) neutrons and protons
41. 4) 21 quarks and 3 leptons
42. 3) $+1.07 \times 10^{-19}$ C
43. 1) $s\bar{c}$
44. 1) charge must be conserved
45. 2) -1e
46. 1) strong
47. up, down, down
48. -1e
49. 1.60×10^{-7} J
50. 3.56×10^{-24} kg
51. 1) uud

THE UNIVERSITY OF THE STATE OF NEW YORK • THE STATE EDUCATION DEPARTMENT • ALBANY, NY 12234

Reference Tables for Physical Setting/PHYSICS

2006 Edition

List of Physical Constants

Name	Symbol	Value
Universal gravitational constant	G	6.67×10^{-11} N•m^2/kg^2
Acceleration due to gravity	g	9.81 m/s^2
Speed of light in a vacuum	c	3.00×10^8 m/s
Speed of sound in air at STP		3.31×10^2 m/s
Mass of Earth		5.98×10^{24} kg
Mass of the Moon		7.35×10^{22} kg
Mean radius of Earth		6.37×10^6 m
Mean radius of the Moon		1.74×10^6 m
Mean distance—Earth to the Moon		3.84×10^8 m
Mean distance—Earth to the Sun		1.50×10^{11} m
Electrostatic constant	k	8.99×10^9 N•m^2/C^2
1 elementary charge	e	1.60×10^{-19} C
1 coulomb (C)		6.25×10^{18} elementary charges
1 electronvolt (eV)		1.60×10^{-19} J
Planck's constant	h	6.63×10^{-34} J•s
1 universal mass unit (u)		9.31×10^2 MeV
Rest mass of the electron	m_e	9.11×10^{-31} kg
Rest mass of the proton	m_p	1.67×10^{-27} kg
Rest mass of the neutron	m_n	1.67×10^{-27} kg

Prefixes for Powers of 10

Prefix	Symbol	Notation
tera	T	10^{12}
giga	G	10^9
mega	M	10^6
kilo	k	10^3
deci	d	10^{-1}
centi	c	10^{-2}
milli	m	10^{-3}
micro	μ	10^{-6}
nano	n	10^{-9}
pico	p	10^{-12}

Approximate Coefficients of Friction

	Kinetic	Static
Rubber on concrete (dry)	0.68	0.90
Rubber on concrete (wet)	0.58	
Rubber on asphalt (dry)	0.67	0.85
Rubber on asphalt (wet)	0.53	
Rubber on ice	0.15	
Waxed ski on snow	0.05	0.14
Wood on wood	0.30	0.42
Steel on steel	0.57	0.74
Copper on steel	0.36	0.53
Teflon on Teflon	0.04	

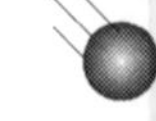

The Electromagnetic Spectrum

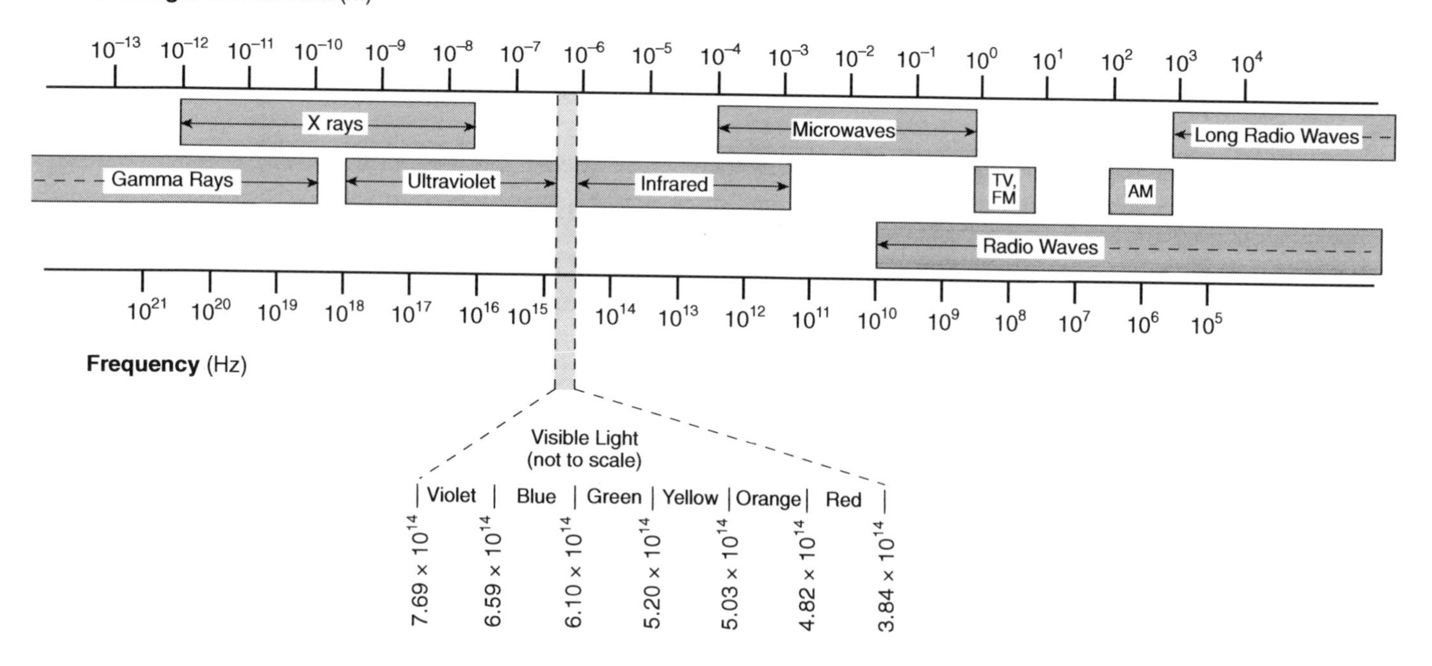

Absolute Indices of Refraction

($f = 5.09 \times 10^{14}$ Hz)

Material	Index
Air	1.00
Corn oil	1.47
Diamond	2.42
Ethyl alcohol	1.36
Glass, crown	1.52
Glass, flint	1.66
Glycerol	1.47
Lucite	1.50
Quartz, fused	1.46
Sodium chloride	1.54
Water	1.33
Zircon	1.92

Energy Level Diagrams

Hydrogen

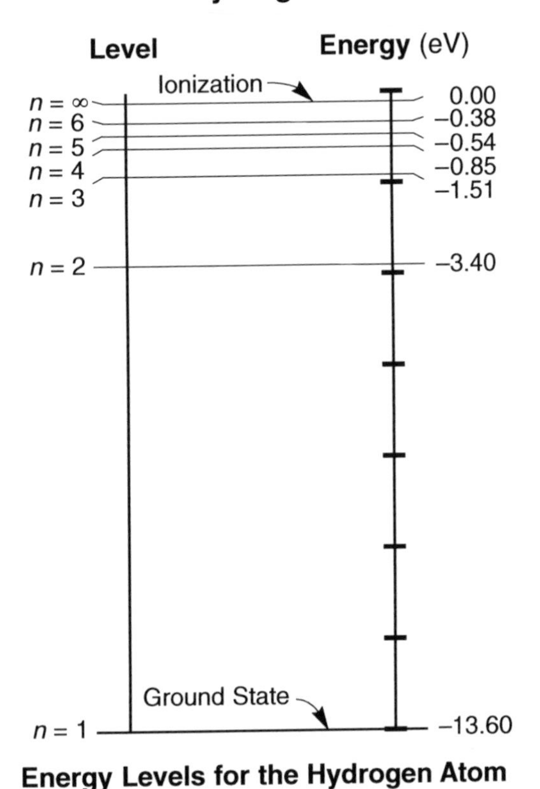

Energy Levels for the Hydrogen Atom

Mercury

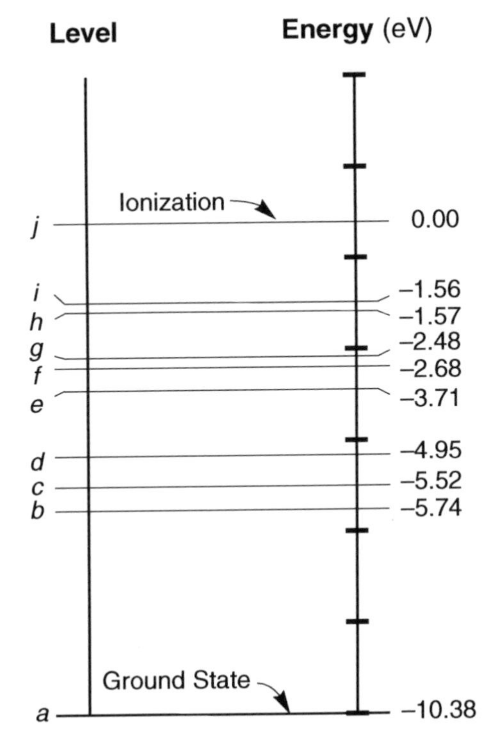

A Few Energy Levels for the Mercury Atom

Classification of Matter

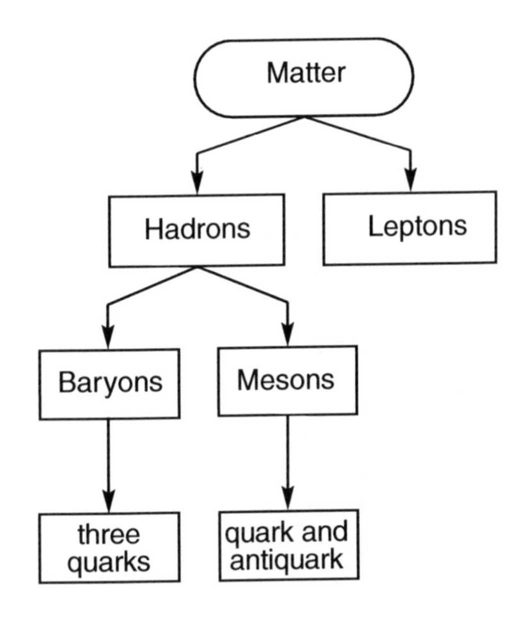

Particles of the Standard Model

Quarks

Name	up	charm	top
Symbol	u	c	t
Charge	$+\frac{2}{3}$e	$+\frac{2}{3}$e	$+\frac{2}{3}$e
	down	strange	bottom
	d	s	b
	$-\frac{1}{3}$e	$-\frac{1}{3}$e	$-\frac{1}{3}$e

Leptons

electron	muon	tau
e	μ	τ
−1e	−1e	−1e
electron neutrino	muon neutrino	tau neutrino
ν_e	ν_μ	ν_τ
0	0	0

Note: For each particle, there is a corresponding antiparticle with a charge opposite that of its associated particle.

Electricity

$$F_e = \frac{kq_1q_2}{r^2}$$

$$E = \frac{F_e}{q}$$

$$V = \frac{W}{q}$$

$$I = \frac{\Delta q}{t}$$

$$R = \frac{V}{I}$$

$$R = \frac{\rho L}{A}$$

$$P = VI = I^2R = \frac{V^2}{R}$$

$$W = Pt = VIt = I^2Rt = \frac{V^2t}{R}$$

A = cross-sectional area
E = electric field strength
F_e = electrostatic force
I = current
k = electrostatic constant
L = length of conductor
P = electrical power
q = charge
R = resistance
R_{eq} = equivalent resistance
r = distance between centers
t = time
V = potential difference
W = work (electrical energy)
Δ = change
ρ = resistivity

Series Circuits

$$I = I_1 = I_2 = I_3 = \ldots$$

$$V = V_1 + V_2 + V_3 + \ldots$$

$$R_{eq} = R_1 + R_2 + R_3 + \ldots$$

Parallel Circuits

$$I = I_1 + I_2 + I_3 + \ldots$$

$$V = V_1 = V_2 = V_3 = \ldots$$

$$\frac{1}{R_{eq}} = \frac{1}{R_1} + \frac{1}{R_2} + \frac{1}{R_3} + \ldots$$

Circuit Symbols

cell
battery
switch
voltmeter
ammeter
resistor
variable resistor
lamp

Resistivities at 20°C	
Material	**Resistivity** (Ω•m)
Aluminum	2.82×10^{-8}
Copper	1.72×10^{-8}
Gold	2.44×10^{-8}
Nichrome	$150. \times 10^{-8}$
Silver	1.59×10^{-8}
Tungsten	5.60×10^{-8}

Waves

$v = f\lambda$

$T = \frac{1}{f}$

$\theta_i = \theta_r$

$n = \frac{c}{v}$

$n_1 \sin \theta_1 = n_2 \sin \theta_2$

$\frac{n_2}{n_1} = \frac{v_1}{v_2} = \frac{\lambda_1}{\lambda_2}$

c = speed of light in a vacuum

f = frequency

n = absolute index of refraction

T = period

v = velocity or speed

λ = wavelength

θ = angle

θ_i = angle of incidence

θ_r = angle of reflection

Modern Physics

$E_{photon} = hf = \frac{hc}{\lambda}$

$E_{photon} = E_i - E_f$

$E = mc^2$

c = speed of light in a vacuum

E = energy

f = frequency

h = Planck's constant

m = mass

λ = wavelength

Geometry and Trigonometry

Rectangle

$A = bh$

Triangle

$A = \frac{1}{2}bh$

Circle

$A = \pi r^2$

$C = 2\pi r$

Right Triangle

$c^2 = a^2 + b^2$

$\sin \theta = \frac{a}{c}$

$\cos \theta = \frac{b}{c}$

$\tan \theta = \frac{a}{b}$

A = area

b = base

C = circumference

h = height

r = radius

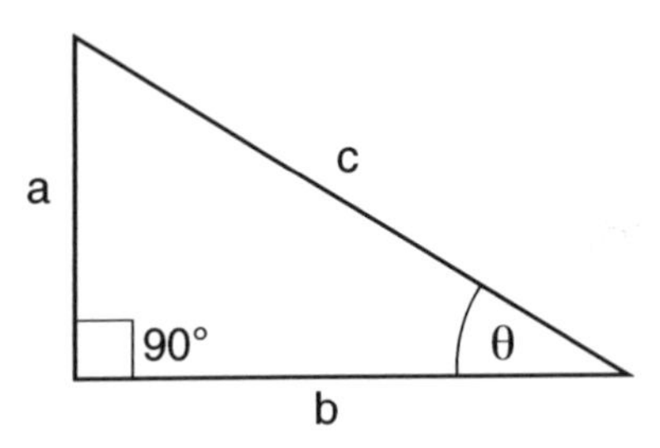

Mechanics

$\bar{v} = \frac{d}{t}$

$a = \frac{\Delta v}{t}$

$v_f = v_i + at$

$d = v_i t + \frac{1}{2}at^2$

$v_f^2 = v_i^2 + 2ad$

$A_y = A \sin \theta$

$A_x = A \cos \theta$

$a = \frac{F_{net}}{m}$

$F_f = \mu F_N$

$F_g = \frac{Gm_1 m_2}{r^2}$

$g = \frac{F_g}{m}$

$p = mv$

$p_{before} = p_{after}$

$J = F_{net}t = \Delta p$

$F_s = kx$

$PE_s = \frac{1}{2}kx^2$

$F_c = ma_c$

$a_c = \frac{v^2}{r}$

$\Delta PE = mg\Delta h$

$KE = \frac{1}{2}mv^2$

$W = Fd = \Delta E_T$

$E_T = PE + KE + Q$

$P = \frac{W}{t} = \frac{Fd}{t} = F\bar{v}$

a = acceleration

a_c = centripetal acceleration

A = any vector quantity

d = displacement or distance

E_T = total energy

F = force

F_c = centripetal force

F_f = force of friction

F_g = weight or force due to gravity

F_N = normal force

F_{net} = net force

F_s = force on a spring

g = acceleration due to gravity or gravitational field strength

G = universal gravitational constant

h = height

J = impulse

k = spring constant

KE = kinetic energy

m = mass

p = momentum

P = power

PE = potential energy

PE_s = potential energy stored in a spring

Q = internal energy

r = radius or distance between centers

t = time interval

v = velocity or speed

$\bar{v}$ = average velocity or average speed

W = work

x = change in spring length from the equilibrium position

Δ = change

θ = angle

μ = coefficient of friction

Made in the USA
Middletown, DE
18 March 2017